Catalytic Transformation of Renewables (Olefin, Bio-Sourced, et. al)

Catalytic Transformation of Renewables (Olefin, Bio-Sourced, et. al)

Editors

Nikolaos Dimitratos
Stefania Albonetti
Tommaso Tabanelli

MDPI • Basel • Beijing • Wuhan • Barcelona • Belgrade • Manchester • Tokyo • Cluj • Tianjin

Editors
Nikolaos Dimitratos
Alma Mater Studiorum-University of Bologna
Italy

Stefania Albonetti
Alma Mater Studiorum-University of Bologna
Italy

Tommaso Tabanelli
Alma Mater Studiorum-University of Bologna
Italy

Editorial Office
MDPI
St. Alban-Anlage 66
4052 Basel, Switzerland

This is a reprint of articles from the Special Issue published online in the open access journal *Catalysts* (ISSN 2073-4344) (available at: https://www.mdpi.com/journal/catalysts/special_issues/ Transformation_Renewables_Olefin_Bio-sourced).

For citation purposes, cite each article independently as indicated on the article page online and as indicated below:

LastName, A.A.; LastName, B.B.; LastName, C.C. Article Title. *Journal Name* **Year**, *Volume Number*, Page Range.

ISBN 978-3-0365-1068-2 (Hbk)
ISBN 978-3-0365-1069-9 (PDF)

© 2021 by the authors. Articles in this book are Open Access and distributed under the Creative Commons Attribution (CC BY) license, which allows users to download, copy and build upon published articles, as long as the author and publisher are properly credited, which ensures maximum dissemination and a wider impact of our publications.

The book as a whole is distributed by MDPI under the terms and conditions of the Creative Commons license CC BY-NC-ND.

Contents

About the Editors vii

Nikolaos Dimitratos, Stefania Albonetti and Tommaso Tabanelli
Catalytic Transformation of Renewables (Olefin, Bio-Sourced, et al.)
Reprinted from: *Catalysts* **2021**, *11*, 364, doi:10.3390/catal11030364 1

Juan Antonio Cecilia, Carmen Pilar Jiménez-Gómez, Virginia Torres-Bujalance, Cristina García-Sancho, Ramón Moreno-Tost and Pedro Maireles-Torres
Oxidative Condensation of Furfural with Ethanol Using Pd-Based Catalysts: Influence of the Support
Reprinted from: *Catalysts* **2020**, *10*, 1309, doi:10.3390/catal10111309 5

Kachaporn Saenluang, Anawat Thivasasith, Pannida Dugkhuntod, Peerapol Pornsetmetakul, Saros Salakhum, Supawadee Namuangruk and Chularat Wattanakit
In Situ Synthesis of Sn-Beta Zeolite Nanocrystals for Glucose to Hydroxymethylfurfural (HMF)
Reprinted from: *Catalysts* **2020**, *10*, 1249, doi:10.3390/catal10111249 25

Oliver Schade, Paolo Dolcet, Alexei Nefedov, Xiaohui Huang, Erisa Saraçi, Christof Wöll and Jan-Dierk Grunwaldt
The Influence of the Gold Particle Size on the Catalytic Oxidation of 5-(Hydroxymethyl)furfural
Reprinted from: *Catalysts* **2020**, *10*, 342, doi:10.3390/catal10030342 37

Mar López, Valentín Santos and Juan Carlos Parajó
Autocatalytic Fractionation of Wood Hemicelluloses: Modeling of Multistage Operation
Reprinted from: *Catalysts* **2020**, *10*, 337, doi:10.3390/catal10030337 51

Camila P. Ferraz, Natalia J. S. Costa, Erico Teixeira-Neto, Ângela A. Teixeira-Neto, Cleber W. Liria, Joëlle Thuriot-Roukos, M. Teresa Machini, Rénato Froidevaux, Franck Dumeignil, Liane M. Rossi and Robert Wojcieszak
5-Hydroxymethylfurfural and Furfural Base-Free Oxidation over AuPd Embedded Bimetallic Nanoparticles
Reprinted from: *Catalysts* **2020**, *10*, 75, doi:10.3390/catal10010075 67

Alessandra Roselli, Yuri Carvalho, Franck Dumeignil, Fabrizio Cavani, Sébastien Paul and Robert Wojcieszak
Liquid Phase Furfural Oxidation under Uncontrolled pH in Batch and Flow Conditions: The Role of In Situ Formed Base
Reprinted from: *Catalysts* **2020**, *10*, 73, doi:10.3390/catal10010073 83

Shahram Alijani, Sofia Capelli, Stefano Cattaneo, Marco Schiavoni, Claudio Evangelisti, Khaled M. H. Mohammed, Peter P. Wells, Francesca Tessore and Alberto Villa
Capping Agent Effect on Pd-Supported Nanoparticles in the Hydrogenation of Furfural
Reprinted from: *Catalysts* **2020**, *10*, 11, doi:10.3390/catal10010011 95

Kristof Van der Borght, Konstantinos Alexopoulos, Kenneth Toch, Joris W. Thybaut, Guy B. Marin and Vladimir V. Galvita
First-Principles-Based Simulation of an Industrial Ethanol Dehydration Reactor
Reprinted from: *Catalysts* **2019**, *9*, 921, doi:10.3390/catal9110921 111

Carlo Lucarelli, Danilo Bonincontro, Yu Zhang, Lorenzo Grazia, Marc Renom-Carrasco, Chloé Thieuleux, Elsje Alessandra Quadrelli, Nikolaos Dimitratos, Fabrizio Cavani and Stefania Albonetti
Tandem Hydrogenation/Hydrogenolysis of Furfural to 2-Methylfuran over a Fe/Mg/O Catalyst: Structure–Activity Relationship
Reprinted from: *Catalysts* **2019**, *9*, 895, doi:10.3390/catal9110895 133

Wipark Anutrasakda, Kanyanok Eiamsantipaisarn, Duangkamon Jiraroj, Apakorn Phasuk, Thawatchai Tuntulani, Haichao Liu and Duangamol Nuntasri Tungasmita
One-Pot Catalytic Conversion of Cellobiose to Sorbitol over Nickel Phosphides Supported on MCM-41 and Al-MCM-41
Reprinted from: *Catalysts* **2019**, *9*, 92, doi:10.3390/catal9010092 149

Xue-Ying Zhang, Zhong-Hua Xu, Min-Hua Zong, Chuan-Fu Wang and Ning Li
Selective Synthesis of Furfuryl Alcohol from Biomass-Derived Furfural Using Immobilized Yeast Cells
Reprinted from: *Catalysts* **2019**, *9*, 70, doi:10.3390/catal9010070 165

Jianguang Zhang, Xiangping Li, Juping Liu and Chuanbin Wang
A Comparative Study of MFI Zeolite Derived from Different Silica Sources: Synthesis, Characterization and Catalytic Performance
Reprinted from: *Catalysts* **2019**, *9*, 13, doi:10.3390/catal9010013 179

Jun Han, Zijiang Xiong, Zelin Zhang, Hongjie Zhang, Peng Zhou and Fei Yu
The Influence of Texture on Co/SBA–15 Catalyst Performance for Fischer–Tropsch Synthesis
Reprinted from: *Catalysts* **2018**, *8*, 661, doi:10.3390/catal8120661 193

Jesús Hidalgo-Carrillo, Almudena Parejas, Manuel Jorge Cuesta-Rioboo, Alberto Marinas and Francisco José Urbano
MPV Reduction of Furfural to Furfuryl Alcohol on Mg, Zr, Ti, Zr–Ti, and Mg–Ti Solids: Influence of Acid–Base Properties
Reprinted from: *Catalysts* **2018**, *8*, 539, doi:10.3390/catal8110539 205

Valentina Verdoliva, Michele Saviano and Stefania De Luca
Zeolites as Acid/Basic Solid Catalysts: Recent Synthetic Developments
Reprinted from: *Catalysts* **2019**, *9*, 248, doi:10.3390/catal9030248 217

About the Editors

Nikolaos Dimitratos obtained his PhD from Liverpool University and is Associate Professor at the University of Bologna, in the Department of Industrial Chemistry "Toso Montarari". He specializes in the development of nanoparticulate materials for a range of catalytic applications.

Stefania Albonetti obtained her PhD from the University of Bologna and is Associate Professor at the University of Bologna, in the Department of Industrial Chemistry "Toso Montarari". She specializes in the development of heterogeneous catalysts for environmentally friendly industrial processes.

Tommaso Tabanelli obtained his PhD from the University of Bologna and is a Young Researcher at the University of Bologna, in the Department of Industrial Chemistry "Toso Montarari". He specializes in the development of heterogeneous catalysts, focusing on biomass-derived chemical processes.

Editorial

Catalytic Transformation of Renewables (Olefin, Bio-Sourced, et al.)

Nikolaos Dimitratos *, Stefania Albonetti and Tommaso Tabanelli

Dipartimento di Chimica Industriale "Toso Montanari", Alma Mater Studiorum Università di Bologna, Viale Risorgimento 4, 40136 Bologna, Italy; stefania.albonetti@unibo.it (S.A.); tommaso.tabanelli@unibo.it (T.T.)
* Correspondence: nikolaos.dimitratos@unibo.it; Tel.: +39-051-2093144

Citation: Dimitratos, N.; Albonetti, S.; Tabanelli, T. Catalytic Transformation of Renewables (Olefin, Bio-Sourced, et al.). *Catalysts* **2021**, *11*, 364. https://doi.org/10.3390/catal11030364

Received: 8 March 2021
Accepted: 9 March 2021
Published: 10 March 2021

Publisher's Note: MDPI stays neutral with regard to jurisdictional claims in published maps and institutional affiliations.

Copyright: © 2021 by the authors. Licensee MDPI, Basel, Switzerland. This article is an open access article distributed under the terms and conditions of the Creative Commons Attribution (CC BY) license (https://creativecommons.org/licenses/by/4.0/).

The objective of this Special Issue is to provide new diverse contributions that can demonstrate recent applications in biomass transformation using heterogeneous catalysts. In recent decades, a wide variety of biomass-derived chemicals have emerged as key platform chemicals for the production of fine chemicals and liquid fuels using heterogeneous catalysts as the preferred option for most of the developed and proposed catalytic processes. A range of heterogeneous catalysts have been evaluated for effective biomass conversion, such as supported metal nanoparticles, mixed metal oxides and zeolites, where the control of particle size, porosity, acid-basic and redox properties is crucial for providing active, stable and selective heterogeneous catalysts. Moreover, the crucial role of the solvent, choice of reactor design and final chemical processes for controlling activity, selectivity and deactivation phenomena has been demonstrated.

In this Special Issue, 14 articles and 1 review are presented. The article by Thivasasith and co-workers demonstrates the efficient conversion of glucose to hydrohymethylfurfural (HMF) by using Sn-Beta zeolites as heterogeneous multifunctional catalysts and demonstrating the role of Sn and beta zeolite for transforming glucose in an efficient and selective way under relatively mild reaction conditions [1]. Grunwaldt and co-workers showed the influence of the gold nanoparticle size for the catalytic oxidation of 5-hydroxymethylfurfural to produce 2,5-furandicarboxylic acid, by synthesising supported gold colloidal nanoparticles, and by optimising the particle size of Au, the yield to FDCA (2,5-furandicarboxylic acid) was successfully increased up to 92% [2]. In another study, Wojcieszak and co-workers investigated the liquid phase oxidation of HMF and furfural using supported bimetallic AuPd nanoparticles and base free conditions [3]. By tuning the Au-Pd atomic ratio, they succeeded in optimising activity and selectivity towards to the desired products. In a following study, the same group reported the liquid phase furfural oxidation in batch and continuous flow conditions using hydrotalcite-supported gold nanoparticles [4]. By tuning Mg:Al molar ratios they succeeded in tuning the acid-base properties of the hydrotalcite-supported nanoparticles and to find the optimum catalyst for transforming furfural to furoic acid with 100% yield and they demonstrated the issues of stability of these catalysts by using continuous flow systems. Villa and co-workers studied the effect of the capping agent of Pd-supported nanoparticles for the hydrogenation of furfural. They demonstrated that by controlling the temperature of the preparation for the preformed Pd colloidal nanoparticles and by altering the stabiliser, the activity and selectivity to the desired products could be controlled. Their studies showed the influence not only of the particle size of Pd but also of the nature of support for affecting the adsorption of the reactant on the Pd active sites [5]. Marinas and co-workers studied the MPV (Meerwein–Ponndorf–Verley) reduction of furfural to furfuryl alcohol using Mg, Zr and Ti mixed systems. Considering the environmentally friendly process of the reduction of a carbonyl compound through hydrogen transfer from a secondary alcohol, the group reported the reduction of furfural to furfuryl alcohol, and it was reported that the presence of Lewis acid sites and especially of acid-base pair sites can favour the MPV reaction [6]. Li and co-workers studied the synthesis of furfuryl alcohol from furfural using a biocatalytic approach. The reduction of furfural was reported in the presence of immobilized

Meyerozyma guilliermondii SC1103 cells. The biocatalytic process was optimised, and 98% conversion of furfural with selectivity over 985 was achieved. Moreover, the scale up process demonstrated the that the furfuryl alcohol productivity can be increased significantly and is among the highest values reported for the biocatalytic synthesis of furfuryl alcohol [7]. Albonetti and co-workers investigated the hydrodeoxygenation of furfural over Fe-containing MgO catalysts using methanol as a hydrogen donor and continuous flow conditions [8]. They reported the production of 2-methylfuran with 92% selectivity, by tuning the content of Fe and, therefore, the content of Lewis acid sites on the basic support. FTIR (Fourier Transform Infrared Spectroscopy) has been used to study the reaction mechanism and especially under which reaction conditions hydrogen transfer reduction or methanol dehydrogenation and/or methanol disproportionation happens during the process. The oxidative condensation of furfural with ethanol and using Pd-based catalysts was studied by Maireles-Torres and co-workers [9]. Particularly, they focused on studying the influence of the support using acidic and basic materials. They concluded that the presence of basic sites can lead to a beneficial effect on the catalytic activity, and the best activity was reported when MgO was the chosen support with 70% yield to furan-2-acrolein. The combination of PdO particles with the high basicity of the MgO was responsible for the improved catalytic behaviour observed. Tungasmita and co-workers synthesised nichel phoshides supported on MCM-41-based materials, and they demonstrated the one pot catalytic conversion of cellobiose to sorbitol, which is one important product from the biomass family [10]. By varying the Ni to P atomic ratio, they optimised the conversion and yield to sorbitol, and by characterisation studies, they reported that the most active phase is in the presence of $Ni_{12}P_5$. The autocatalytic fractionation of wood hemicellulose was reported by Parajo and co-workers [11]. *Eucalyptus globulus* wood samples were treated with hot, compressed water (autohydrolysis) in consecutive stages under non-isothermal conditions in order to convert the hemicellulose fraction into soluble compounds through reactions catalysed by in situ generated acids. The concentration profiles determined for the soluble saccharides, acids, and furans present in the liquid phases from the diverse crossflow stages were employed for kinetic modelling, based on pseudohomogeneous reactions and Arrhenius-type dependence of the kinetic coefficients on temperature. Galvita and co-workers reported the design and optimization of an industrial ethanol dehydration by employing a multiscale model ranging from nano- and micro-, to macroscale [12]. The intrinsic kinetics of the elementary steps was quantified through ab initio obtained rate and equilibrium coefficients. Heat and mass transfer limitations for the industrial design case were assessed via literature correlations. The developed industrial reactor model indicated that it is not advantageous to utilize feeds with high concentrations of ethanol, since a lower ethanol conversion and ethene yield were observed. Li and co-workers reported the synthesis, characterisation and catalytic performance of MFI zeolite derived from different silica sources [13]. The dry gel conversion (DGC) method was used to synthesize silicalite-1 and ZSM-5 with MFI structures. From the characterization results, it was reported that the high-quality coffin-like silicalite-1 was synthesized using silica spheres with a particle size of 300 nm as a silica source. The performance of aqueous phase eugenol hydrodeoxygenation over the Pd/C-ZSM-5 catalyst was evaluated, and it was shown that high hydrocarbon selectivity up to 74% could be achieved with eugenol conversion of 97%. Yu and co-workers showed the influence of the Co/SBA–15 catalyst texture, such as pore size and pore length, on Fischer–Tropsch (FT) synthesis [14]. The authors reported that the increase in pore size could improve the activity of the Co/SBA–15 catalyst until a certain value. Moreover, it was also found that the pore length of the Co/SBA–15 catalyst played a key role in the catalytic activity. CO_2 and C4+ selectivity was 2.0% and 74%, respectively, during the simulated syngas (64% H_2: 32% CO: balanced N_2) FT over the Co/SBA–15 catalysts, and CO conversion rate and CH_4 selectivity were 10.8% and 15.7%, respectively, after 100 h time on stream. Finally, one review is presented in this Special Issue, focusing on the usage of zeolites as acid/basic solid catalysts and on recent synthetic developments. De Luca and co-workers present, in this review, the key properties of zeolites as acid, both

Lewis and Brønsted, and basic solid support [15]. Moreover, their application as catalysts is discussed by reviewing published works, and their still unexplored potential as a green, mild and selective catalyst is also reported.

Funding: This research received no external funding.

Conflicts of Interest: The authors declare no conflict of interest.

References

1. Saenluang, K.; Thivasasith, A.; Dugkhuntod, P.; Pornsetmetakul, P.; Salakhum, S.; Namuangruk, S.; Wattanakit, C. In Situ Synthesis of Sn-Beta Zeolite Nanocrystals for Glucose to Hydroxymethylfurfural (HMF). *Catalysts* **2020**, *10*, 1249. [CrossRef]
2. Schade, O.; Dolcet, P.; Nefedov, A.; Huang, X.; Saraçi, E.; Wöll, C.; Grunwaldt, J.-D. The Influence of the Gold Particle Size on the Catalytic Oxidation of 5-(Hydroxymethyl)furfural. *Catalysts* **2020**, *10*, 342. [CrossRef]
3. Ferraz, C.P.; Costa, N.J.S.; Teixeira-Neto, E.; Teixeira-Neto, Â.A.; Liria, C.W.; Thuriot-Roukos, J.; Machini, M.T.; Froidevaux, R.; Dumeignil, F.; Rossi, L.M.; et al. 5-Hydroxymethylfurfural and Furfural Base-Free Oxidation over AuPd Embedded Bimetallic Nanoparticles. *Catalysts* **2020**, *10*, 75. [CrossRef]
4. Roselli, A.; Carvalho, Y.; Dumeignil, F.; Cavani, F.; Paul, S.; Wojcieszak, R. Liquid Phase Furfural Oxidation under Uncontrolled pH in Batch and Flow Conditions: The Role of In Situ Formed Base. *Catalysts* **2020**, *10*, 73. [CrossRef]
5. Alijani, S.; Capelli, S.; Cattaneo, S.; Schiavoni, M.; Evangelisti, C.; Mohammed, K.M.H.; Wells, P.P.; Tessore, F.; Villa, A. Capping Agent Effect on Pd-Supported Nanoparticles in the Hydrogenation of Furfural. *Catalysts* **2020**, *10*, 11. [CrossRef]
6. Hidalgo-Carrillo, J.; Parejas, A.; Cuesta-Rioboo, M.J.; Marinas, A.; Urbano, F.J. MPV Reduction of Furfural to Furfuryl Alcohol on Mg, Zr, Ti, Zr–Ti, and Mg–Ti Solids: Influence of Acid–Base Properties. *Catalysts* **2018**, *8*, 539. [CrossRef]
7. Zhang, X.-Y.; Xu, Z.-H.; Zong, M.-H.; Wang, C.-F.; Li, N. Selective Synthesis of Furfuryl Alcohol from Biomass-Derived Furfural Using Immobilized Yeast Cells. *Catalysts* **2019**, *9*, 70. [CrossRef]
8. Lucarelli, C.; Bonincontro, D.; Zhang, Y.; Grazia, L.; Renom-Carrasco, M.; Thieuleux, C.; Quadrelli, E.A.; Dimitratos, N.; Cavani, F.; Albonetti, S. Tandem Hydrogenation/Hydrogenolysis of Furfural to 2-Methylfuran over a Fe/Mg/O Catalyst: Structure–Activity Relationship. *Catalysts* **2019**, *9*, 895. [CrossRef]
9. Cecilia, J.A.; Jiménez-Gómez, C.P.; Torres-Bujalance, V.; García-Sancho, C.; Moreno-Tost, R.; Maireles-Torres, P. Oxidative Condensation of Furfural with Ethanol Using Pd-Based Catalysts: Influence of the Support. *Catalysts* **2020**, *10*, 1309. [CrossRef]
10. Anutrasakda, W.; Eiamsantipaisarn, K.; Jiraroj, D.; Phasuk, A.; Tuntulani, T.; Liu, H.; Tungasmita, D.N. One-Pot Catalytic Conversion of Cellobiose to Sorbitol over Nickel Phosphides Supported on MCM-41 and Al-MCM-41. *Catalysts* **2019**, *9*, 92. [CrossRef]
11. López, M.; Santos, V.; Parajó, J.C. Autocatalytic Fractionation of Wood Hemicelluloses: Modeling of Multistage Operation. *Catalysts* **2020**, *10*, 337. [CrossRef]
12. Van der Borght, K.; Alexopoulos, K.; Toch, K.; Thybaut, J.W.; Marin, G.B.; Galvita, V.V. First-Principles-Based Simulation of an Industrial Ethanol Dehydration Reactor. *Catalysts* **2019**, *9*, 921. [CrossRef]
13. Zhang, J.; Li, X.; Liu, J.; Wang, C. A Comparative Study of MFI Zeolite Derived from Different Silica Sources: Synthesis, Characterization and Catalytic Performance. *Catalysts* **2019**, *9*, 13. [CrossRef]
14. Han, J.; Xiong, Z.; Zhang, Z.; Zhang, H.; Zhou, P.; Yu, F. The Influence of Texture on Co/SBA–15 Catalyst Performance for Fischer–Tropsch Synthesis. *Catalysts* **2019**, *9*, 444. [CrossRef]
15. Verdoliva, V.; Saviano, M.; De Luca, S. Zeolites as Acid/Basic Solid Catalysts: Recent Synthetic Developments. *Catalysts* **2019**, *9*, 248. [CrossRef]

Article

Oxidative Condensation of Furfural with Ethanol Using Pd-Based Catalysts: Influence of the Support

Juan Antonio Cecilia, Carmen Pilar Jiménez-Gómez, Virginia Torres-Bujalance, Cristina García-Sancho, Ramón Moreno-Tost and Pedro Maireles-Torres *

Departamento de Química Inorgánica, Cristalografía y Mineralogía (Unidad Asociada al ICP-CSIC), Facultad de Ciencias, Campus de Teatinos, Universidad de Málaga, 29071 Málaga, Spain; jacecilia@uma.es (J.A.C.); carmenpjg@uma.es (C.P.J.-G.); vtorresbujalance@gmail.com (V.T.-B.); cristinags@uma.es (C.G.-S.); rmtost@uma.es (R.M.-T.)
* Correspondence: maireles@uma.es; Tel.: +34-952137534

Received: 19 October 2020; Accepted: 8 November 2020; Published: 12 November 2020

Abstract: PdO nanoparticles were deposited on several supports (β-zeolite, Al_2O_3, Fe_2O_3, MgO, and SiO_2), which displayed different crystallinity, textural properties, and amount of acid and basic sites. These catalysts were characterized by X-ray diffraction (XRD), transmission electron microscopy (TEM), N_2 adsorption–desorption isotherms at −196 °C, NH_3 and CO_2 thermoprogrammed desorption analyses (NH_3- and CO_2-TPD, and X-ray photoelectron spectroscopy (XPS). Pd-based catalysts were tested in the oxidative condensation of furfural with ethanol to obtain value-added chemicals. The catalytic results revealed high conversion values, although the presence of a high proportion of carbonaceous deposits, mainly in the case of the PdO supported on β-zeolite and Al_2O_3, is also noteworthy. The presence of basic sites led to a beneficial effect on the catalytic behavior, since the formation of carbonaceous deposits was minimized. Thus, the 2Pd-MgO (2 wt.% Pd) catalyst reached the highest yield of furan-2-acrolein (70%) after 3 h of reaction at 170 °C. This better catalytic performance can be explained by the high basicity of MgO, used as support, together with the large amount of available PdO, as inferred from XPS.

Keywords: furfural; oxidative condensation; furan-2-acrolein; Pd-based catalysts

1. Introduction

The increase in the world population has caused a progressive decline of fossil fuels. This fact has led to the quest of alternative sources, which can provide both energy and chemicals. A large variety of energy sources are now available to be used synergistically and responsibly to supply the world's population, although most of them are only useful for obtaining energy. Among them, biomass is the only source that allows the production of both energy and chemicals, in such a way that it is the unique source that could widely replace conventional fossil feedstocks [1].

Biomass is worldwide distributed, which together its high availability, could favor the energy self-sufficiency of countries, but it must be sustainable, without competing with the food supply. In this sense, lignocellulosic biomass is attracting the interest of the scientific community, as it can be obtained from low-cost agricultural wastes [2,3]. Lignocellulose is mainly formed by cellulose (40–50%), hemicellulose (20–35%), and lignin (15–25%), which can be easily fractionated through physical and chemical treatments [4,5]. In the case of hemicellulose, the hemicellulose fibers embedded in the cell walls of plants are extracted by using alkaline hydroperoxide [6,7], obtaining xylans as the main products [4–7]. Then, xylans can be hydrolyzed in their respective monomers by a mild acid treatment, giving rise mainly to xylose [8], which in turn, can be dehydrated through acid catalysis to obtain furfural (FUR) [8].

After bioethanol, FUR is the second most produced chemical in the sugar platform [8]. The high interest of this building block molecule is ascribed to its chemical structure (an aldehyde group linked to a furan ring with α-β unsaturations), which confers it high reactivity for undergoing a large spectrum of chemical reactions [8]. Thus, it has been reported that FUR can react through hydrogenation, decarbonylation, alkylation, condensation, oxidation, or opening-ring reactions to form many valuable products [9,10]. Focusing on the oxidation reactions, it is well known that FUR can suffer a Cannizzaro process giving rise to furfuryl alcohol (FOL) and furoic acid (FURAc), through homogeneous catalysis with NaOH; however, the yield of this reaction is only limited to 50% [11,12]. The direct oxidation of FUR to furoic acid requires strong oxidants such as chlorite, permanganate, or chromate species [13,14], in such a way the reaction medium is highly toxic to health and the environment. Taking into account the environmental awareness of recent decades, the scientific community is developing more environmentally friendly catalysts. In this sense, the use of noble metals (Au, Pt, Pd, Ru) has emerged as a sustainable alternative for FUR oxidation processes [12]. These catalysts are prone to fast deactivation due to the strong interaction between the furoate intermediates and the active phase, although this deactivation can be minimized by the addition of a base into the reaction medium [15].

The reaction medium plays a key role in determining the selectivity pattern. Thus, the oxidation of FUR leads to furoic acid in aqueous medium [12], whereas the use of short-chain alcohols, such as methanol, favors the oxidative esterification, obtaining methyl furoate as product [12], when Au-based catalysts are used [16–19]. The use of longer-chain alcohols with H in the C_α position causes a competitive reaction between the oxidative esterification and oxidative condensation [20–22], although oxidative condensation is mainly favored [12]. The influence of several experimental parameters has been evaluated, and it has been found that ethanol and propanol are the most reactive to carry out the oxidative condensation process [23]. In the same way, the strength of the base also exerts an important role, since strong bases also favor oxidative condensation [21], although the basicity must be modulated because side reactions may appear due to uncontrolled condensation, thus decreasing the yield [17]. In the oxidative condensation of FUR, the most widely used catalysts have been based on Au [20–23], although other noble metals, such as Pt [24,25] or Pd [26] dispersed on different metal oxides have also been studied. However, recently, much interest is being paid to the development of non-noble and abundant-metal-based catalysts, since noble metals are quite expensive and scarce. In this sense, Cu [27] and Fe complexes [28], as well as CuO/CeO_2 [29] and Co_xO_y-based catalysts [23], have shown activity in the oxidative condensation of FUR, although more severe conditions are needed. This catalytic process gives rise to furan-2-acrolein (F2A), where the increase in the hydrocarbon chain is an appropriate strategy to diminish the volatility of liquid fuels, which can be obtained after subsequent hydrogenation of F2A [30,31]. In addition, F2A is used as an intermediate in the synthesis of drugs and in the fragrance industry due to its pleasant smell [32].

Taking into account that oxidative condensation of FUR using Pd-based catalysts has hardly been studied in the literature, the aim of the present work was to evaluate the catalytic activity of Pd-based catalysts in oxidative condensation. Moreover, it is also intended to evaluate the influence of the support on the catalytic performance, taking into account the amount of acid and basic centers, as well as their strength. Furthermore, the effect of Pd loading and the reaction time will also be studied.

2. Characterization of Catalysts

Before evaluating the catalytic behavior of Pd-based catalysts in FUR oxidation reactions, these samples were characterized by different physico-chemical techniques in order to get insights into structural, textural and acidic properties.

The analysis of crystalline phases present in the Pd-based catalysts was studied by X-ray diffraction (Figure 1). Since the catalysts possess a low Pd loading, which renders it difficult to observe the corresponding diffraction peaks, only the characterization data of catalysts with the highest Pd content, i.e., 2 wt.%, will be shown.

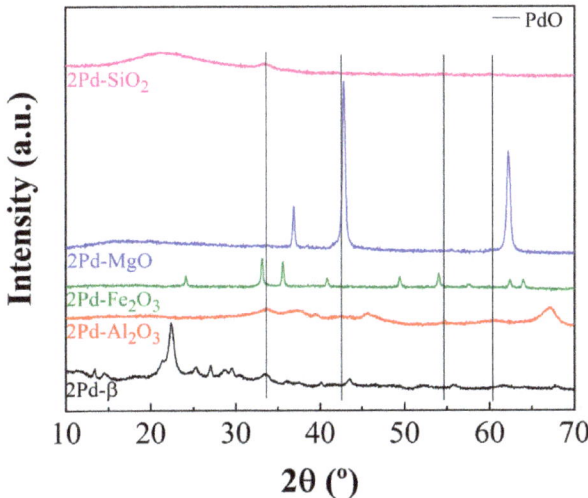

Figure 1. X-ray diffraction patterns of Pd supported on β-zeolite, Al$_2$O$_3$, Fe$_2$O$_3$, MgO, and SiO$_2$ (Pd content: 2 wt.%; vertical lines indicate the position of PdO signals).

The diffractogram of 2Pd-β shows a large number of peaks of variable intensity. Among them, the most intense peaks appear at 2θ (°) = 21.4, 22.4, 25.3, 27.1, 28.7, 29.5, and 32.6, which are typical of this microporous aluminosilicate (PDF: 00-048-0074). In the case of 2Pd-Al$_2$O$_3$, the diffraction peaks are broader and less intense. These diffraction peaks are located at 2θ (°) = 37.4, 39.6, 45.8, 60.3, and 67.4, which are ascribed to the existence of γ-Al$_2$O$_3$ (PDF: 00-001-1303). The diffraction profile of 2Pd-Fe$_2$O$_3$ exhibits well-defined and narrower diffraction peaks, located at 2θ (°) = 24.2, 33.1, 35.7, 40.9, 49.4, 54.1, 57.6, 62.5, and 64.1, associated to Fe$_2$O$_3$ crystals of α-hematite (PDF: 01-089-2810). Similarly, 2Pd-MgO also displays well-defined diffraction peaks located at 2θ (°) = 37.0, 42.9, and 62.2, corresponding to MgO crystals with a periclase structure. Finally, 2Pd-SiO$_2$ shows a broad band, whose maximum is located at 2θ (°) = 21.6, characteristic of amorphous silica.

Considering the low Pd loading (2 wt.%), diffraction peaks attributed to Pd species were hardly detected. Thus, it is only noticeable the presence of a small diffraction peak at 2θ (°) = 33.8, which can be assigned to PdO (PDF: 00-048-0074), difficult to distinguish in the case of 2Pd-MgO and 2Pd-Fe$_2$O$_3$ due to the high crystallinity of these supports.

The average crystal size of metal oxides, used as supports, was determined by the Williamson–Hall equation [33]. Different sizes can be observed, from 7 nm for Al$_2$O$_3$ to 35 and 50 nm for MgO and Fe$_2$O$_3$, respectively. In the case of β-zeolite, the determination of the crystal size is very complex, since the diffraction pattern depends on the Si/Al molar ratio, while silica has an amorphous structure.

The morphology of Pd-based catalysts was evaluated by transmission electron microscopy (TEM) (Figure 2). From these micrographs, the particle sizes of supports corroborated the XRD data, and homogeneous distributions of PdO particles smaller than 15 nm were observed in all cases.

The textural properties were determined from N$_2$ adsorption–desorption isotherms at −196 °C. The shape of isotherms (Figure 3) is very different, depending on the type of support. Thus, according to the International Union of Pure and Applied Chemistry (IUPAC) classification, the isotherm of 2Pd-SiO$_2$ can be considered as Type II [34], associated to macroporous materials with an important N$_2$ adsorption at high relative pressure due to interparticle voids between adjacent SiO$_2$ particles. The catalysts 2Pd-MgO and 2Pd-Fe$_2$O$_3$ also exhibit Type II isotherms. The low N$_2$ adsorption at low relative pressures indicates the absence of micropores. These data seem to be in agreement with the presence of bigger crystals of MgO and Fe_2O_3 in comparison to other supports (Figures 1 and 2). The adsorption in the 2Pd-Fe_2O_3 at high relative pressure suggests the presence of some macroporosity. 2Pd-β is mainly microporous, as inferred from the raise of adsorption at low relative pressure. In the case of 2Pd-Al_2O_3, the profile can be fitted as Type IV (a) [34], which is typical of mesoporous materials, where the presence of hysteresis loops suggests the presence of pores wider than 4 nm.

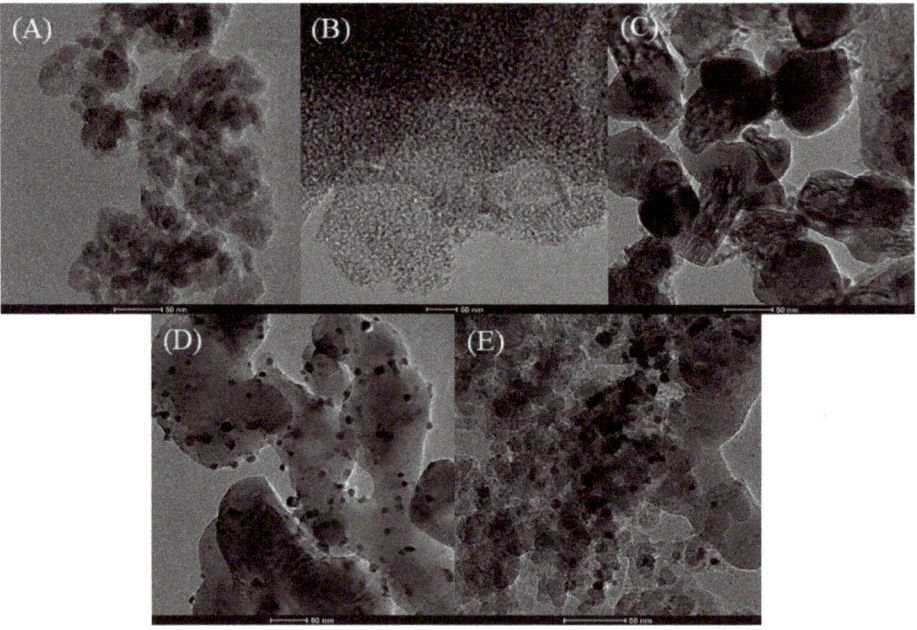

Figure 2. TEM micrographs of 2Pd-β (**A**), 2Pd-Al_2O_3 (**B**), 2Pd-Fe_2O_3 (**C**), 2Pd-MgO (**D**), and 2Pd-SiO_2 (**E**) catalysts.

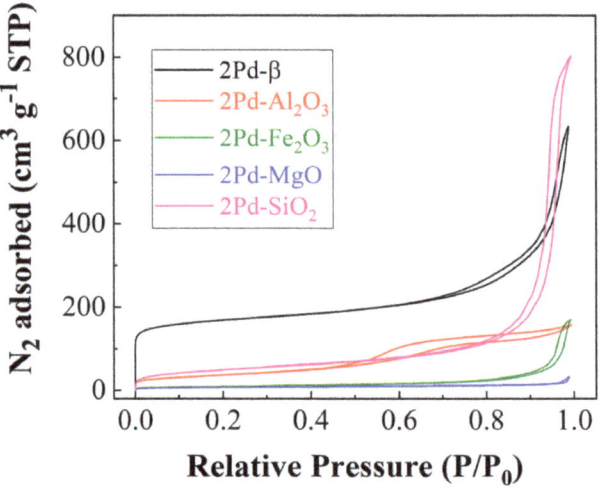

Figure 3. N_2 adsorption–desorption isotherm at −196 °C of Pd-based catalysts (Pd content: 2 wt.%).

The analysis of the hysteresis loops (Figure 3) reveals that 2Pd-β, 2Pd-Fe$_2$O$_3$, and 2Pd-SiO$_2$ catalysts show a Type H3 loop [34], which is given by non-rigid aggregates of plate-like particles. In the case of the 2Pd-Al$_2$O$_3$, the loop can be adjusted to H2 (b) [34], being associated with pore blocking, but the size distribution of neck widths is large.

The specific surface area was determined by using the Brunauer–Emmett–Teller equation (S_{BET}) [35] (Table 1). As expected due to its microporous character, 2Pd-β displays the highest S_{BET} (774 m^2 g^{-1}), mainly ascribed to its high microporosity, as inferred from the *t*-plot data [36] (587 m^2 g^{-1}). The catalysts 2Pd-SiO$_2$ and 2Pd-Al$_2$O$_3$ have intermediate S_{BET} values (177 and 132 m^2 g^{-1}, respectively), while the microporosity is very similar in both cases, although much lower than that shown by 2Pd-β. The catalysts 2Pd-MgO and 2Pd-Fe$_2$O$_3$ exhibit the poorest textural properties, with the lowest S_{BET} values and microporosity, as a consequence of the formation of bigger MgO or Fe$_2$O$_3$ particles, which gives rise to lower interparticle voids.

Table 1. Textural properties of Pd-based catalysts (Pd content: 2 wt.%).

Sample	Surface Area (m^2 g^{-1})	t-Plot (m^2 g^{-1})	V_P (cm^3 g^{-1})	V_{MP} (cm^3 g^{-1})
2Pd-β	774	587	0.614	0.178
2Pd-Al$_2$O$_3$	132	26	0.210	0.010
2Pd-Fe$_2$O$_3$	33	3	0.087	0.001
2Pd-MgO	24	12	0.024	0.006
2Pd-SiO$_2$	177	28	0.484	0.010

The pore size distributions, compiled in Figure 4, were estimated by the density functional theory (DFT) [37]. As expected, 2Pd-β displays microporous, while the catalysts with the largest pore sizes are both 2Pd-β and 2Pd-SiO$_2$. It is also noteworthy that the catalyst with the narrowest pore size distribution is 2Pd-Al$_2$O$_3$, with a pore diameter between 3.6 and 11.0 nm.

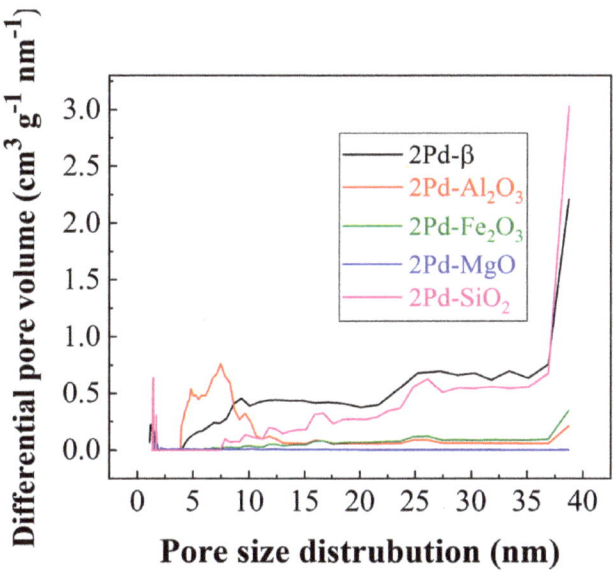

Figure 4. Pore size distribution (estimated from the density functional theory (DFT) method) of Pd supported on β-zeolite, Al$_2$O$_3$, Fe$_2$O$_3$, MgO, and SiO$_2$ catalysts (Pd content: 2 wt.%).

Once textural properties were determined, the quantification of the total amount of acid and basic sites was accomplished. The estimation of acid sites from NH$_3$-TPD analysis (Figure 5A and Table 2) reveals that the catalyst with the best textural properties, 2Pd-β, also displays the highest amount of acid sites (455 µmol g^{-1}), as expected, whereas the 2Pd-SiO$_2$ catalyst exhibited the lowest acidity values (25 µmol g^{-1}), even lower than that shown by a basic support like MgO. In this latter case, its acidity can be associated to structural defects in the metal oxide network.

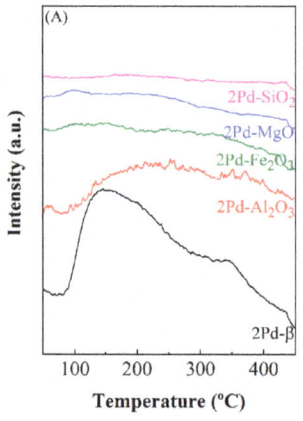 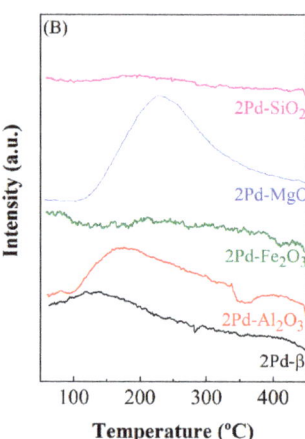

Figure 5. NH$_3$-TPD (**A**) and CO$_2$-TPD (**B**) for Pd-based catalysts with 2 wt.% Pd.

Table 2. Amount of acid sites, estimated by NH$_3$-TPD, and basic sites, deduced from CO$_2$-TPD, of Pd-based catalysts with 2 wt.% Pd.

Sample	Total Amount of Acid Sites (µmol g^{-1})	Total Amount of Basic Sites (µmol g^{-1})
2Pd-β	455	85
2Pd-Al$_2$O$_3$	235	104
2Pd-Fe$_2$O$_3$	90	15
2Pd-MgO	105	221
2Pd-SiO$_2$	25	26

The total amount of basic sites was determined from CO$_2$-TPD (Figure 5B and Table 2). Generally, the total amount of basic sites is less than that shown for acid sites ones. From CO$_2$-TPD profiles, it can be observed how, in spite of its poor textural properties, 2Pd-MgO reaches the highest amount of basic sites (221 µmol g^{-1}). In the case of 2Pd-Al$_2$O$_3$ and 2Pd-β, the amount of basic sites are 104 and 85 µmol g^{-1}, respectively. The lowest amount of basic sites was found for 2Pd-SiO$_2$ and 2Pd-Fe$_2$O$_3$ catalysts.

The surface chemical analysis of catalysts was performed by X-ray photoelectron spectroscopy (XPS), and the data are shown in Table 3. The electronic charge effects were corrected by using adventitious carbon at 284.8 eV as reference [38]. Together with this adventitious carbon peak, C 1s core level spectra also display another contribution at 288–289 eV for 2Pd-Al$_2$O$_3$, 2Pd-Fe$_2$O$_3$, and 2Pd-MgO. This contribution is ascribed to the presence of carbonate species, since basic sites present in basic or amphoteric metal oxides tend to react with atmospheric carbon dioxide to form the corresponding metal carbonates. However, crystalline metal carbonates were not observed in the XRD patterns, pointing out that carbonation only takes place on the catalyst surface.

Table 3. XPS data of Pd-based catalysts (Pd content: 2 wt.%).

Sample	Binding Energy, eV (Atomic Concentration, %)								Pd/M Atomic Ratio
	C 1s		O 1s		Si 2p	Al 2p	Fe 2p	Mg 2p	Pd 3d	
2Pd-β	284.8 (12.3)	-	532.6 (61.9)	-	103.7 (24.4)	75.1 (1.2)	-	-	336.8 (0.1)	0.004
2Pd-Al$_2$O$_3$	284.8 (10.0)	288.4 (0.5)	531.1 (59.8)	-	-	74.1 (29.6)	-	-	336.7 (0.2)	0.007
2Pd-Fe$_2$O$_3$	284.8 (9.7)	288.4 (1.0)	530.0 (49.0)	531.9 (14.1)	-	-	710.7 (25.5)	-	336.7 (0.7)	0.027
2Pd-MgO	284.8 (11.6)	288.6 (2.7)	529.4 (32.5)	531.8 (16.7)	-	-	-	48.9 (35.7)	336.7 (0.8)	0.022
2Pd-SiO$_2$	284.8 (4.5)	-	532.8 (67.9)	-	103.6 (27.4)	-	-	-	336.7 (0.2)	0.007

The analysis of the O 1s core level spectra showed a main contribution at 529.4–532.8 eV, attributed to oxide species. Catalysts prepared with less acidic supports (Pd-Fe$_2$O$_3$ and Pd-MgO) presented another contribution located at higher binding energy (531.8–531.9 eV), which is ascribed to both hydroxide and carbonate species, according to the C 1s core level data, previously noted.

The analysis of the Si 2p, Al 2p, Fe 2p, and Mg 2p core level spectra leads to single contributions in each region, which are attributed to their respective oxide species [38]. Finally, the study of the Pd 3d core level spectra revealed the existence of a band located at 336.7 eV (Figure 6) [38], being ascribed to Pd as PdO, confirming XRD data (Figure 1). In this sense, it is noteworthy that those catalysts whose surface Pd content is lower (2Pd-Al$_2$O$_3$, 2Pd-β, and 2Pd-SiO$_2$) are those where the main diffraction peak of PdO was detected. This fact would imply that the dispersion of PdO in these three catalysts must be worse than in 2Pd-MgO and 2Pd-Fe$_2$O$_3$ catalysts.

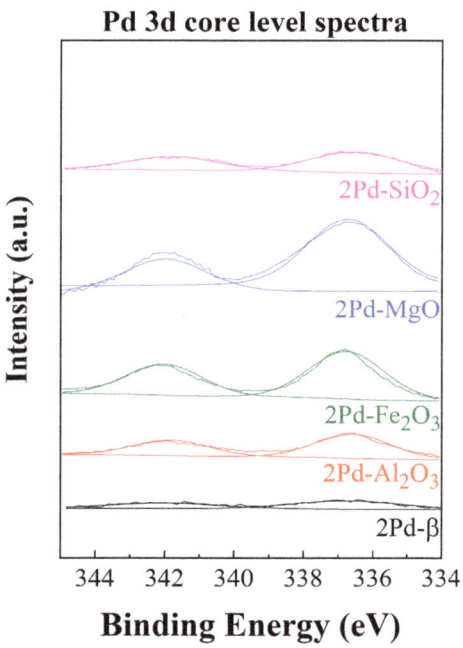

Figure 6. Pd 3d core level spectra of Pd-based catalysts (Pd content: 2 wt.%).

Regarding the Pd/M (M = Si, Al, Mg, Fe) atomic ratio (Table 3), the highest values are found for 2Pd-Fe$_2$O$_3$ and 2Pd-MgO, thus indicating that Pd dispersion is better on these two supports, and hence, more Pd species are available. In the case of 2Pd-β, its higher specific surface area, and mainly its microporous character (Table 1), together with the lower Pd loading, could make the detection of Pd by XPS difficult, since Pd nanoparticles would be located in the inner of zeolite particles. It must be taken into account that this surface technique only allows us to analyze a depth of 2–3 nm.

3. Catalytic Results

After the physico-chemical characterization of Pd-based catalysts, these were tested in FUR oxidation using H$_2$O$_2$ as oxidant and ethanol, which simultaneously acts as solvent and reagent (Figure 7).

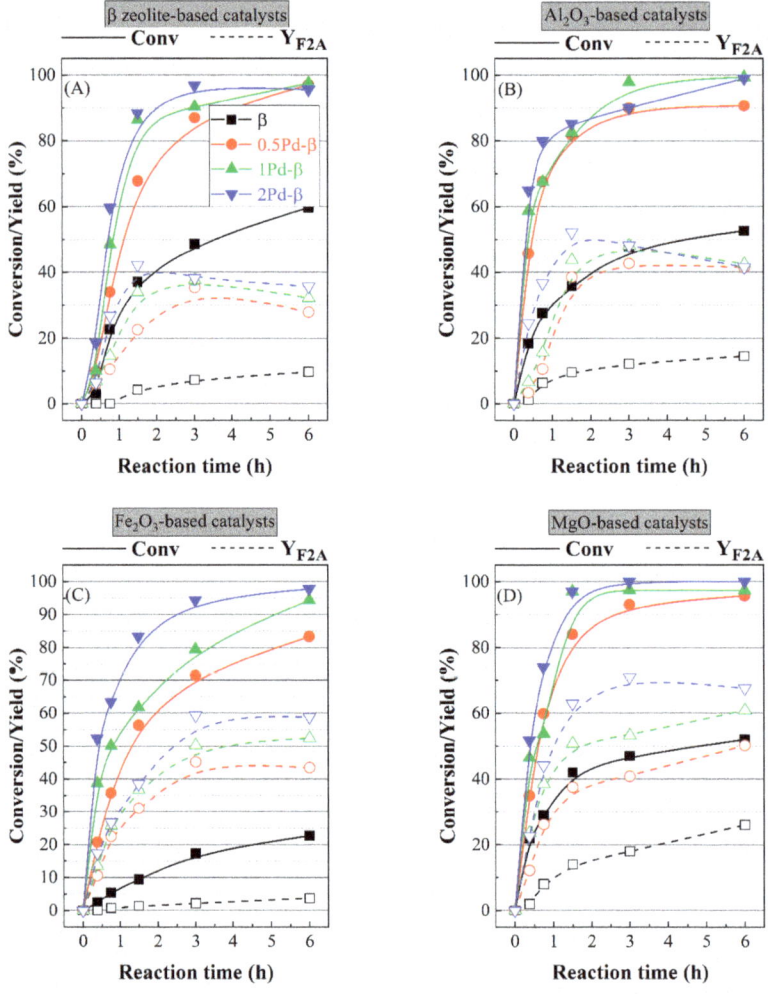

Figure 7. *Cont.*

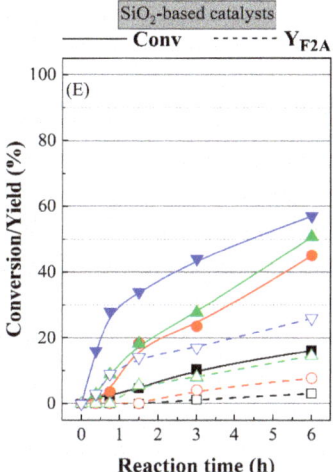

Figure 7. Furfural conversion and F2A yield in the oxidative condensation of FUR using Pd supported on β-zeolite (**A**), Al$_2$O$_3$ (**B**), Fe$_2$O$_3$ (**C**), MgO (**D**), and SiO$_2$ (**E**). Catalyst without Pd (—), catalyst with 0.5 wt.% of Pd (—), catalyst with 1 wt.% of Pd (—) and catalyst with 2 wt.% of Pd (—). Experimental conditions: temperature: 170 °C; catalyst loading: 0.05 g; V (ethanol): 6 mL; mass of Na$_2$CO$_3$: 0.10 g; V (H$_2$O$_2$, 30 wt.%): 0.5 mL.

The catalytic study of Pd-based catalysts supported on β-zeolite reveals that FUR conversion rises with the Pd loading, reaching almost full conversion after 3 h of reaction with the highest loading (2 wt%). It is striking that the F2A yield is much lower than FUR conversion in all cases. Thus, the catalytic results show that the yield towards F2A reaches a maximum value of 42% for 2Pd-β after 1.5 h of reaction. Moreover, zeolite-supported catalysts suffer a decrease in F2A yield at longer reaction times. This fact is attributed to FUR is highly reactive in an oxidizing medium, leading to the formation of soluble and insoluble polymers and humins, as a consequence of polymerization processes. It must be taken into account that FUR molecule presents a carbonyl group and unsaturations in α–β position that confer its great reactivity, and these functional groups are also present in F2A molecule. These data are in agreement with others reported in the literature for the oxidation of furfural in aqueous H$_2$O$_2$, catalyzed by titanium silicalite, to maleic acid, with carbon balances of about 50–60% for all catalysts [39,40]. In addition, these authors also pointed out that the poor carbon balances could be ascribed to the adsorption of FUR over the internal and external surface of zeolite, or the formation of undetectable products. In the present work, the decay of the F2A observed at longer reaction time (> 3 h) is accompanied by the formation of small proportion of 2,2′-difuryl methane (DFM), although its yield is below 8%, as was reported by Fang et al. using Au/NiO catalysts [41]. The formation of DFM agrees with previous findings about Pd-based catalysts as active in several organic synthesis, such as oxidation and C–C coupling [42,43]. Considering that the amount of DFM is approximately similar to the amount of F2A lost after 1.5 h of reaction, it can be expected that F2A is not very prone to generate carbonaceous residues, as easily as it takes place in the case of FUR. The high amount of non-detected products and the decay of the F2A yield with Pd-β could be ascribed to its textural properties, since both FUR and reaction products can be retained in the micropores of β-zeolite, and its higher acidity could also favor side reactions.

In the case of Pd-Al$_2$O$_3$, a similar trend to that of Pd-β was observed, reaching almost full FUR conversion after 6 h. In spite of displaying a lower PdO dispersion than that observed in Pd-β, the samples supported on Al$_2$O$_3$ attain a higher conversion and F2A yield values at shorter reaction times. In the same way, the carbon balances also improve slightly in comparison to catalysts supported on zeolite. This fact could be ascribed to the textural properties of the support, since the catalysts supported on Al_2O_3 show higher meso- and macroporosity, which means easier access of the FUR molecules and subsequent desorption of reaction products. Similarly to Pd-β catalysts, the decrease in the F2A at longer reaction times is accompanied by the formation of DFM as a consequence of a coupling reaction between two molecules of F2A.

For Pd-Fe_2O_3, the increase in FUR conversion with the reaction time is more gradual than those observed with catalysts supported on β-zeolite and Al_2O_3. In addition, it is also noteworthy that the carbon balance improves, even more than with previous catalysts. With regard to the detected products, F2A is the only product, attaining a maximum yield of 59% at a longer reaction time, whereas the formation of DFM was not observed. The catalysts that give rise to the formation of DFM are those with the highest acid values (2Pd-Al_2O_3 and 2Pd-β, Table 2). This fact would demonstrate that the coupling reaction between two F2A molecules exhibits slower kinetic and requires the participation of acidic sites on the support. However, the catalysts supported on Fe_2O_3 reached high F2A yields. From these data, it can be inferred that the use of supports with high acidity are not suitable to obtain F2A, since these catalysts tend to form a higher proportion of carbonaceous deposits, as well as favor side reactions such as C–C coupling [41].

Concerning the catalytic performance of the Pd-MgO catalysts, the FUR conversion also increased very fast at shorter reaction times, similar to that observed for Pd-Al_2O_3, with almost full FUR conversion after only 1.5 h of reaction. It deserves to be noted that the carbon balance improved notably in comparison to the other catalysts, mainly those catalysts that displayed a higher proportion of acid centers (Pd-β and Pd-Al_2O_3). The 2Pd-MgO catalyst also allowed us to attain the highest F2A yield: 70% after 3 h of reaction. On the other hand, the absence of DFM with Pd-MgO catalysts, similar to Pd-Fe_2O_3, is also noticeable, confirming that C–C coupling reactions are favored using acidic supports. On the other hand, the presence of a higher amount of basic sites seems to show beneficial effects on the catalytic behavior, probably due to this catalyst providing additional basic sites (together with Na_2CO_3) in the reaction. If the textural properties are considered, the catalytic behavior of Pd-MgO is even more remarkable, since the specific surface area is lower than those shown for other catalysts, thus demonstrating the key role of basicity. However, the amount of Pd detected by XPS is the highest for the 2Pd-MgO catalyst.

Finally, Pd supported on SiO_2 catalysts were also evaluated. Similarly, FUR conversion increases with the Pd content, although the activity is the lowest of all series of catalysts. Again, the main product is F2A, although the yield is more limited due to the poorest activity, reaching a F2A yield of 25% for 2Pd-SiO_2 after 6 h of reaction at 170 °C. Therefore, the presence of both acid and basic sites should be playing an important role in the condensation reaction. In the case of basic supports, mainly MgO, the catalytic results are very interesting, since it can act as a heterogeneous basic catalyst that can have a synergistic effect with Na_2CO_3.

Several possible routes have been proposed in the oxidative condensation of FUR [20–22]. According to the obtained products, the most reliable route involved is, firstly, the oxidation of ethanol to acetaldehyde by H_2O_2, and the subsequent base subtraction of an H from C_α position of the acetaldehyde, occurring an aldol condensation reaction, giving rise to F2A as product. The absence of furoic acid among the reaction products discarded the direct oxidation of FUR, while the absence of furfuryl alcohol (FOL) ruled out a catalytic transfer hydrogenation (CTH) process, or the Cannizzaro reaction. In fact, when FOL was fed instead of FUR, no F2A was detected, although its concentration slightly decreased, probably due to polymerization of FOL. Then, F2A is undergone to coupling reaction to form DFM, although longer reaction times and an acid support are required.

In order to evaluate the efficiency of the Pd-based catalyst, the next study was focused on the analysis of the amount of F2A obtained per gram of catalyst and active site (productivity). This study was carried out assuming that the oxidative condensation and oxidative esterification are considered as pseudo-first-order kinetics [44,45]. In addition, short reaction times, which implies low FUR conversion and the formation of F2A as only product, were chosen so that the logarithmic scale fit was linear

(Figure 8). From the obtained data, it can be inferred that 2Pd-MgO is the catalyst that produces the highest amount of F2A per gram and per active phase. Next, catalysts with an acid character such as 2Pd-Al$_2$O$_3$ and 2Pd-Fe$_2$O$_3$ also present active sites in the oxidative condensation of FUR to F2A, although their activity seems to be a bit far from that obtained by the 2Pd-MgO catalyst. With regard to the 2Pd-β, the activity of the active sites of this acidic catalyst is even lower, probably due to this support being very microporous, as previously noted. The poorest behavior was observed for 2Pd-SiO$_2$, since F2A productivity can be considered as negligible in comparison to other catalysts.

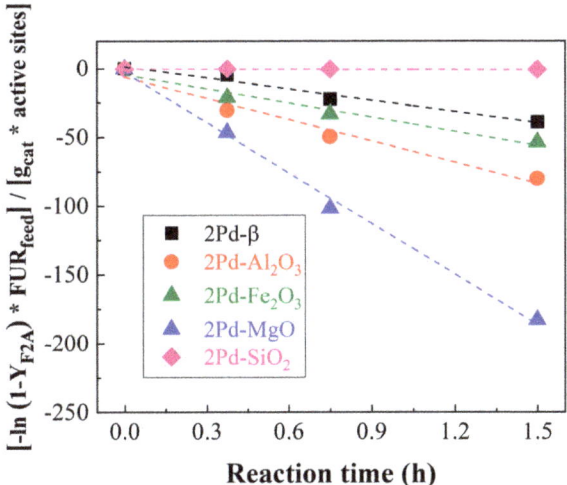

Figure 8. Kinetic study in the formation of F2A with Pd-based catalysts. Experimental conditions: temperature: 170 °C, catalyst loading: 0.05 g, V$_{(ethanol)}$: 6 mL, amount of base (Na$_2$CO$_3$): 0.10 g, V (H$_2$O$_2$, 30 wt%): 0.5 mL.

From the analysis of the slope of F2A productivity, it is possible to establish a kinetic constant of F2A formation, denoted as k$_{F2A}$ (Figure 9). These data reveal the highest k$_{F2A}$ (122 h^{-1}) for 2Pd-MgO, while, in the opposite case, 2Pd-SiO$_2$ displays k$_{F2A}$ close to zero.

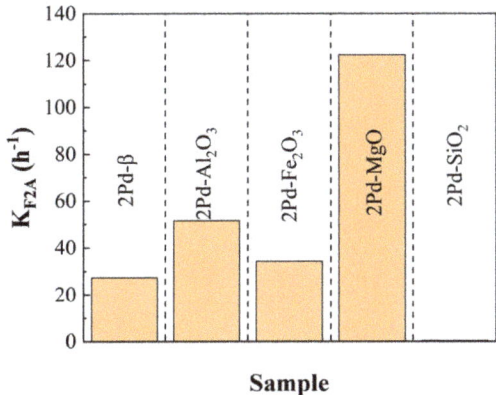

Figure 9. Kinetic constant of F2A formation (K$_{F2A}$) for Pd-based catalysts. Experimental conditions: temperature: 170 °C, time: 3 h, catalyst loading: 0.05 g, V$_{(ethanol)}$: 6 mL, amount of base (Na$_2$CO$_3$): 0.10 g, V (H$_2$O$_2$, 30 wt%): 0.5 mL.

As Pd-based catalysts are quite expensive, the goal must be the development of catalytic systems that have low Pd content and, in turn, are reusable to extend their life cycles to obtain economically sustainable and competitive catalysts. Thus, the following study was focused on the catalyst evaluation during several reaction cycles. Between each cycle, the catalyst was washed with water and dried at 80 °C to be able to eliminate the excess Na_2CO_3 existing in the reaction medium. The catalytic results, compiled in Figure 10, show how the FUR conversion decreases after each run in all cases. This decay is more pronounced in the case of the 2Pd-β catalyst, where the conversion decreased from 97 to 47% after only three cycles. This decrease also takes place in other catalyst such as 2Pd-Al_2O_3 and 2Pd-Fe_2O_3 although it was more gradual, reaching conversions of 64 and 75%, respectively, after three cycles. The less prone catalyst to suffer deactivation was 2Pd-MgO, since the FUR conversion only diminishes from 100 to 81% after three cycles. In the case of 2Pd-SiO_2, in spite of this catalyst displays lower conversion values, FUR conversion follows the same pattern to that observed for the other catalysts, diminishing from 44 to 28%. With regard to the obtained products, it can be observed a high amount of non-detected products in all cases, although it seems that the proportion of undetected products, as well as the deactivation degree, are related to the amount of acid sites. Thus, the 2Pd-β catalyst is the most prone to suffer deactivation as a consequence of the high proportion of non-detected products, which must adsorb on the catalyst surface, in such a way that the active sites are partially blocked, diminishing the amount of available Pd sites. The decrease in catalytic activity could also be ascribed to the leaching of Pd species. However, the amount of Pd in the reaction medium was quantified by ICP-MS for the catalyst with stronger deactivation (2Pd-β), being negligible, with only 0.0009% of leached Pd species.

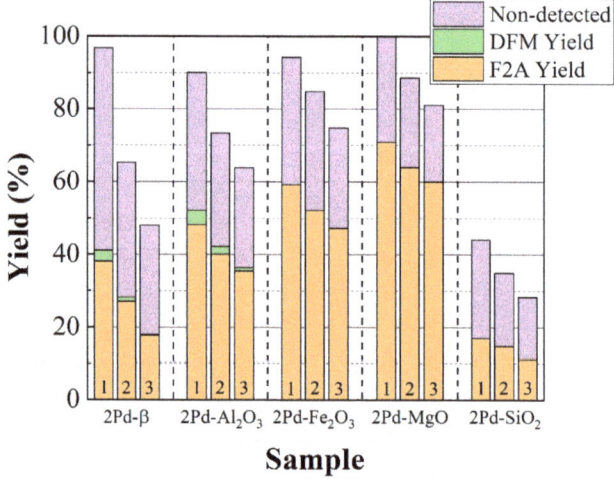

Figure 10. Reuse for Pd-based catalysts supported on β-zeolite, Al_2O_3, Fe_2O_3, MgO, and SiO_2. Experimental conditions: temperature: 170 °C, time: 3 h, catalyst loading: 0.05 g, $V_{(ethanol)}$: 6 mL, amount of base (Na_2CO_3): 0.10 g, V (H_2O_2, 30 wt%): 0.5 mL.

The analysis of reaction products shows that F2A is the only detected product in the case of 2Pd-MgO, 2Pd-SiO_2, and 2Pd-Fe_2O_3, while it was the main product for 2Pd-β and 2Pd-Al_2O_3. F2A yield follows the same trend to that observed for FUR conversion in all catalysts. F2A yield in the most active catalyst (2Pd-MgO) decreases from 71 and 60% (170 °C and 3 h) after three cycles. In the case of the catalysts with higher acidity, a more pronounced decay of the F2A with the reusing is observed. In addition, these catalysts with higher acidity also show the presence of small proportions of DFM, coming from C–C coupling reaction of F2A, although this product tends to disappear after several runs.

The effect of the Na$_2$CO$_3$ addition was also evaluated (Figure 11), and an important decrease in the yield of target product (F2A) was observed. Moreover, the trend concerning the influence of the acid-base properties of the support was also confirmed. The diminishing of non-detected products with a lower amount of added base is also noteworthy. Thus, for example, with 0.02 g of Na$_2$CO$_3$, the use of an acid support, such as β-zeolite, resulted in a yield of F2A similar to that achieved with MgO, but with a higher proportion of non-detected products. In the absence of catalyst, neither F2A nor DFM were produced, although furfural polymerization could take place, as would be inferred from a furfural conversion of 45%.

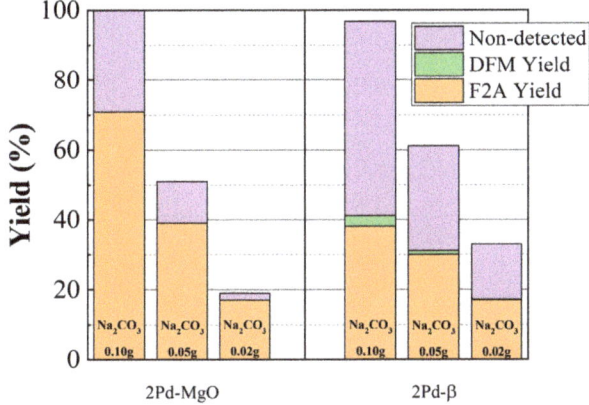

Figure 11. Influence of the Na$_2$CO$_3$ addition on Pd-based catalysts supported on MgO and β-zeolite. Experimental conditions: temperature: 170 °C, time: 3 h, catalyst loading: 0.05 g, V$_{(ethanol)}$: 6 mL, V (H$_2$O$_2$, 30 wt%): 0.5 mL.

In order to study the evolution of the active phase during the oxidative condensation of FUR, catalysts were collected after 3 h of reaction at 170 °C and analyzed by XPS (Table 4, Figures 12 and 13). In all cases, an increase in surface C content is observed. In fact, a correlation between the surface carbon deposition and deactivation after several reaction cycles seems to exist (Figure 12). Moreover, the C 1s region is composed by a main band located at 284.8 eV, which is assigned to C–C and C–H bonds, and other much less intense contributions are located at 286.2 and 287.8 eV, due to –C–O–C– and –C=O, respectively [46,47]. This could point out the deposition of FUR, as well as some products obtained from the FUR polymerization due to the presence of an oxidizing reaction medium. With regards to the other core level spectra (Si 2p, Al 2p, Mg 2p, and Fe 2p), these remained unchanged after the reaction, confirming that the supports are stable under these experimental conditions. However, the atomic concentrations are lower than those observed for the catalysts before the reaction, as a consequence of the formation of carbonaceous deposits on the catalyst surface.

Table 4. Atomic concentration data, obtained from XPS, for Pd-based catalysts (Pd content: 2 wt.%) after FUR oxidative condensation reaction. Experimental conditions: temperature: 170 °C, time: 3 h, catalyst loading: 0.05 g, V$_{(ethanol)}$: 6 mL, amount of base (Na$_2$CO$_3$): 0.10 g, V (H$_2$O$_2$, 30 wt%): 0.5 mL.

Sample	Atomic Concentration, %						
	C 1s	O 1s	Si 2p	Al 2p	Fe 2p	Mg 2p	Pd 3d
2Pd-β-u	67.31	22.58	10.13	0.79	-	-	0.03
2Pd-Al$_2$O$_3$-u	62.70	27.00	-	10.08	-	-	0.06
2Pd-Fe$_2$O$_3$-u	48.68	43.22	-	-	7.88		0.20
2Pd-MgO-u	44.23	42.20	-	-	-	13.62	0.17
2Pd-SiO$_2$-u	29.51	48.46	22.04	-	-	-	0.19

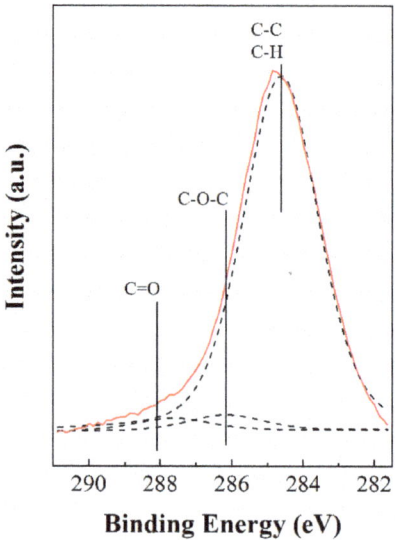

Figure 12. C 1s core level spectra for 2Pd-β catalyst, after FUR oxidative condensation reaction. Experimental conditions: temperature: 170 °C, time: 3 h, catalyst loading: 0.05 g, $V_{(ethanol)}$: 6 mL, amount of base (Na_2CO_3): 0.10 g, V (H_2O_2, 30 wt%): 0.5 mL.

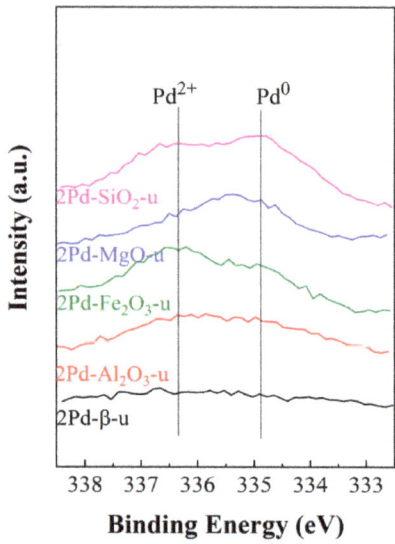

Figure 13. Pd 3d core level spectra for Pd-based catalysts, after FUR oxidative condensation reaction. Experimental conditions: temperature: 170 °C, time: 3 h, catalyst loading: 0.05 g, $V_{(ethanol)}$: 6 mL, amount of base (Na_2CO_3): 0.10 g, V (H_2O_2, 30 wt%): 0.5 mL.

The Pd 3d core level spectra (Figure 13) suffer modifications in comparison to those obtained before the reaction, since the presence of Pd^0 and Pd^{2+} can be observed. In this sense, it is difficult to elucidate the phenomena involved in this process, since, on the one hand, the reaction medium is oxidant. This fact should allow the Pd^{2+} to remain oxidized. On the other hand, the solvent/reagent used in the reaction is ethanol, and it is well-known that alcohols are used as reducing agents to reduce transition metals ($M^{x+} \rightarrow M^0$). In addition, the X-ray irradiation of the XPS can also cause the photo-reduction in the Pd^{2+} species, although, in any case, this last suggestion must be ruled out because no photo-reduction phenomena were observed in the catalysts before the oxidative condensation reaction. Therefore, the only explanation for the presence of Pd(0) would be the reduction due to the alcohol. Moreover, the decrease in surface Pd content can be explained by the deposition of carbonaceous species on the catalyst surface.

The carbon content of used Pd-based catalysts was also analyzed by thermogravimetric (TG) analysis (Figure 14). In all cases, a broad band between 250 °C and 450 °C can be observed, which could be ascribed to the overlapping of the desorption of chemisorbed FUR [48] and the combustion of the carbonaceous deposits [46]. Thermogravimetric (TG) measurements of the used catalysts follow the same trend to that observed by XPS (Table 4), where the catalysts with higher proportion of mass loss also are those with higher proportion of C content on their surface. In the same way, the mass loss is also directly related to the amount acid sites, as a consequence of the stronger interaction with the FUR molecules, as well as the formation of carbonaceous species.

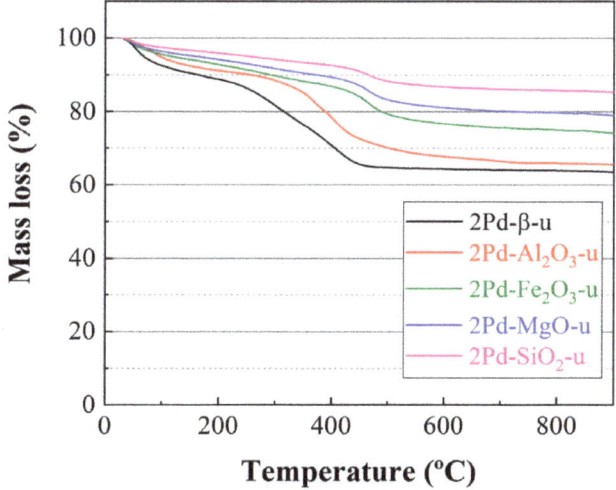

Figure 14. Thermogravimetric analysis of Pd-based catalysts, after FUR oxidative condensation reaction. Experimental conditions: temperature: 170 °C, time: 3 h, catalyst loading: 0.05 g, $V_{(ethanol)}$: 6 mL, amount of base (Na_2CO_3): 0.10 g, V (H_2O_2, 30 wt%): 0.5 mL.

4. Materials and Methods

4.1. Synthesis of the Catalysts

Pd-based catalysts were synthesized from $PdCl_2$ (Pressure Chemical CO., Smallman St., Pittsburgh, PA, USA), which was dissolved in HCl (VWR, Fontenay sous Bois, France) to obtain a solution of H_2PdCl_4 0.01 M. Then, this acid was used to incorporate Pd species on several supports, such as SiO_2, Al_2O_3, MgO, Fe_2O_3, and β-H-zeolite with a SiO_2/Al_2O_3 molar ratio of 25, by incipient wetness impregnation, incorporating a Pd content ranging between 0.5 and 2 wt.% for all supports. SiO_2 (Sigma-Aldrich, Saint Louis, MO, USA), Al_2O_3 (VWR, West Chester, PA, USA), and MgO (Sigma-Aldrich, Saint Louis, MO, USA) were employed without previous treatment; however, β-H-zeolite was synthesized from its NH_4 form (β-NH_4^+-zeolite) by calcination at 400 °C with a rate of 2 °C min^{-1}. In the case of Fe_2O_3, this support was synthesized from a solution of $FeCl_3$ (0.2M) (Sigma-Aldrich, Saint Louis, MO, USA), where Fe^{3+} species have been precipitated by increasing the pH with a 0.5 M NaOH aqueous solution. Then, the obtained solid was filtered, dried overnight at 80 °C, and calcined at 400 °C with a rate of 2 °C min^{-1} for 2 h.

After incipient wetness impregnation, the catalysts were dried at 80 °C overnight and calcined at 400 °C with a ramp of 2 °C min^{-1}, maintaining this temperature for 2 h. Catalysts were labeled as xPd support, where x is the weight percentage of Pd incorporated to each support.

4.2. Characterization of the Catalysts

The crystallinity of the Pd-based catalysts was determined by X-ray diffraction by using a PANalytical X'Pert PRO diffractometer (Bruker, Rheinstetten, Germany), using the Cu Kα (1.5406 Å) radiation and a germanium monochromator, over a 2θ range with Bragg–Brentano geometry. Average crystallite size and lattice strain values were determined by using the Williamson–Hall equation: B·cosθ = (K·λ/D) + (2·ε·sinθ), where θ is the Bragg angle, B is the full width at half maximum (FWHM) of the XRD peak, K is the Scherrer constant, λ is the wavelength of the X-ray, and ε the lattice strain [33].

The morphology of the catalysts was evaluated by transmission electronic microscopy (TEM), using a Philips CCM 200 Supertwin DX4 high-resolution (Thermo Fisher Scientific, Waltham, MA, USA). Microanalysis was performed with an EDX Super-X system, with 4 X-ray detectors and an X-FEG beam.

The textural properties of the Pd-based catalysts were determined from their N_2 adsorption–desorption isotherms at −196 °C with an ASAP 2020 model of Micromeritics Inc. (Micrometrics, Norcross, GA, USA). Prior to N_2 adsorption, the sample was outgassed at 200 °C overnight. Specific surface areas were obtained by using the Brunauer–Emmett–Teller (BET) equation, with a cross-section of 16.2 Å^2 for the N_2 molecule [35]. Pore size distribution profiles were deduced from the density functional theory model (DFT) [37], and total pore volume from N_2 adsorbed at P/P_0 = 0.996.

The chemical composition on the catalyst surface, including the oxidation state of Pd species, was determined by X-ray photoelectron spectroscopy (XPS). A Physical Electronics PHI5700 spectrometer (Physical Electronics, Eden Prairie, MN, USA), with non-monochromatic Mg Kα radiation (300 W, 15 kV and 1253.6 eV) with a multichannel detector, was employed. Spectra were recorded in the constant pass energy mode, at 39.35 eV using a 720 μm diameter analysis area. Charge referencing was measured against adventitious carbon (C 1 s at 284.8 eV). Acquisition and data analysis were performed with a PHI ACCESS ESCA-V6.0F software package (Eden Prairie, Minnesota, MN, USA), subtracting a Shirley-type background from signals. Gaussian–Lorentzian curves were used for fitting recorded spectra to more accurately determine the binding energies of the different element core levels.

As Pd-based catalysts have been synthesized using a wide variety of supports, the total amount of acid and basic sites was also determined. Thus, thermo-programmed desorption of ammonia (NH_3-TPD) was carried out to determinate the total amount of acid sites. For each experiment, 0.08 g of catalyst was placed in a quartz reactor. Prior to the analysis, the sample was cleaned using a He flow of 40 mL·min^{-1} from room temperature to 400 °C with a heating rate of 10 °C·min^{-1}. Then, the catalyst was cooled until 100 °C under the same He flow, and once the temperature is stabilized at 100 °C, the catalyst was saturated with NH_3 for 5 min and later physisorbed NH_3 was removed under He flow. NH_3 desorption was performed by heating the sample from 100 to 400 °C, with a rate of 10 °C·min^{-1}, registering the signal using a Shimadzu GC-14B instrument (Shimadzu, Kioto, Japan) equipped with a thermal conductivity detector (TCD). In the case of the amount of basic sites, the quantification was carried out by thermoprogrammed desorption of CO_2 (CO_2-TPD). In each analysis, 0.03 g of catalyst was pretreated under a He flow (40 mL min^{-1}) at 400 °C, maintaining this temperature for 15 min (10 °C·min^{-1}). Later, the sample was cooled to 100 °C and a pure CO_2 stream, with a flow of 60 mL min^{-1}, was introduced into the quartz reactor for 30 min. Finally, the amount of CO_2 evolved was analyzed using signal using a Shimadzu GC-14B instrument equipped with a thermal conductivity detector (TCD) between 100 and 600 °C with a rate of 10 °C·min^{-1} under He flow.

Pd leached along the reaction was determined by ICP-MS on Perkin Elmer spectrophotometer (NexION 300D, Waltham, MA, USA), after digestion of samples in an Anton Paar device (Multiwave 3000, Graz, Austria) by using HNO_3, HCl and HF.

The TG-DSC data were registered with a Mettler-Toledo (TGA/DSC-1) instrument (Columbus, OH, USA), equipped with a MX5 microbalance, by varying the temperature from room temperature to 900 °C, at a heating rate of 5 °C min^{-1}, under an air flow. Samples were put in open platinum crucibles.

4.3. Catalytic Reaction

The oxidation reaction was performed in glass pressure reactors with thread bushing (Ace, 15 mL) in a temperature-controlled aluminum block under magnetic stirring. In each experiment, 0.05 g of catalyst was mixed with 6 mL of ethanol as alcohol, 0.10 g of base (Na_2CO_3), 0.5 mL of H_2O_2 as oxidant, and 0.02 mL of o-xylene as internal standard. Once the reactions finished, the reactors were removed from the aluminum block and submerged in cool water to stop the catalytic process. Reaction products were microfiltered and analyzed by gas chromatography (Shimadzu GC model 14A, Shimadzu, Kioto, Japan), equipped with a flame ionization detector and a TBR-14 capillary column. FUR conversion, selectivity, and yield were calculated as follows (Equations (1)–(3)):

$$\text{Conversion (\%)} = \frac{\text{mol of FUR converted}}{\text{mol of FUR fed}} \times 100 \quad (1)$$

$$\text{Selectivity (\%)} = \frac{\text{mol of product}}{\text{mol of FUR converted}} \times 100 \quad (2)$$

$$\text{Yield (\%)} = \frac{\text{mol of product}}{\text{mol of FUR fed}} \times 100 \quad (3)$$

5. Conclusions

Pd species were supported on several supports with different textural properties, crystallinity, and amount of acid/basic sites and tested in the FUR oxidative condensation with ethanol.

The characterization of catalysts revealed that the Pd species are in the form of PdO. These particles are well dispersed on the catalyst surface, mainly when supported on β-zeolite and SiO_2. On the other hand, the analysis of the textural properties showed that Fe_2O_3 and MgO are the supports with poorer textural properties.

The analysis of the catalytic behavior of the Pd-based catalysts in the FUR oxidative condensation revealed that all catalysts were active, except 2Pd-SiO_2 which reached the lowest conversion values. In all cases, the main product is F2A, although the carbon balance also reported the presence of non-detected products, probably by the formation of carbonaceous deposits on the surface of the catalysts. These carbonaceous deposits were more evident in the case of catalysts with higher amount of acid sites, i.e., those supported on Al_2O_3 and β-zeolite. In the same way, these acid sites seem to favor C–C coupling reaction, although high reaction times are required to observe small proportions of DFM. In contrast, the presence of basic sites exerts a beneficial effect on the catalytic behavior, since the formation of carbonaceous deposits is minimized. In addition, these basic centers can collaborate with Na_2CO_3 in condensation reactions. Thus, the maximum F2A yield was reached with 2Pd-MgO (70%), the catalyst prepared with the most basic support, after 3 h of reaction at 170 °C.

Author Contributions: Conceptualization; J.A.C. and P.M.-T.; methodology: J.A.C., C.P.J.-G. and V.T.-B.; validation: J.A.C., C.G.-S., R.M.-T. and P.M.-T.; formal analysis, J.A.C., C.P.J.-G. and V.T.-B.; investigation, J.A.C., C.P.J.G. and V.T.B.; data curation: J.A.C., C.G.-S., R.M.-T. and P.M.-T.; writing—original draft preparation: J.A.C; writing—review and editing: C.G.-S., R.M.-T. and P.M.-T.; visualization: P.M.-T.; supervision: J.A.C. and P.M.-T.; project administration: P.M.-T.; funding acquisition: P.M.-T. and C.G.-S. All authors have read and agreed to the published version of the manuscript.

Funding: The authors are grateful to financial support from the Spanish Ministry of Innovation, Science and Universities (Project RTI2018-094918-B-C44) and FEDER (European Union) funds, and University of Málaga (UMA18-FEDERJA-171).

Acknowledgments: J.A.C. and C.P.J.-G. thank University of Malaga for contracts of PhD incorporation. C.G.-S. acknowledges FEDER funds for a postdoctoral contract (UMA18-FEDERJA-171).

Conflicts of Interest: The authors declare that they have no known competing financial interests or personal relationships that could have appeared to influence the work reported in this paper.

References

1. Corma, A.; Iborra, S.; Velty, A. Chemical routes for the transformation of biomass into chemicals. *Chem. Rev.* **2007**, *107*, 2411–2502. [CrossRef] [PubMed]
2. Binder, J.B.; Raines, R.T. Simple chemical transformation of lignocellulosic biomass into furans for fuels and chemicals. *J. Am. Chem. Soc.* **2009**, *131*, 1979–1985. [CrossRef]
3. Zhou, C.H.; Xia, X.; Li, C.C.; Tong, D.S.; Beltramini, J. Catalytic conversion of lignocellulosic biomass to fine chemicals and fuels. *Chem. Soc. Rev.* **2011**, *40*, 5588–5617. [CrossRef]
4. Kumar, A.K.; Sharma, S. Recent updates on different methods of pretreatment of lignocellulosic feedstocks: A review. *Bioresour. Bioprocess.* **2017**, *4*, 7. [CrossRef]
5. Hendricks, A.T.W.M.; Zeeman, G. Pretreatments to enhance the digestibility of lignocellulosic biomass. *Bioresour. Technol.* **2009**, *100*, 10–18. [CrossRef]
6. Doner, L.W.; Hicks, K.B. Isolation of hemicellulose from corn fiber by alkaline hydrogen peroxide extraction. *Cereal Chem.* **1997**, *74*, 176–181. [CrossRef]
7. Sun, R.C.; Tomkinson, J.; Wang, Y.X.; Xiao, B. Physico-chemical and structural characterization of hemicelluloses from wheat straw by alkaline peroxide extraction. *Polymer* **2000**, *41*, 2647–2656. [CrossRef]
8. Mariscal, R.; Maireles-Torres, P.; Ojeda, M.; Sádaba, I.; López Granados, M. Furfural: A renewable and versatile platform molecule for the synthesis of chemicals and fuels. *Energy Environ. Sci.* **2016**, *9*, 1144–1189. [CrossRef]
9. Lange, J.P.; van der Heide, E.; van Buijtenen, J.; Price, R. Furfural—A promising platform for lignocellulosic biofuels. *ChemSusChem* **2012**, *5*, 150–166. [CrossRef]
10. Li, X.; Jia, P.; Wang, T. Furfural: A promising platform compound for sustainable production of C4 and C5 chemicals. *ACS Catal.* **2016**, *6*, 7621–7640. [CrossRef]
11. Hurd, C.D.; Garrett, J.W.; Osborne, E.N. Furan reactions. IV. Furoic acid from furfural. *J. Am. Chem. Soc.* **1933**, *55*, 1082–1084. [CrossRef]
12. Arias, P.L.; Cecilia, J.A.; Gandarias, I.; Iglesias, J.; López Granados, M.; Mariscal, R.; Morales, G.; Moreno-Tost, R.; Maireles-Torres, P. Oxidation of lignocellulosic platform molecules to value-added chemicals using heterogeneous catalytic technologies. *Catal. Sci. Technol.* **2020**, *10*, 2721–2757. [CrossRef]
13. Dalcanale, E.; Montanari, F. Selective oxidation of aldehydes to carboxylic acids with sodium chlorite-hydrogen peroxide. *J. Org. Chem.* **1986**, *51*, 567–569. [CrossRef]
14. Sekar, K.G. Oxidation of furfural by imidazolium dichromate in acid medium. *Int. J. Chem. Sci.* **2003**, *1*, 227–232.
15. Douthwaite, M.; Huang, X.; Iqbal, S.; Miedziak, P.J.; Brett, G.L.; Kondrat, S.; Edwards, J.K.; Sankar, M.; Knight, D.W.; Bethell, D.; et al. The controlled catalytic oxidation of furfural to furoic acid using AuPd/Mg(OH)$_2$. *Catal. Sci. Technol.* **2017**, *7*, 5284–5293. [CrossRef]
16. Signoretto, M.; Menegazzo, F.; Contessotto, L.; Pinna, M.; Manzoli, F.; Boccuzzi, F. Au/ZrO$_2$: An efficient and reusable catalyst for the oxidative esterification of renewable furfural. *Appl. Catal. B* **2013**, *129*, 287–293. [CrossRef]
17. Menegazzo, F.; Signoretto, M.; Pinna, F.; Manzoli, M.; Aina, V.; Cerrato, G.; Boccuzzi, F. Oxidative esterification of renewable furfural on gold-based catalysts: Which is the best support? *J. Catal.* **2014**, *309*, 241–247. [CrossRef]
18. Ampelli, C.; Barbera, K.; Centi, G.; Genovese, C.; Papanikolaou, G.; Perathoner, S.; Schouten, K.J.; van der Waal, J.K. On the nature of the active sites in the selective oxidative esterification of furfural on Au/ZrO$_2$ catalysts. *Catal. Today* **2016**, *278*, 56–65. [CrossRef]
19. Menegazzo, F.; Signoretto, M.; Fantinel, T.; Manzoli, M. Sol-immobilized vs deposited-precipitated Au nanoparticles supported on CeO$_2$ for furfural oxidative esterification. *J. Chem. Technol. Biotechnol.* **2017**, *92*, 2196–2205. [CrossRef]

20. Tong, X.; Liu, Z.; Yu, L.; Li, Y. A Tunable Process: Catalytic transformation of renewable furfural with aliphatic alcohols in the presence of molecular oxygen. *Chem. Commun.* **2015**, *51*, 3674–3677. [CrossRef]
21. Tong, X.; Liu, Z.; Hu, J.; Liao, S. Au-catalyzed oxidative condensation of renewable furfural and ethanol to produce furan-2-acrolein in the presence of molecular oxygen. *Appl. Catal. A* **2016**, *510*, 196–203. [CrossRef]
22. Ning, L.; Liao, S.; Liu, X.; Guo, P.; Zhang, Z.; Zhang, H.; Tong, X. A regulatable oxidative valorization of furfural with aliphatic alcohols catalyzed by functionalized metal-organic frameworks-supported Au nanoparticles. *J. Catal.* **2018**, *364*, 1–13. [CrossRef]
23. Yu, L.; Liao, S.; Ning, L.; Xue, S.; Liu, Z.; Tong, X. Sustainable and cost-effective protocol for cascade oxidative condensation of furfural with aliphatic alcohols. *ACS Sustain. Chem. Eng.* **2016**, *4*, 1894–1898. [CrossRef]
24. Liu, Z.; Tong, X.; Liu, J.; Xue, S.; Sci, C.; Liu, Z.; Tong, X.; Liu, J.; Xue, S. A smart catalyst system for the valorization of renewable furfural in aliphatic alcohols. *Catal. Sci. Technol.* **2016**, *6*, 1214–1221. [CrossRef]
25. Ning, L.; Liao, S.; Cui, H.; Tong, X. Selective conversion of renewable furfural with ethanol to produce furan-2-acrolein mediated by Pt@MOF-5. *ACS Sustain. Chem. Eng.* **2018**, *6*, 135–142. [CrossRef]
26. Tong, X.; Zhang, Z.; Gao, Y.; Zhang, Y.; Yu, L.; Li, Y. Selective carbon-chain increasing of renewable furfural utilizing oxidative condensation reaction catalyzed by mono-dispersed palladium oxide. *Mol. Catal.* **2019**, *477*, 110545. [CrossRef]
27. Cui, H.; Tong, X.; Yu, L.; Zhang, M.; Yan, Y.; Zhuang, X. A catalytic oxidative valorization of biomass-derived furfural with ethanol by copper/azodicarboxylate system. *Catal. Today* **2019**, *319*, 100–104. [CrossRef]
28. Zhang, Z.; Tong, X.; Zhang, H.; Li, Y. Versatile catalysis of Iron: Tunable and selective transformation of biomass-derived furfural in aliphatic alcohol. *Green Chem.* **2018**, *20*, 3092–3100. [CrossRef]
29. Tong, X.; Yu, L.; Luo, X.; Zhuang, X.; Liao, S.; Xue, S. Efficient and selective transformation of biomass-derived furfural with aliphatic alcohol catalyzed by binary Cu-Ce oxide. *Catal. Today* **2017**, *298*, 175–186. [CrossRef]
30. West, R.M.; Liu, Z.Y.; Peter, M.; Dumesic, J.A. Liquid alkanes with targeted molecular weights from biomass-derived carbohydrates. *ChemSusChem* **2008**, *1*, 417–424. [CrossRef]
31. Dutta, S.; De, S.; Saha, B.; Alam, M.D.I. Advances in conversion of hemicellulosic biomass to furfural and upgrading to biofuels. *Catal. Sci. Technol.* **2012**, *2*, 2025–2036. [CrossRef]
32. Zeitsch, K.J. *The Chemistry and Technology of Furfural and Its Many By-Products*, 1st ed.; Elsevier Ldt.: Amsterdam, The Netherlands, 2000; Volume 13.
33. Williamson, G.K.; Hall, W.H. X-Ray line broadening from filed aluminium and wolfram. *Acta Metall.* **1953**, *1*, 22–31. [CrossRef]
34. Thommes, M.; Kaneko, K.; Neimark, A.V.; Oliver, J.P.; Rodríguez-Reinoso, F.; Rouquerol, J.; Sing, K.S.W. Physisorption of gases, with special reference to the evaluation of surface area and pore size distribution (IUPAC Technical Report). *Pure Appl. Chem.* **2015**, *87*, 1051–1069. [CrossRef]
35. Brunauer, S.; Emmett, P.H.; Teller, E. Adsorption of gases in multimolecular layers. *J. Am. Chem. Soc.* **1938**, *60*, 309–319. [CrossRef]
36. de Boer, J.H.; Lippens, B.C.; Linsen, B.G.; Broekhoff, J.C.P.; van den Heuvel, A.; Osinga, T.J. The t-curve of multimolecular N_2-adsorption. *J. Colloid Interface Sci.* **1966**, *21*, 405–414. [CrossRef]
37. Landers, J.; Gor, G.Y.; Neimark, A.V. Density functional theory methods for characterization of porous materials. *Colloids Surf. A Physicochem. Eng. Asp.* **2013**, *437*, 3–32. [CrossRef]
38. Wagner, C.D.; Moulder, J.F.; Davis, L.E.; Riggs, W.M. *Handbook of X-ray Photoelectron Spectroscopy*; Perkin Elmer Corporation: Eden Prairie, MN, USA, 1992.
39. Alba-Rubio, A.C.; Fierro, J.L.G.; León-Reina, L.; Mariscal, R.; Dumesic, J.A.; López-Granados, M. Oxidation of furfural in aqueous H_2O_2 catalysed by titanium silicalite: Deactivation processes and role of extraframework Ti oxides. *Appl. Catal. B Environ.* **2017**, *202*, 269–280. [CrossRef]
40. Alonso-Fagúndez, N.; Agirrezabal-Telleria, I.; Arias, P.L.; Fierro, J.L.G.; Mariscal, R.; López-Granados, M. Aqueous-phase catalytic oxidation of furfural with H_2O_2: High yield of maleic acid by using titanium silicalite-1. *RSC Adv.* **2014**, *4*, 54960–54972. [CrossRef]
41. Fang, Q.; Qin, Z.X.; Shi, Y.; Liu, F.; Barkaoui, S.; Abroshan, H.; Li, G. Au/NiO composite: A catalyst for one-pot cascade conversion of furfural. *ACS Appl. Energy Mater.* **2019**, *2*, 2654–2661. [CrossRef]
42. Liu, J.; Peng, X.; Sun, W.; Zhao, Y.; Xia, C. Magnetically separable Pd catalyst for carbonylative Sonogashira coupling reactions for the synthesis of α,β-alkynyl ketones. *Org. Lett.* **2008**, *10*, 3933–3936. [CrossRef]

43. Verma, S.; Verma, D.; Sinha, A.K.; Jain, S.L. Palladium complex immobilized on graphene oxide-magnetic nanoparticle composites for ester synthesis by aerobic oxidative esterification of alcohols. *Appl. Catal. A Gen.* **2015**, *489*, 17–23. [CrossRef]
44. Manzoli, M.; Menegazzo, F.; Signoretto, M.; Marchese, D. Biomass derived chemicals: Furfural oxidative esterification to methyl-2-furoate over gold catalysts. *Catalysts* **2016**, *6*, 107. [CrossRef]
45. Desai, D.S.; Yadav, G.D. Green synthesis of furfural acetone by solvent-free aldol condensation of furfural with acetone over La_2O_3–MgO mixed oxide catalyst. *Ind. Eng. Chem. Res.* **2019**, *58*, 16096–16105. [CrossRef]
46. Jiménez-Gómez, C.P.; Cecilia, J.A.; Moreno-Tost, R.; Maireles-Torres, P. Selective production of 2-Methylfuran by gas-phase hydrogenation of furfural on copper incorporated by complexation in mesoporous silica catalysts. *ChemSusChem* **2017**, *10*, 1448–1459. [CrossRef]
47. Liu, D.; Zemlyanov, D.; Wu, T.; Lobo-Lapidus, R.J.; Dumesic, J.A.; Miller, J.T.; Marshall, C.L. Deactivation mechanistic studies of copper chromite catalyst for selective hydrogenation of 2-furfuraldehyde. *J. Catal.* **2013**, *299*, 336–345. [CrossRef]
48. Dimas-Rivera, G.L.; de la Rosa, J.R.; Lucio-Ortiz, C.J.; de los Reyes-Heredia, J.A.; González-González, V.; Hernández, T. Desorption of furfural from bimetallic Pt-Fe oxides/alumina catalysts. *Materials* **2014**, *7*, 527–541. [CrossRef]

Publisher's Note: MDPI stays neutral with regard to jurisdictional claims in published maps and institutional affiliations.

© 2020 by the authors. Licensee MDPI, Basel, Switzerland. This article is an open access article distributed under the terms and conditions of the Creative Commons Attribution (CC BY) license (http://creativecommons.org/licenses/by/4.0/).

Article

In Situ Synthesis of Sn-Beta Zeolite Nanocrystals for Glucose to Hydroxymethylfurfural (HMF)

Kachaporn Saenluang [1], Anawat Thivasasith [1,*], Pannida Dugkhuntod [1], Peerapol Pornsetmetakul [1], Saros Salakhum [1], Supawadee Namuangruk [2] and Chularat Wattanakit [1]

[1] Department of Chemical and Biomolecular Engineering, Vidyasirimedhi Institute of Science and Technology, School of Energy Science and Engineering, Rayong 21210, Thailand; kachaporn.s_s16@vistec.ac.th (K.S.); s15_pannida.d@vistec.ac.th (P.D.); Peerapol.p_s18@vistec.ac.th (P.P.); s15_saros.s@vistec.ac.th (S.S.); chularat.w@vistec.ac.th (C.W.)

[2] National Nanotechnology Center (NANOTEC), National Science and Technology Development Agency, Pathum Thai 12120, Thailand; supawadee@nanotec.or.th

* Correspondence: anawat.t@vistec.ac.th; Tel.: +66-3-301-4262

Received: 30 September 2020; Accepted: 12 October 2020; Published: 28 October 2020

Abstract: The Sn substituted Beta nanocrystals have been successfully synthesized by in-situ hydrothermal process with the aid of cyclic diquaternary ammonium (CDM) as the structure-directing agent (SDA). This catalyst exhibits a bifunctional catalytic capability for the conversion of glucose to hydroxymethylfurfural (HMF). The incorporated Sn acting as Lewis acid sites can catalyze the isomerization of glucose to fructose. Subsequently, the Brønsted acid function can convert fructose to HMF via dehydration. The effects of Sn amount, zeolite type, reaction time, reaction temperature, and solvent on the catalytic performances of glucose to HMF, were also investigated in the detail. Interestingly, the conversion of glucose and the HMF yield over 0.4 wt% Sn-Beta zeolite nanocrystals using dioxane/water as a solvent at 120 °C for 24 h are 98.4% and 42.0%, respectively. This example illustrates the benefit of the in-situ synthesized Sn-Beta zeolite nanocrystals in the potential application in the field of biomass conversion.

Keywords: in-situ synthesis; Sn-Beta zeolite; isomorphous substitution; glucose; HMF

1. Introduction

In recent years, the conversion of biomass to high value-added chemicals, for example, lactic acid, formic acid, levulinic acid, and 5-hydroxymethylfurfural (5-HMF), has been extensively studied [1,2]. In particular, the HMF product has been widely used in the synthesis of many useful compounds, novel polymer materials, plastic resins, and diesel fuel additives [3]. Typically, one of the most promising alternative feedstocks of biomass to produce HMF is glucose, because it is the most abundant monosaccharide and the cheapest hexose, making it as a promising candidate to produce fructose, and subsequently HMF [4]. In a typical procedure, the conversion of glucose to HMF requires the following two main steps: (i) isomerization of glucose to fructose; (ii) the dehydration of fructose to HMF, in which bifunctional catalysts composed of Lewis acid sites and Brønsted acid function are responsible for these two steps, respectively [5,6].

Indeed, there are many types of catalysts that have been used for the conversion of glucose to HMF, such as metal chloride salts ($MgCl_2$, $SnCl_4$) [7], metal oxides (TiO_2, ZrO_2, Nb_2O_5) [8,9], metal organic frameworks (ZIF-8) [10], zeolites [11–16], and Sn, Ti and Zr-containing zeolites [17,18]. Among them, the metal incorporated in a zeolite is one of the most important candidates for this reaction because of high metal dispersion [19], high surface area [20], unique shape selective properties [21], suitable acid properties [22] and high thermal/hydrothermal stability [23], for example.

Various catalysts, such as Sn, Ti and Zr-containing zeolites and especially Sn-incorporated in Beta zeolite framework (Sn-Beta), have been extensively used in the conversion of glucose to fructose [24]. The Sn-Beta zeolite exhibits unique Lewis acidity [17], which is used for the isomerization of glucose to fructose [25]. In addition, by combining with Brønsted acid sites, the dehydration of fructose to HMF has been further proceeded, and therefore, the bifunctional catalysts containing Sn as Lewis acid sites together with Brønsted acid zeolites play an important role in these reaction pathways [26]. Typically, the Sn incorporated Beta zeolite can be produced by following two major strategies: (i) bottom-up approaches, in which the Sn sources are directly added to the zeolite precursor by an in-situ or hydrothermal synthesis process, resulting in Sn being tetrahedrally built into the silica framework of the BEA topology; (ii) top-down approaches, in which the Sn species is deposited on the zeolite surfaces by a post synthesis method via either a wet-impregnation method or an ion-exchange method [27]. However, it seems that the Sn-Beta zeolite obtained via an in-situ or hydrothermal synthesis method has played a very important role because it is a simple method, well-controlled metal dispersion, and is convenient with respect to other methods.

Over the past decade, the Sn-Beta zeolite has been utilized in many catalytic reactions, such as Baeyer-Villiger oxidations [28] and Meerwein-Ponndorf-Verley oxido-reductions [29]. Recently, the Sn-Beta zeolite has been successfully applied in several carbohydrate-related reactions like glucose isomerization to fructose, and subsequently fructose dehydration to HMF [30]. The HMF formation from glucose requires a catalyst containing both Lewis acid function and Brønsted acids site. For example, a physical mixture of Sn Lewis acids and Brønsted acid catalysts can be used as a catalyst [31,32]. It was found that high amount of glucose can be converted to a high yield of HMF in a one-pot reaction system. In addition, Qiang Guo et al. [23] reported that the Amberlyst-131 can act as a Brønsted acid catalyst, and Sn incorporated in Beta zeolite can act as Lewis acid sites. The reaction was carried out in the biphasic phase of dioxane with 5% water content when using fructose or glucose as reactants. It was found that the yield of 5-HMF is up to 74% and 56% from fructose and glucose as reactants, respectively. Moreover, Mark E. Davis et al. [33] also reported that the Sn-Beta zeolite can be used to produce HMF from glucose by a one-pot biphasic water/tetrahydrofuran (THF) reaction system, eventually resulting in achieving the glucose conversions of 79% and HMF selectivity of 72% at 180 °C of the reaction temperature.

As mentioned above, the addition of an organic solvent needs to be considered in the proper selection of HMF extracting solvent (e.g., MIBK, dioxane). Because an organic solvent is an extracting phase in which HMF preferably dissolves with respect to the aqueous solution, the selectivity towards HMF is improved significantly by controlling the production of humins and other by-products during the reaction [34,35]. Simona M et al. [36] also reported that HMF selectivity of 84.3% can be achieved by using Nb-Beta in a biphasic (H_2O/tetrahydrofuran (THF)) system. However, without THF as a solvent, the HMF selectivity only reaches 41.7% under an aqueous condition.

In the present study, we reported the facile preparation of Sn incorporated in Brønsted acid Beta nanocrystals obtained via an in-situ hydrothermal synthesis process with the aid of cyclic diquaternary ammonium (CDM) as the structure-directing agent (SDA). By combining the incorporated Sn as Lewis acid sites together with the hierarchical structure of Brønsted acid Beta, it significantly improves the catalytic performance in terms of conversion and selectivity under the aqueous and biphasic phase (dioxane/H_2O) for the one-pot synthesis of HMF from glucose. In addition, the effects of Sn amount, zeolite type, reaction time, reaction temperature, and solvent on the conversion catalytic performances of glucose to HMF were also investigated and further discussed in more detail.

2. Results and Discussion

2.1. Characterization of Synthesized Sn-Beta Nanocrystals

The Sn-incorporated Beta nanocrystals with 0.4 wt% of Sn (0.4 wt% Sn-Beta) have been successfully synthesized by the in-situ hydrothermal process following a modified literature procedure [37,38].

The XRD patterns of all the synthesized samples were used to check the crystalline structure (characteristic) of all catalysts, as shown in Figure 1. Compared with the commercial Beta zeolite (Beta-COM), which was supplied from the Zeolyst International company, the intensities of XRD peaks of the synthesized Sn-Beta nanocrystals are lower than those of the synthesized bare Beta and the commercial Beta zeolite (Beta-COM), indicating that the crystallinity of Beta zeolite nanocrystals was decreased after adding Sn in the zeolite framework [39]. In addition, the XRD pattern has no diffraction peak of Sn due to the small content of Sn (0.4 wt%) or the highly dispersed species inside the Beta zeolite nanocrystal framework [40].

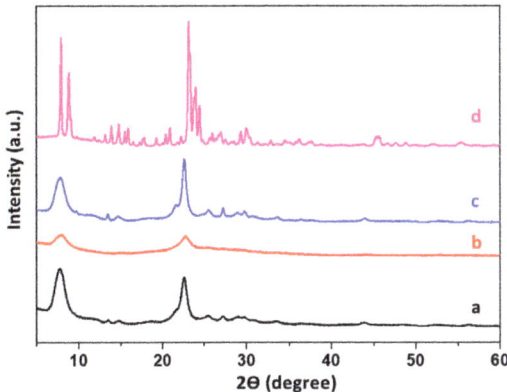

Figure 1. XRD patterns of: (**a**) the bare Beta, (**b**) the 0.4 wt% Sn-Beta, (**c**) the commercial Beta (Beta-COM), and (**d**) the conventional ZSM-5 (ZSM-5-CON).

However, the plate-shaped morphology of the 0.4 wt% Sn-Beta sample was not affected by the incorporation of Sn [41] as can be seen in SEM and TEM images (Figure 2A–C,E–G). Obviously, the plate-shaped particle size of the 0.4 wt% Sn-Beta sample and the synthesized bare Beta is approximately 25.3 ± 12 and 21.7 ± 7 nm, respectively. In strong contrast to this, the plate-shaped particle size of the commercial Beta (Beta-COM) is a little bit larger, with the size of 38.3 nm. However, other zeolite frameworks, such as ZSM-5, have also been used for the comparison. In the case of the conventional ZSM-5 sample (ZSM-5-CON), the coffin-shaped crystal is very large, with a particle size of approximately 2500 ± 200 nm. The particle size distribution of all samples is shown in Figure S1. Moreover, to confirm the existence of Sn in the 0.4 wt% Sn-Beta sample, the EDS elemental mapping of Sn was used to observe the Sn component, as shown in Figure 2I.

To further investigate the Sn species of the 0.4 wt% Sn-Beta, the X-ray photoelectron spectroscopy (XPS) was employed as shown in Figure 3. There were three distinct peaks that appeared at 484.4, 486.5, and 487.5 eV, which contributed to Sn^0, SnO_2, and Sn isomorphous substitution in the zeolite framework, respectively [42,43]. These observations reveal that the 0.4 wt% Sn-Beta obtained by an in-situ hydrothermal process can successfully incorporate Sn active sites into the zeolite framework.

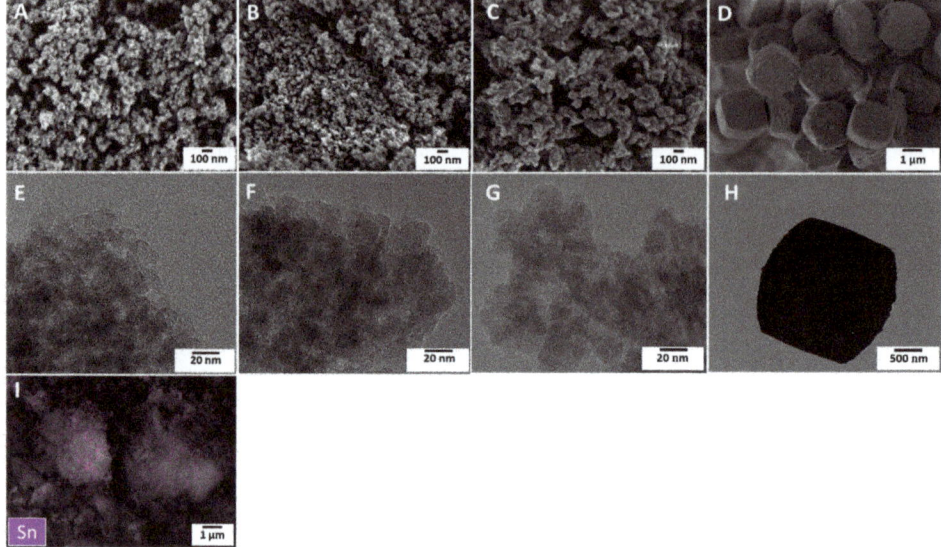

Figure 2. SEM and TEM images of: (**A,E**) the bare Beta, (**B,F**) the 0.4 wt% Sn-Beta, (**C,G**) the commercial Beta (Beta-COM), (**D,H**) the conventional ZSM-5 (ZSM-5-CON), and (**I**) the EDS Sn mapping of the in-situ synthesized Sn incorporated Beta (0.4 wt% Sn-Beta).

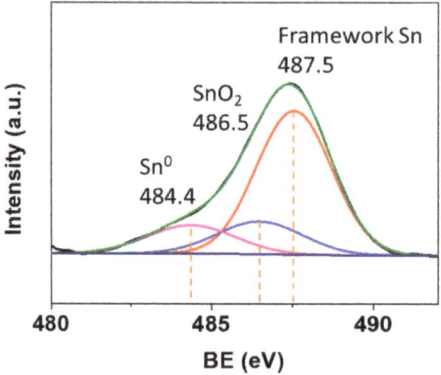

Figure 3. XPS spectrum of the 0.4 wt% Sn-Beta representing Sn^0, SnO_2, and Sn isomorphous substitution in the zeolite framework at 484.4, 486.5, and 487.5 eV, respectively.

To confirm the Si/Al ratio of all samples, the XRF technique was used and it was found that the Si/Al ratios of the bare Beta and 0.4 wt% Sn-Beta are 9.0, and 9.4, respectively. These observations obviously demonstrate that Al sites in the Beta zeolite framework were replaced by Sn, and therefore the Si/Al ratio of the 0.4 wt% Sn-Beta sample is slightly higher than that of the bare Beta [44]. Moreover, the Si/Sn ratio of the 0.4 wt% Sn-Beta sample is approximately 56. Furthermore, the Si/Al ratio of the commercial Beta (Beta-COM) and the conventional ZSM-5 (ZSM-5-CON) was observed as 11.7 and 11.5, respectively as can be seen in Table S1.

To investigate the textural properties of the bare Beta, and the 0.4 wt% Sn-Beta, N_2 sorption isotherms and the summarized data are shown in Figure S2 and Table 1, respectively. Obviously, the 0.4 wt% Sn-Beta exhibits a significantly lower BET surface area compared with the bare Beta, while total pore volume and external pore volume of the 0.4 wt% Sn-Beta is higher than those of the Beta zeolite [41]. These observations also relate to the fact that the low crystallinity of 0.4 wt% Sn-Beta leads to the production of the high external pore volume and the increased total pore volume compared with the synthesized bare Beta. This is similar to what has been described previously [40,45].

Table 1. Textural properties of the bare Beta and the 0.4 wt% Sn-Beta.

Sample	S_{BET} [a] (m^2/g)	S_{micro} [b] (m^2/g)	S_{ext} [c] (m^2/g)	V_{total} [d] (cm^3/g)	V_{micro} [e] (cm^3/g)	V_{ext} [f] (cm^3/g)	V_{ext}/V_{total} [g]
Bare Beta	711	670	41	1.49	0.16	1.33	0.89
0.4 wt% Sn-Beta	492	470	22	2.05	0.10	1.95	0.95

[a] S_{BET}: BET specific surface area. [b] S_{micro}: microporous surface area. [c] S_{ext}: external surface area. [d] V_{total}: total pore volume. [e] V_{micro}: micropore volume, [f] $V_{ext} = V_{total} - V_{micro}$; all surface areas and pore volumes are in the units of $m^2\ g^{-1}$ and $cm^3\ g^{-1}$, respectively. [g] Fraction of external volume.

To evaluate the acid properties, the ammonia temperature-programmed desorption (NH$_3$-TPD) profiles were presented in Figure S3 and the analyzed data are summarized in Table 2. The synthesized bare Beta, the in-situ synthesized Sn incorporated Beta nanocrystals (0.4 wt% Sn-Beta), and the commercial Beta (Beta-COM) demonstrate the similar trend of acid profiles composing of weak (0.237–0.340 mmol g^{-1}) and strong acid sites (0.424–0.449 mmol g^{-1}) appeared at the temperature in the range of 180, and 300–550 °C, respectively, eventually leading to a similar total acid density in the range of 0.661–0.789 mmol g^{-1}. In addition, the conventional ZSM-5 (ZSM-5-CON) demonstrates a slight difference in the densities of weak acid sites, strong acid sites, and total acid sites of 0.440, 0.423 mmol g^{-1}, and 0.863 mmol g^{-1}, respectively.

Furthermore, to identify the type of acid sites containing Brønsted acid sites (BAS) and Lewis acid sites (LAS) of the 0.4 wt% Sn-Beta and the bare Beta, the adsorption of pyridine monitored by FTIR spectroscopy was performed at 150 °C. It was found that the 0.4 wt%Sn-Beta shows the significant presence of both characteristic signals for Brønsted acid sites (ν = 1545 cm^{-1}), and Lewis acid (ν = 1455 cm^{-1}) sites. In addition, the band at 1490 cm^{-1} is attributed to pyridine adsorbed on both Brønsted acid sites and Lewis acid sites (BAS+LAS) as shown in Figure 4. Compared with the bare Beta, the B/L ratio of the 0.4 wt% Sn-Beta is significantly decreased, implying that Lewis acid sites increase after Sn insertion into the Beta framework (Table S2). This makes it clear that the 0.4 wt% Sn-Beta composes of both Brønsted acid sites and Lewis acid functions simultaneously and it should be suitable as a catalyst for glucose conversion to HMF, which typically requires the bifunctional catalyst containing Brønsted acid and Lewis acid sites.

Table 2. Acid sites density of all samples determined via the ammonia temperature-programmed desorption (NH$_3$-TPD).

Samples/T_{max} (°C)	Acid Site Density (mmol g^{-1}) [a]		
	Weak (180 °C)	Strong (300–550 °C)	Total
Bare Beta	0.318	0.441	0.759
0.4 wt% Sn-Beta	0.237	0.424	0.661
Commercial Beta (Beta-COM)	0.340	0.449	0.789
Conventional ZSM-5 (ZSM-5-CON)	0.440	0.423	0.863

[a] The number of acid sites measured by NH3-TPD and analyzed by Gaussian deconvolution.

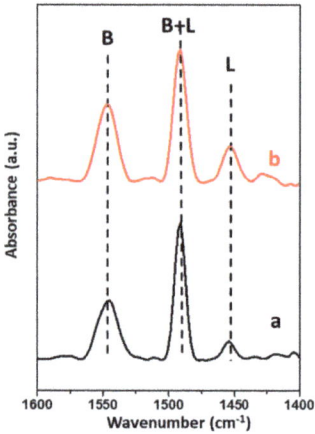

Figure 4. FTIR spectra of pyridine adsorbed on: (**a**) the bare Beta, and (**b**) the 0.4 wt% Sn-Beta at 150 °C representing in the adsorption region of Brønsted and Lewis acid sites.

2.2. Catalytic Test in the Glucose Conversion to 5-Hydroxymethylfurfural (5-HMF)

To illustrate the beneficial effect of the 0.4 wt% Sn-Beta, the conversion of glucose to 5-HMF in the aqueous phase was initially performed at various reaction times and reaction temperatures, as can be seen in Table 3. Although the addition of Sn into the Beta framework can obviously improve the glucose conversion with respect to the bare Beta, the 0.7 wt% Sn-Beta, does not provide significantly improved catalytic performances with respect to the one with lower Sn loading (0.4 wt%). For example, the glucose conversion and HMF yield over the 0.7 wt% Sn-Beta are 73.8% and 21.5%, respectively, while the glucose conversion and HMF yield over the 0.4 wt% Sn-Beta are 72.0% and 19.1%, respectively.

To further investigate the effect of the reaction time and temperature, the in-situ synthesized Sn-incorporated Beta nanocrystals with 0.4 wt% Sn loading (0.4 wt% Sn-Beta) were used to study in the next step. Obviously, when the reaction time was decreased from 48 h to 24 h, the glucose conversion and HMF yield is decreased from 72.0% to 57.5% and from 19.1% to 15.1%, respectively. As can be seen when decreasing the reaction time from 48 h to 24 h, the HMF yield was decreased only 4.0%, and therefore the 24 h of reaction time is more suitable to leave for a shorter global reaction time. To evaluate the effect of reaction temperature, when increasing the reaction temperature from 120 to 140 °C at 24 h using the 0.4 wt% Sn-Beta catalyst, the glucose conversion was significantly increased from 63.1 to 94.1%. However, the HMF product cannot be produced under this condition because it can be further rehydrated to form formic acid (FA), and levulinic acid (LA) [20,24,31,41].

Table 3. Catalytic performances in glucose conversion to 5-hydroxymethylfurfural (5-HMF) in the aqueous phase over the in-situ synthesized Sn-incorporated Beta nanocrystals with various Sn contents obtained at different reaction times and reaction temperatures.

Samples	Time (h)	T (°C)	Conversion (%) Glu [a]	Product Selectivity (%) Fru [b]	Lac [c]	FA [d]	LA [e]	HMF [f]	Yield (%) [g]	MB (%) [h]
Bare Beta	48	120	61.1	36.1	16.4	13.9	12.0	21.6	13.2	86
0.7 wt% Sn-Beta	48	120	73.8	33.5	6.2	5.8	25.4	29.1	21.5	68
0.4 wt% Sn-Beta	48	120	72.0	37.9	5.1	5.6	24.9	26.5	19.1	74
0.4 wt% Sn-Beta	24	120	57.5	38.8	4.8	13.7	16.5	26.2	15.1	82
0.4 wt% Sn-Beta	24	140	94.1	0.7	16.4	42.9	40	0.0	0.0	67

[a] Glucose, [b] Fructose, [c] Lactose, [d] Formic acid, [e] Levulinic acid, [f] Hydroxymethylfurfural, [g] Yield of HMF (%), [h] Mass balance (%).

To further study the effect of solvent on the conversion of glucose to HMF over the 0.4 wt% Sn-Beta, the biphasic phase containing the mixture of dioxane and H_2O was used (Table 4). Interestingly, the catalytic conversion of glucose to HMF over the 0.4 wt% Sn-Beta using a biphasic system at 120 °C and a reaction time of 24 h is significantly improved with the glucose conversion and HMF yield of 98.4% and 42.0%, respectively. Even though the glucose conversion to HMF at a much shorter reaction time is also improved, the glucose conversion and HMF yield at 120 °C for 4 h of the reaction time is approximately 77.3% and 25.0%, respectively. Interestingly, when the glucose conversion to HMF was tested at the lower temperature of 100 °C for 24 h, the glucose conversion and HMF yield were slightly decreased, with the values of 80.3% and 29.0%, respectively, with respect to the one obtained at the reaction temperature of 120 °C for 24 h. However, the glucose conversion to HMF was tested at 100 °C for 4 h and it was found that the glucose conversion and HMF yields are significantly lower with respect to the reaction condition at 100 °C for 24 h. These observations clearly demonstrate that the suitable condition for converting glucose to HMF over the 0.4 wt% Sn-Beta in a biphasic system is the reaction temperature of 100 to 120 °C and a reaction time of 24 h.

Table 4. Catalytic performances in glucose conversion to 5-hydroxymethylfurfural (5-HMF) in the biphasic phase (dioxane/water) over different catalysts at different reaction times and reaction temperatures.

Samples	Phase [a]	Time (h)	T (°C)	Conversion (%) Glu [b]	Product Selectivity (%) Fru [c]	Lac [d]	FA [e]	LA [f]	HMF [g]	Yield (%) [h]	MB (%) [i]
0.4 wt% Sn-Beta	aq.	24	120	57.5	38.8	4.8	13.7	16.5	26.2	15.1	82
0.4 wt% Sn-Beta	bi.	24	120	98.4	5.9	42.0	8.8	0.0	42.7	42.0	61
0.4 wt% Sn-Beta	bi.	4	120	77.3	24.3	36.3	6.6	0.5	32.3	25.0	81
0.4 wt% Sn-Beta	bi.	24	100	80.3	22.5	31.2	9.8	0.4	36.1	29.0	68
0.4 wt% Sn-Beta	bi.	4	100	30.1	75.3	8.2	3.1	0	13.4	4.1	93
Beta-COM	bi.	24	100	44.9	45.8	19.3	4.6	0	30.3	13.6	89
ZSM-5-CON	bi.	24	100	42.7	45.9	16.2	4.3	0	33.6	14.4	91
ZSM-5-COM	bi	24	100	59.4	40.5	21.4	9.7	0.3	28.1	16.7	81

[a] Phase: aqueous phase and biphasic phase represent by aq and bi., respectively, [b] Glucose, [c] Fructose, [d] Lactose, [e] Formic acid, [f] Levulinic acid, [g] Hydroxymethylfurfural, [h] Yield of HMF (%), [i] Mass balance (%).

To compare the catalytic performances in the glucose conversion to HMF over the different zeolite frameworks and the commercial Beta, the glucose conversion is significantly lower when using the commercial Beta (Beta-COM), the conventional ZSM-5 (ZSM-5-CON), and the commercial ZSM-5 (ZSM-5-COM, as the characterization of this sample as shown in Figure S4 and Table S3) (44.9%, 42.7% and 59.4% for Beta-COM, ZSM-5-CON and ZSM-5-COM, respectively, and the HMF yield is 13.6%, 14.4%, and 16.7% for Beta-COM, ZSM-5-CON, and ZSM-5-COM, respectively). The decrease in the catalytic activity confirms the two effects: (i) the increase in the catalytic performance when using the bare Beta nanocrystals with respect to the commercial one, (ii) the incorporation of Sn acting as the additional Lewis active site. These finding open up the perspectives to design the hierarchical Beta nanocrystals with the incorporation of Sn in the framework, which can greatly improve the catalytic conversion of glucose to HMF in a biphasic system.

3. Experimental Section

3.1. Chemicals and Materials

α,α'-dichloro-p-xylene (98%, TCI), N,N,N′,N′-tetramethyl-1,6-hexanediamine (98%, TCI), sodium silicate ($Na_2Si_3O_7$: 26.5 wt% SiO_2, and 10.6 wt% Na_2O, Merck), aluminum sulphate ($Al_2(SO_4)_3 \cdot 18H_2O$, Univar, Ajax Finechem), sulfuric acid (H_2SO_4: 96%, RCI Labscan), tin(IV) chloride pentahydrate ($SnCl_4 5H_2O$, Sigma-Aldrich,), and sodium hydroxide (NaOH: 98%, Carlo Erba) were used as starting materials for the synthesis of Sn-incorporated Beta nanocrystals. Glucose (>97%, TCI), hydrochloric acid (HCl: 37%, Merck), and dioxane ($C_4H_8O_2$, Merck) were used for the catalytic activity testing without any further purification. ZSM-5-CON zeolites were prepared by a one-pot hydrothermal process following the procedure reported in the literature [38], the commercial Beta zeolite was supplied from Zeolyst International ((NH_4)BEA, Si/Al 12.5, CP814P) and the commercial ZSM-5 zeolite was supplied from Zeolyst International ((NH_4)MFI, Si/Al 15, 3024E).

3.2. In Situ Synthesis of Sn-Beta Zeolite Nanocrystals

Firstly, the cyclic diquaternary ammonium (CDM) as a structure-directing agent (SDA) was synthesized according to the literature methods with some modification [37]. Briefly, N,N,N′,N′-tetramethyl-1,6-hexanediamine was mixed with an equimolar amount of α,α′-dichloro-p-xylene in acetonitrile. The mixture was synthesized under 60 °C for 3 h in a round-bottom flask, controlling the temperature by oil bath. Subsequently, the solid powder was collected by filtering and washing with acetonitrile followed by diethyl ether. Finally, the solid CDM was dried in an oven at 100 °C for 12 h.

Secondly, the Sn-incorporated Beta zeolite nanocrystals were synthesized by using a one-pot hydrothermal process, in which sodium silicate and aluminum sulphate were used as silica and alumina sources, respectively. Meanwhile, the CDM was used as the structure-directing agent (SDA) to control the structure of Beta zeolite, and $SnCl_4$ was used as an Sn precursor. The molar composition was $xSnCl_4$: $30Na_2O$:$2.5Al_2O_3$:$100SiO_2$:$10CDM$:$15H_2SO_4$:$6000H_2O$ [46]. In a typical procedure, the Sn-Beta samples were prepared from two solutions. The first solution was 1 g of sodium silicate in a sodium hydroxide solution. To prepare the second solution, 0.13 g of aluminum sulphate in H_2SO_4 solution was obtained. Subsequently, the second solution was added dropwise to the first solution under vigorous stirring. Before stirring at room temperature for 1 h, the CDM and $SnCl_4$ were added into the mixed solution. Afterwards, the resultant mixture was transferred to a Teflon lined stainless steel autoclave for hydrothermal treatment at 170 °C for 24 h. The obtained product was filtered and washed with deionized water, dried at 100 °C overnight, and calcined at 550 °C for 6 h. The solid products were transformed to protonated form by ion exchanged with 1 M ammonium chloride solution at 80 °C for 2 h, repeated for three times. Then, the products were washed with deionized water, dried at 100 °C overnight, and calcined at 550 °C for 6 h. The samples are denoted as x%Sn-Beta where x corresponds to the Sn loading (0, 0.4 and 0.7 wt%).

3.3. Characterization of Catalysts

The powder X-ray diffraction (XRD) patterns were used to investigate the crystalline structure of catalysts by using a Bruker D8 ADVANCE instrument (Billerica, Massachusetts,. United States) with 0.02° step sizes and a scan rate of 1° min^{-1} in the 2θ range of 5–60°. Wavelength-dispersive X-ray fluorescence spectrometry (WDXRF) was used to analyze the elemental composition of the catalysts and was performed using a Bruker S8 Tiger ECO instrument (Billerica, Massachusetts,. United States). Scanning electron microscopy (SEM) and transmission electron microscopy (TEM) were used to observe the surface morphology and topology of catalysts by performing on a JEOL JSM-7610F microscope (Tokyo, Japan) with an acceleration of 1 kV and a JEOL-JEM-ARM2000F microscope (Tokyo, Japan) operating at 200 kV, respectively. Energy dispersive spectroscopy (EDS) was used to measure the elemental distribution of catalysts by using a JEOL JSM-7610F microscope (Tokyo, Japan) operating at 15 kV. The textural properties of catalysts were determined via an N_2 adsorption/desorption technique at −196 °C, performed on a BELL-MAX analyzer (Tokyo, Japan). Prior to measurement, the catalysts were carefully degassed at 300 °C for 24 h with the temperature ramp rate of 20 °C min^{-1} under the vacuum system (10^{-1}–10^{-2} kPa). The specific surface area (S_{BET}) was calculated by using the Brunauer-Emmett-Teller (BET) theory. The total pore volume (V_{total}) was calculated at P/P_0 = 0.99. The t-plot method was used to estimate the micropore volume (V_{micro}), and the external volume (V_{ext}). The profiles of temperature-programmed desorption of ammonia (NH_3-TPD) using a BELL-CAT II analyzer (Tokyo, Japan) were performed to observe the acid properties (acid density and acid strength) of catalysts. The NH_3-TPD profiles were monitored in the temperature range of 100 to 700 °C with a heating rate of 10 °C min^{-1} in the flow of He. The X-ray photoelectron spectroscopy (XPS) depth profiles were recorded using a JEOL JPS-9010 (Tokyo, Japan) equipped with nonmonochromatic Mg K X-rays (1486.6 eV). An argon ion gun was used to etch the samples with the etching rate of 0.5 nm s^{-1}, and the XPS spectra were obtained at approximately 20 nm depth intervals. FTIR spectra of pyridine adsorption were recorded in the range of 4000–600 cm^{-1} performed on a Bruker Invenio R (Ettlingen, Germany) instrument equipped with an MCT detector. The spectra were gained at a resolution of 4 cm^{-1} with 64 scans. Before measuring spectra, the zeolite sample was pretreated by 10 v/v% of N$_2$ at 500 °C for 1 h with the temperature ramp rate of 10 °C min^{-1}. Subsequently, pyridine was introduced into the chamber at its vapor pressure at 40 °C for 1 h. After removing the physically adsorbed pyridine under vacuum for 1 h, the spectra were recorded at 150 °C [47].

3.4. Catalytic Testing

The synthesized catalysts were tested on glucose conversion under an aqueous and biphasic system (water/dioxane) with a small amount of HCl (pH = 1) performed in a high-pressure batch reactor. The reaction phase consists of a glucose (0.54 g, 3.6 wt%), catalyst (0.15 g), and deionized water (15 mL). In the case of the biphasic system, the solvent is the mixture of 13 mL of dioxane, and 2 mL of deionized water (deionized water phase was perfectly dissolved in organic phase). The experiment was carried out at different temperatures (100, 120, and 140 °C), autogenous pressures, and different time durations (4, 24, and 48 h). After the reaction, the liquid phases were separated from the solid phases (catalyst and any solid products formed) using a 0.45 µm membrane filter. High performance liquid chromatography (HPLC) with a Refractive index detector (RID) detector was used to analyze liquid products. Glucose and other products were monitored with an SH1011 sugar Shodex column, using MilliQ water (0.5 mM of H$_2$SO$_4$) as the mobile phase at a flow rate of 0.5 mL/min and a column temperature of 60 °C. The glucose conversion, selectivity, and yield of products are defined as follows:

$$Glucose\ conversion\ (\%) = \frac{mole\ of\ converted\ glucose}{mole\ of\ starting\ glucose} \times 100$$

$$Selectivity\ of\ products\ (\%) = \frac{mole\ of\ product}{total\ mole\ of\ products} \times 100$$

$$Yield\ of\ products\ (\%) = \frac{conversion \times selectivity}{100}$$

The mass balance of all experiments was calculated in the range of 61–93%.

To analyze the product in the organic phase, gas chromatography (GC) (Agilent, GC system 7890 B) (Agilent Technologies, Palo Alto, CA, USA) equipped with mass spectrometer (Agilent, system 5977A MSD) (Agilent Technologies, Palo Alto, CA, USA) with an HP-5MS capillary column (30 m, 0.32 mm i.d., stationary phase thickness of 0.25 µm) was performed to analyze the products of the organic phase from the reaction.

The Sn-Beta nanocrystals exhibit a bifunctional catalytic capability for the conversion of glucose to HMF. The incorporated Sn acting as Lewis acid sites can catalyze the isomerization of glucose to fructose. Subsequently, the Brønsted acid function can convert fructose to HMF via dehydration, as shown in Scheme 1.

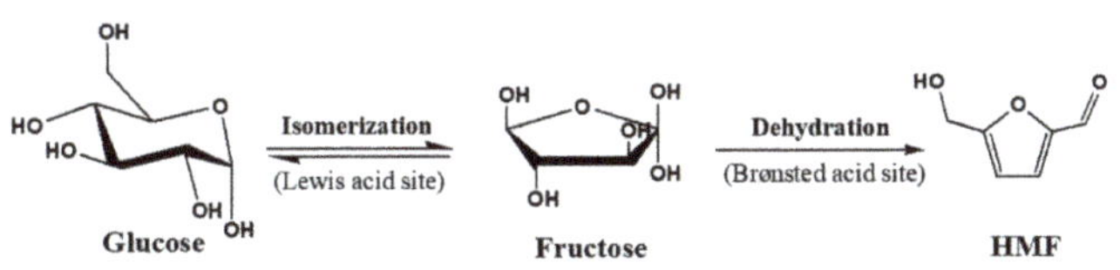

Scheme 1. The conversion of glucose to HMF through the glucose isomerization and subsequent dehydration on bifunctional catalyst.

4. Conclusions

In the present study, the Sn-incorporated Beta nanocrystals have been successfully synthesized via an in-situ hydrothermal process. The Sn species in the Beta zeolite nanocrystals is an isomorphous species, which can be incorporated inside the zeolite framework. In addition, the synthesized Sn-Beta nanocrystals have been applied as bifunctional catalysts for converting glucose to HMF. These synthesized materials composed of Lewis and Brønsted acid sites generated by the incorporated Sn and Brønsted acid sites, respectively, providing the glucose isomerization to fructose on the Sn active sites and the dehydration of fructose to HMF on Brønsted acid sites. Furthermore, the effects of various parameters including Sn content, zeolite type, reaction time, reaction temperature, and solvent on the glucose conversion to HMF were investigated in the detail. It was found that using the 0.4 wt% Sn-Beta zeolite nanocrystals in a dioxane/H_2O system at 120 °C for 24 h exhibited the greatest performance for producing HMF, with a yield of 42.0%. These findings open up the perspectives for the development of Sn incorporated in the Beta zeolite nanocrystals via an in-situ hydrothermal synthesis for the biomass conversion to high value-added chemical products.

Supplementary Materials: The following are available online at http://www.mdpi.com/2073-4344/10/11/1249/s1, Figure S1: Particle size distribution of: (a) the synthesized bare Beta, (b) the in-situ synthesized Sn incorporated Beta (0.4 wt% Sn-Beta), (c) the commercial Beta (Beta-COM), and (d) the conventional ZSM-5 (ZSM-5-CON), Figure S2: N_2 adsorption/desorption isotherms of (a) the synthesized bare Beta, and (b) the in-situ synthesized Sn incorporated Beta (0.4 wt% Sn-Beta), Figure S3: NH_3-TPD profiles of (a) the synthesized bare Beta, (b) the in-situ synthesized Sn incorporated Beta (0.4 wt% Sn-Beta), (c) the commercial Beta (Beta-COM), and (d) the conventional ZSM-5 (ZSM-5-CON), Figure S4: (A) XRD pattern (B) SEM image, (C) Particle size distribution and (D) NH_3-TPD profile of the commercial ZSM-5 (ZSM-5-COM) zeolite. Table S1: Chemical compositions analyzed by XRF of the synthesized bare Beta, the in-situ synthesized Sn incorporated Beta (0.4 wt% Sn-Beta), the commercial Beta (Beta-COM), and the conventional ZSM-5 (ZSM-5-CON), Table S2: Brønsted/Lewis acid site ratio was calculated by integrated area of main peaks, Table S3: Acid sites density of all samples determined via the ammonia temperature-programmed desorption (NH_3-TPD).

Author Contributions: Conceptualization, C.W.; Methodology, K.S., A.T., P.D., S.S., P.P.; Formal Analysis, K.S., A.T.; Investigation, S.N.; Writing—Original Draft Preparation, K.S., A.T.; Writing—Review and Editing, A.T., C.W. All authors have read and agreed to the published version of the manuscript.

Funding: This work was financially supported by the Vidyasirimedhi Institute of Science and Technology, and the National Research Council of Thailand (NRCT: Mid-Career Research Grant 2020). This research also received financial support from TSRI.

Conflicts of Interest: The authors declare no conflict of interest.

References

1. Kohli, K.; Prajapati, R.; Sharma, B. Bio-based chemicals from renewable biomass for integrated biorefineries. *Energies* **2019**, *12*, 233. [CrossRef]
2. Gallo, J.M.R.; Trapp, A.M. The chemical conversion of biomass-derived saccharides: An overview. *J. Braz. Chem. Soc.* **2017**, *28*, 1586607. [CrossRef]
3. Kucherov, F.A.; Romashov, L.V.; Galkin, K.I.; Ananikov, V.P. Chemical transformations of biomass-derived C_6-furanic platform chemicals for sustainable energy research, materials science, and synthetic building blocks. *ACS Sustain. Chem. Eng.* **2018**, *6*, 8064–8092. [CrossRef]
4. Menegazzo, F.; Ghedini, E.; Signoretto, M. 5-hydroxymethylfurfural (HMF) production from real biomasses. *Molecules* **2018**, *23*, 2201. [CrossRef]
5. Li, M.; Li, W.; Lu, Y.; Jameel, H.; Chang, H.M.; Ma, L. High conversion of glucose to 5-hydroxymethylfurfural using hydrochloric acid as a catalyst and sodium chloride as a promoter in a water/γ-valerolactone system. *RSC Adv.* **2017**, *7*, 14330–14336. [CrossRef]
6. Torres-Olea, B.; Mérida-Morales, S.; García-Sancho, C.; Cecilia, J.A.; Maireles-Torres, P. Catalytic activity of mixed Al_2O_3-ZrO_2 oxides for glucose conversion into 5-hydroxymethylfurfural. *Catalysts* **2020**, *10*, 878. [CrossRef]
7. Tempelman, C.; Jacobs, U.; Hut, T.; Pereira de Pina, E.; Van Munster, M.; Cherkasov, N.; Degirmenci, V. Sn exchanged acidic ion exchange resin for the stable and continuous production of 5-HMF from glucose at low temperature. *Appl. Catal. A* **2019**, *588*, 117267. [CrossRef]

8. Zhang, M.; Su, K.; Song, H.; Li, Z.; Cheng, B. The excellent performance of amorphous Cr_2O_3, SnO_2, SrO and graphene oxide-ferric oxide in glucose conversion into 5-HMF. *Catal. Commun.* **2015**, *69*, 76–80. [CrossRef]
9. Silahua-Pavón, A.A.; Espinosa-González, C.G.; Ortiz-Chi, F.; Pacheco-Sosa, J.G.; Pérez-Vidal, H.; Arévalo-Pérez, J.C.; Godavarthi, S.; Torres-Torres, J.G. Production of 5-HMF from glucose using TiO_2-ZrO_2 catalysts: Effect of the sol-gel synthesis additive. *Catal. Commun.* **2019**, *129*, 105723. [CrossRef]
10. Oozeerally, R.; Ramkhelawan, S.D.K.; Burnett, D.L.; Tempelman, C.H.L.; Degirmenci, V. ZIF-8 metal organic framework for the conversion of glucose to fructose and 5-hydroxymethyl furfural. *Catalysts* **2019**, *9*, 812. [CrossRef]
11. Dapsens, P.Y.; Mondelli, C.; Jagielski, J.; Hauert, R.; Pérez-Ramírez, J. Hierarchical Sn-MFI zeolites prepared by facile top-down methods for sugar isomerisation. *Catal. Sci. Technol.* **2014**, *4*, 2302–2311. [CrossRef]
12. Gardner, D.W.; Huo, J.; Hoff, T.C.; Johnson, R.L.; Shanks, B.H.; Tessonnier, J.-P. Insights into the hydrothermal stability of ZSM-5 under relevant biomass conversion reaction conditions. *ACS Catal.* **2015**, *5*, 4418–4422. [CrossRef]
13. Xue, Z.; Ma, M.-G.; Li, Z.; Mu, T. Advances in the conversion of glucose and cellulose to 5-hydroxymethylfurfural over heterogeneous catalysts. *RSC Adv.* **2016**, *6*, 98874–98892. [CrossRef]
14. Fan, X.; Jiao, Y. Chapter 5—Porous materials for catalysis: Toward sustainable synthesis and applications of zeolites. *Sustain. Nanoscale Eng.* **2020**, 115–137. [CrossRef]
15. Khan, W.; Jia, X.; Wu, Z.; Choi, J.; Yip, A. Incorporating hierarchy into conventional zeolites for catalytic biomass conversions: A review. *Catalysts* **2019**, *9*, 127. [CrossRef]
16. Abildstrøm, J.O.; Ali, Z.N.; Mentzel, U.V.; Mielby, J.; Kegnæs, S.; Kegnæs, M. Mesoporous MEL, BEA, and FAU zeolite crystals obtained by in situ formation of carbon template over metal nanoparticles. *New J. Chem.* **2016**, *40*, 4223–4227. [CrossRef]
17. Siqueira, B.G.; Silva, M.A.P.; Moraes, C. Synthesis of HMF from glucose in aqueous medium using niobium and titanium oxides. *Braz. J. Petroleum Gas* **2013**, *7*, 71–82. [CrossRef]
18. Tamura, M.; Chaikittisilp, W.; Yokoi, T.; Okubo, T. Incorporation process of Ti species into the framework of MFI type zeolite. *Micropor. Mesopor. Mater.* **2008**, *112*, 202–210. [CrossRef]
19. Bayu, A.; Karnjanakom, S.; Kusakabe, K.; Abudula, A.; Guan, G. Preparation of Sn-β-zeolite via immobilization of Sn/choline chloride complex for glucose-fructose isomerization reaction. *Chin. J. Catal.* **2017**, *38*, 426–433. [CrossRef]
20. Sun, Y.; Shi, L.; Wang, H.; Miao, G.; Kong, L.; Li, S.; Sun, Y. Efficient production of lactic acid from sugars over Sn-Beta zeolite in water: Catalytic performance and mechanistic insights. *Sustain. Ener. Fuels* **2019**, *3*, 1163–1171. [CrossRef]
21. Yang, G.; Wang, C.; Lyu, G.; Lucia, L.A.; Chen, J. Catalysis of glucose to 5-hydroxymethylfurfural using Sn-Beta zeolites and a Brønsted acid in biphasic systems. *BioResources* **2015**, *10*, 5863–5875. [CrossRef]
22. Moliner, M.; Román-Leshkov, Y.; Davis, M.E. Tin-containing zeolites are highly active catalysts for the isomerization of glucose in water. *Proc. Natl. Acad. Sci. USA* **2010**, *107*, 6164–6168. [CrossRef] [PubMed]
23. Guo, Q.; Ren, L.; Alhassan, S.M.; Tsapatsis, M. Glucose isomerization in dioxane/water with Sn-β catalyst: Improved catalyst stability and use for HMF production. *Chem. Commun.* **2019**, *55*, 14942–14945. [CrossRef] [PubMed]
24. Li, L.; Ding, J.; Jiang, J.-G.; Zhu, Z.; Wu, P. One-pot synthesis of 5-hydroxymethylfurfural from glucose using bifunctional [Sn,Al]-Beta catalysts. *Chin. J. Catal.* **2015**, *36*, 820–828. [CrossRef]
25. Li, Y.-P.; Head-Gordon, M.; Bell, A.T. Analysis of the reaction mechanism and catalytic activity of metal-substituted Beta zeolite for the isomerization of glucose to fructose. *ACS Catal.* **2014**, *4*, 1537–1545. [CrossRef]
26. Wang, L.; Guo, H.; Xie, Q.; Wang, J.; Hou, B.; Jia, L.; Cui, J.; Li, D. Conversion of fructose into furfural or 5-hydroxymethylfurfural over HY zeolites selectively in γ-butyrolactone. *Appl. Catal. A* **2019**, *572*, 51–60. [CrossRef]
27. Meng, Q.; Liu, J.; Xiong, G.; Li, X.; Liu, L.; Guo, H. The synthesis of hierarchical Sn-Beta zeolite via aerosol-assisted hydrothermal method combined with a mild base treatment. *Micropor. Mesopor. Mater.* **2019**, *287*, 85–92. [CrossRef]
28. Corma, A.; Nemeth, L.T.; Renz, M.; Valencia, S. Sn-zeolite beta as a heterogeneous chemoselective catalyst for Baeyer-Villiger oxidations. *Nature* **2001**, *412*, 423–425. [CrossRef]

29. Dijkmans, J.; Schutyser, W.; Dusselier, M.; Sels, B.F. Snβ-zeolite catalyzed oxido-reduction cascade chemistry with biomass-derived molecules. *Chem. Commun.* **2016**, *52*, 6712–6715. [CrossRef]
30. Wang, T.; Nolte, M.W.; Shanks, B.H. Catalytic dehydration of C_6 carbohydrates for the production of hydroxymethylfurfural (HMF) as a versatile platform chemical. *Green Chem.* **2014**, *16*, 548–572. [CrossRef]
31. Takagaki, A.; Ohara, M.; Nishimura, S.; Ebitani, K. A one-pot reaction for biorefinery: Combination of solid acid and base catalysts for direct production of 5-hydroxymethylfurfural from saccharides. *Chem. Commun.* **2009**, *41*, 6276–6278. [CrossRef] [PubMed]
32. Ohara, M.; Takagaki, A.; Nishimura, S.; Ebitani, K. Syntheses of 5-hydroxymethylfurfural and levoglucosan by selective dehydration of glucose using solid acid and base catalysts. *Appl. Catal. A* **2010**, *383*, 149–155. [CrossRef]
33. Nikolla, E.; Román-Leshkov, Y.; Moliner, M.; Davis, M.E. "One-Pot" synthesis of 5-(hydroxymethyl)furfural from carbohydrates using tin-Beta zeolite. *ACS Catal.* **2011**, *1*, 408–410. [CrossRef]
34. Roman-Leshkov, Y.; Dumesic, J.A. Solvent effects on fructose dehydration to 5-hydroxymethylfurfural in biphasic systems saturated with inorganic salts. *Top Catal.* **2009**, *52*, 297–303. [CrossRef]
35. Patil, S.K.R.; Lund, C.R.F. Formation and growth of humins via aldol addition and condensation during acid-catalyzed conversion of 5-hydroxymethylfurfural. *Energy Fuels* **2011**, *25*, 4745–4755. [CrossRef]
36. Candu, N.; El Fergani, M.; Verziu, M.; Cojocaru, B.; Jurca, B.; Apostol, N.; Teodorescu, C.; Parvulescu, V.I.; Coman, S.M. Efficient glucose dehydration to HMF onto Nb-BEA catalysts. *Catal. Today* **2018**, *325*, 109–116. [CrossRef]
37. Na, K.; Choi, M.; Ryoo, R. Cyclic diquaternary ammoniums for nanocrystalline BEA, MTW and MFI zeolites with intercrystalline mesoporosity. *J. Mater. Chem.* **2009**, *19*, 6713–6719. [CrossRef]
38. Rodaum, C.; Thivasasith, A.; Suttipat, D.; Witoon, T.; Pengpanich, S.; Wattanakit, C. Modified acid-base ZSM-5 derived from core-shell ZSM-5@aqueous miscible organic-layered double hydroxides for catalytic cracking of n-pentane to light olefins. *ChemCatChem* **2020**, *12*, 4288–4296. [CrossRef]
39. Dijkmans, J.; Gabriëls, D.; Dusselier, M.; de Clippel, F.; Vanelderen, P.; Houthoofd, K.; Malfliet, A.; Pontikes, Y.; Sels, B.F. Productive sugar isomerization with highly active Sn in dealuminated β zeolites. *Green Chem.* **2013**, *15*, 2777–2785. [CrossRef]
40. Yang, X.; Liu, Y.; Li, X.; Ren, J.; Zhou, L.; Lu, T.; Su, Y. Synthesis of Sn-containing nanosized Beta zeolite as efficient catalyst for transformation of glucose to methyl lactate. *ACS Sustain. Chem. Eng.* **2018**, *6*, 8256–8265. [CrossRef]
41. Van der Graaff, W.N.P.; Tempelman, C.H.L.; Pidko, E.A.; Hensen, E.J.M. Influence of pore topology on synthesis and reactivity of Sn-modified zeolite catalysts for carbohydrate conversions. *Catal. Sci. Tech.* **2017**, *7*, 3151–3162. [CrossRef]
42. Jiao, L.; Sun, S.; Meng, X.; Ji, P. Sn-based porous coordination polymer synthesized with two ligands for tandem catalysis producing 5-hydroxymethylfurfural. *Catalysts* **2019**, *739*, 1–13. [CrossRef]
43. Dai, W.; Wang, C.; Tang, B.; Wu, G.; Guan, N.; Xie, Z.; Hunger, M.; Li, L. Lewis acid catalysis confined in zeolite cages as a strategy for sustainable heterogeneous hydration of epoxides. *ACS Catal.* **2016**, *6*, 2955–2964. [CrossRef]
44. Liu, M.; Jia, S.; Li, C.; Zhang, A.; Song, C.; Guo, X. Facile preparation of Sn-β zeolites by post-synthesis (isomorphous substitution) method for isomerization of glucose to fructose. *Chin. J. Catal.* **2014**, *35*, 723–732. [CrossRef]
45. Van der Graaff, W.N.P.; Li, G.; Mezari, B.; Pidko, E.A.; Hensen, E.J.M. Synthesis of Sn-Beta with exclusive and high framework Sn content. *ChemCatChem* **2015**, *7*, 1152–1160. [CrossRef]
46. Choi, M.; Na, K.; Ryoo, R. The synthesis of a hierarchically porous BEA zeolite via pseudomorphic crystallization. *ChemComm.* **2009**, *20*, 2845–2847.
47. Saenluang, K.; Imyen, T.; Wannapakdee, W.; Suttipat, D.; Dugkhuntod, P.; Ketkaew, M.; Thivasasith, A.; Wattanakit, C. Hierarchical nanospherical ZSM-5 nanosheets with uniform Al distribution for alkylation of benzene with ethanol. *ACS Appl. Nano Mater.* **2020**, *3*, 3252–3263. [CrossRef]

Publisher's Note: MDPI stays neutral with regard to jurisdictional claims in published maps and institutional affiliations.

© 2020 by the authors. Licensee MDPI, Basel, Switzerland. This article is an open access article distributed under the terms and conditions of the Creative Commons Attribution (CC BY) license (http://creativecommons.org/licenses/by/4.0/).

Article

The Influence of the Gold Particle Size on the Catalytic Oxidation of 5-(Hydroxymethyl)furfural

Oliver Schade [1,2], Paolo Dolcet [1], Alexei Nefedov [3], Xiaohui Huang [4], Erisa Saraçi [2], Christof Wöll [3] and Jan-Dierk Grunwaldt [1,2,*]

1. Institute for Chemical Technology and Polymer Chemistry, Karlsruhe Institute of Technology (KIT), D-76131 Karlsruhe, Germany; oliver.schade@kit.edu (O.S.); paolo.dolcet@kit.edu (P.D.)
2. Institute of Catalysis Research and Technology, Karlsruhe Institute of Technology (KIT), D-76344 Eggenstein-Leopoldshafen, Germany; erisa.saraci@kit.edu
3. Institute of Functional Interfaces, Karlsruhe Institute of Technology (KIT), D-76344 Eggenstein-Leopoldshafen, Germany; alexei.nefedov@kit.edu (A.N.); christof.woell@kit.edu (C.W.)
4. Institute of Nanotechnology, Karlsruhe Institute of Technology (KIT), D-76344 Eggenstein-Leopoldshafen, Germany; xiaohui.huang@partner.kit.edu
* Correspondence: grunwaldt@kit.edu; Tel.: +49-721-608-42120

Received: 24 February 2020; Accepted: 16 March 2020; Published: 19 March 2020

Abstract: For the production of chemicals from biomass, new selective processes are required. The selective oxidation of 5-(Hydroxymethyl)furfural (HMF), a promising platform molecule in fine chemistry, to 2,5-furandicarboxylic acid (FDCA) is considered a promising approach and requires the oxidation of two functional groups. In this study, Au/ZrO$_2$ catalysts with different mean particle sizes were prepared by a chemical reduction method using tetrakis(hydroxymethyl)phosphonium chloride (THPC) and tested in HMF oxidation. The catalyst with the smallest mean particle size (2.1 nm) and the narrowest particle size distribution was highly active in the oxidation of the aldehyde moiety of HMF, but less active in alcohol oxidation. On the other hand, increased activity in FDCA synthesis up to 92% yield was observed over catalysts with a larger mean particle size (2.7 nm), which had a large fraction of small and some larger particles. A decreasing FDCA yield over the catalyst with the largest mean particle size (2.9 nm) indicates that the oxidation of both functional groups require different particle sizes and hint at the presence of an optimal particle size for both oxidation steps. The activity of Au particles seems to be influenced by surface steps and H bonding strength, the latter particularly in aldehyde oxidation. Therefore, the presence of both small and some larger Au particles seem to give catalysts with the highest catalytic activity.

Keywords: gold catalysis; selective oxidation; colloidal synthesis; 5-(hydroxymethyl)furfural; 2,5-furandicarboxylic acid; particle size; biomass conversion

---

1. Introduction

The depletion of fossil resources and the ever-growing demand for energy and chemical products drive the need to exploit renewable carbon sources such as biomass [1,2]. Inspired by current value chains that are based on a small number of platform chemicals like ethylene or propylene, the production of platform molecules through the partial fragmentation of biomass is one promising approach [3]. 5-(Hydroxymethyl)furfural (HMF) is considered one of the most versatile platform molecules that can be produced from carbohydrate biomass via dehydration [4–6]. More specifically, HMF is obtained from hexoses and can therefore also be produced from non-edible polymeric carbohydrates like cellulose [7]. As a platform molecule, HMF can undergo a variety of reactions like hydrogenation [8], dehydrogenation [9], or hydrodeoxygenation [10]. Furthermore, the selective oxidation of HMF has gained great interest in recent years. Selective oxidation of HMF can for example give 5-hydroxymethyl-2-furancarboxylic acid (HFCA) or 2,5-diformylfuran. One of the most important oxidation products is 2,5-furandicarboxylic acid (FDCA), [11,12] which is produced via the oxidation of both functional groups of HMF (Scheme 1). The structural similarity of FDCA to terephthalic acid led to its consideration as one of the twelve important molecules that can be produced from sugar-containing biomass feedstock [11,12]. Therefore, FDCA may be used in future bio-based polymers thus overcoming the need of petrochemical-based terephthalic acid [13,14]. Synthetic approaches for FDCA synthesis include stoichiometric oxidation reactions [15] as well as catalytic routes with mostly molecular oxygen. The latter range from bio- [16,17] to electro- [18–20] and metal catalysis. Current FDCA production is based on the homogeneously catalyzed AMOCO process, which is carried out in acetic acid solvent at 125 °C and 70 bar air in the presence of a Cu/Mn/Br catalyst [21]. In addition, heterogeneous metal catalysts have been applied in HMF oxidation. Especially supported noble metals like Pt [22–24] or Pd [25–27] are catalytically active in this reaction. Among others, supported gold-based catalysts show high activity in the oxidation of HMF with oxygen to FDCA [28–35]. For example, Au supported on CeO_2 nanoparticles allowed the production of FDCA in 96% yield after 5 h at 130 °C [28].

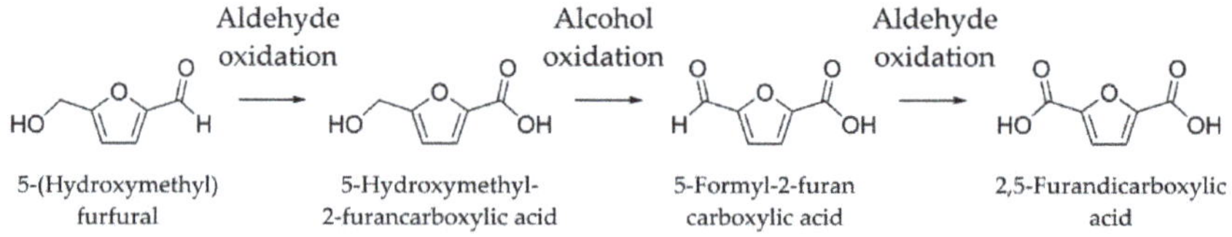

Scheme 1. Possible products and involved oxidation steps of 5-(Hydroxymethyl)furfural (HMF) oxidation over noble metal catalysts in alkaline solution.

Considering the chemical inertness of the bulk metal, gold has in early times been regarded to be catalytically inactive as well. However, small Au particles of few nanometers in size show catalytic activities in a variety of both gas and liquid phase reactions. Triggered by pioneering studies on gold-catalyzed oxidation of CO [36], gold catalysts have been used in a variety of oxidation reactions [37]. A special feature of gold catalysts is a strong dependence of catalytic activities on the gold particle size. For example, small particles of 2 nm in size were first found most active in the oxidation of CO [38] and propylene [39,40]. Later studies found that small clusters rather than nanoparticles are the catalytically active species [41], but such clusters may require high surface areas of the support material [42]. In contrast, slightly larger particles in the nanometer range (≈7 nm) are most active in alcohol oxidation [43,44].

Given that HMF bears two different functional groups, one alcohol and one aldehyde, and the strong influence of the Au particle size on the oxidation of different groups, an optimal Au particle size is needed to oxidize each group in high yields. In this study, we prepared gold catalysts of four different Au mean particle sizes (ranging from 2.1 to 2.9 nm) using a chemical reduction method with THPC. The catalysts were characterized and tested in the batch-wise oxidation of HMF, in which the influence of reaction temperature, amount of added base and air pressure were investigated.

2. Results

2.1. Catalyst Characterization

The starting gold colloids, prior to deposition on the ZrO_2 support, were analyzed by UV-Vis spectroscopy (Figure S1). The spectra showed very weak surface plasmon resonance (SPR) bands, indicating that the particle size of the Au nanoparticles is smaller than ≈3 nm. According to literature [38,45,46], below this value the SPR is strongly suppressed and surface scattering is the factor dominating the UV-Vis spectrum. The colloid with the expected larger size presents additionally a faint band centered at about 525 nm, indicating the possible presence of particles of up to 20 nm size.

The supported catalysts are labelled as AuZX, where Z represents the ZrO_2 support material and X represents the mean particle size by number based on scanning transmission electron microscopy (STEM, see text below and Figure 1). The metal loading (nominal loading: 2 wt.%) and the specific surface area of the four catalysts used in this study are listed in Table 1. Deposition of the gold colloids did not affect the specific surface area of the used ZrO_2 support material showing that the Au colloids or reductant did not block the pores of the support material. The support was added to the colloidal suspension to achieve a Au loading of 2 wt.%, however lower metal loadings show an incomplete deposition of the formed gold colloids on the support materials.

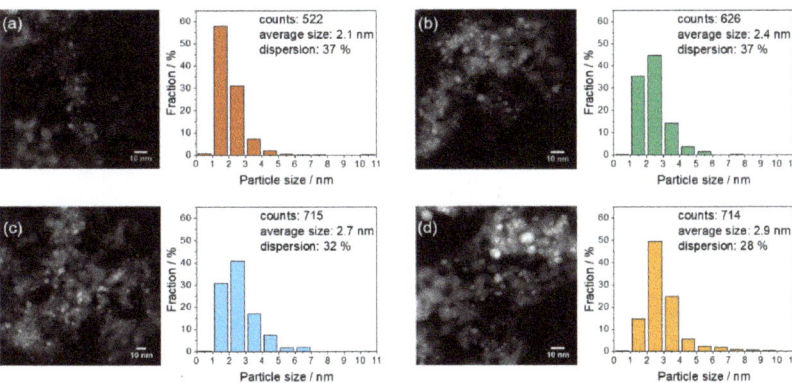

Figure 1. Representative STEM images and the corresponding particle size distributions by number of the four freshly prepared catalysts. The catalysts were labelled according to their mean particle sizes (**a**) AuZ2.1, (**b**) AuZ2.4, (**c**) AuZ2.7, and (**d**) AuZ2.9.

Table 1. Metal loading and specific surface area of the Au/ZrO_2 catalysts and the ZrO_2 support material.

Entry	Support/Catalyst	Au Loading/wt.%	Specific Surface Area/m^2 g^{-1}
1	ZrO_2	-	99
2	AuZ2.1	1.4	97
3	AuZ2.4	1.8	98
4	AuZ2.7	1.8	100
5	AuZ2.9	1.8	102

For none of the catalysts, were any reflections of gold observed in the powder X-ray diffraction (XRD) patterns (Figure S2). The absence of Au reflections is in good agreement with the Au loading and the expected particle sizes, which are below the detection limit of XRD [47]. In addition, similar diffraction patterns of the catalysts to the pure support material in all cases evidenced that ZrO_2 was stable in the colloidal suspension.

To gain insight into the particle sizes and size distributions of the catalysts, STEM images were recorded (Figure 1). AuZ2.1 (Figure 1a) shows a narrow particle size distribution with an average particle size of 2.1 nm at 37% dispersion. For this catalyst, 90% of all counted particles have a size of less than 3 nm with 58% of all particles being in the range of 1–2 nm. The average particle size is slightly larger than it was previously reported using this preparation procedure [43]. In contrast to the previous report, the increase in the particle size was not that pronounced for the four catalysts with average sizes of 2.4 nm, 2.7 nm, and 2.9 nm, respectively (Figure 1b–d). The increasing average particle size can be attributed to a broadening of the particle size distributions in all cases. This means, that for the catalysts b-d most particles are in the range of 2–3 nm (b: 45%, c: 41%, d: 50%), however, the fraction of large particles increases, which can both be seen in the representative TEM images as well as the particle size distributions (Figure 1). The difference in particle size distributions compared to literature might be explained by the fact that we used another support material and chose to characterize the final supported catalysts. Possibly, larger particles were not adsorbed on the support material, which might also explain the slightly lower metal loading. Note that the trend in particle size prevails also when normalized to the surface atoms instead of the number of particles.

IR spectra were recorded for all the supported catalysts to investigate whether remains of the reducing agent THPC were still present in the catalysts (Figure 2). In contrast to the pure support material, all catalysts show two bands at 1130 cm^{-1} and 1052 cm^{-1}. In the basic solution (see details in the Experimental Section), bis(hydroxymethyl)phosphinic acid is formed from THPC [43]. Therefore, the bands can be assigned to the P = O (1130 cm^{-1}) and P-OH (1052 cm^{-1}) vibrations, respectively. This shows, that all catalysts still contained residues of THPC even after a thorough washing procedure.

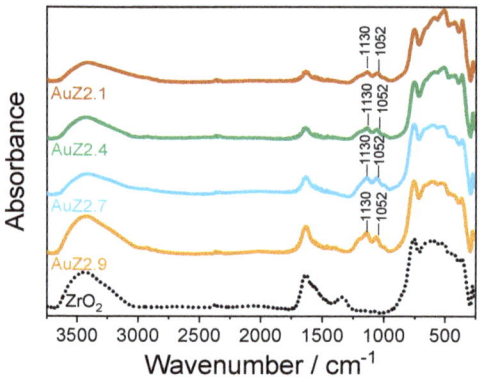

Figure 2. IR spectra of the supported as prepared catalysts and the pure ZrO$_2$ support (dotted line).

In order to quantify the surface contamination with remains of THPC, the catalysts were additionally characterized using X-ray photoelectron spectroscopy (XPS, Table 2, Figures S3 and S4). For all of them, the phosphorous content is indeed limited, and found in the range of 1–2 at%, with similar percentages also for gold. Only in the case of the sample with the largest average size the determined gold content is lower than 1 at% and therefore the P/Au ratio is four-fold that of the other samples. The samples contain also relevant amount of chlorine, originating both from the THPC and the gold precursor itself (tetrachloroauric acid). The latter is the most likely source, given the low amount of phosphorous present.

Table 2. Atomic percentages of the elements composing the surface of the samples, determined by XPS.

Entry	Catalyst	C/at%	O/at%	P/at%	Au/at%	Zr/at%	Cl/at%
1	AuZ2.1	49.4	33.1	1.4	1.0	6.9	8.2
2	AuZ2.4	42.9	38.9	0.9	1.0	8.4	7.9
3	AuZ2.7	43.0	37.6	1.8	1.6	7.4	8.6
4	AuZ2.9	35.5	38.3	1.6	0.4	12.0	11.0

The detailed analysis of the Au4f regions (Figure S4) showed that Au is mainly in metallic state (Au4f$_{7/2}$ ≈ 83.5 eV), while oxidized components (Au$^+$, Au4f$_{7/2}$ ≈ 86.7 eV) are limited to 2–5%. Although cationic Au species also correlated with HMF oxidation activity for Au/CeO$_2$ catalysts [48], their effect in this study is most likely not pronounced due to their low quantity, comparable among all the samples.

2.2. Catalytic Activity

The catalytic activity of all catalysts in HMF oxidation was studied under the same reaction conditions of 100 °C, 10 bar air pressure in the presence of four equivalents of NaOH (Table 3), based on a thorough literature review (e.g., refs. [28,30,31,48,49]). A catalyst prepared by deposition-precipitation with a mean particle size of 3.7 nm (31% dispersion) was tested as a reference [50]. The generally lower product yields compared to Au/ZrO$_2$ prepared via deposition-precipitation (Table 3, entry 1) might be explained by the preparation procedure, since the reference sample was calcined (350 °C, 4 h) prior to testing while the samples discussed here were used as prepared to avoid particle growth. A possible influence could arise from the presence of the P and Cl contaminants, which might block some of the active sites of the gold nanoparticles. On the other hand, the trend in activity does not follow the variations seen for P and Cl contents since, e.g., AuZ2.9 with the highest relative contaminations has a productivity similar to that of AuZ2.4, which has the least amount of impurities. Hence, the difference in catalytic activity can be rationalized in terms of differences in noble metal particle size. Attributing differences in catalytic activity to the nanoparticle size rather than small Au clusters seems also reasonable, as the reaction is carried out in liquid phase and particles in the nanometer range are the ones active in alcohol oxidation [43,44]. Accordingly, nanoparticles rather than clusters will provide active sites. Furthermore, no correlation of the catalytic activity to the electronic structure of Au can be found. AuZ2.1 and AuZ2.9 are almost similar in terms of electronic structure based on XPS, but differ widely in their catalytic activity. This also underlines that the activity differences arise from the particle sizes.

Table 3. Catalyst screening of Au/ZrO$_2$ catalysts with different mean particle sizes in the oxidation of HMF. Reaction conditions: 100 °C, 10 bar air pressure, 4 equivalents of NaOH, 5 h reaction time, 1 mmol HMF in 10 mL H$_2$O, 98.5 mg catalyst.

Entry	Catalyst	HMF Conversion/%	Yield/%		C-Balance/%	Productivity [a]/mol$_{FDCA}$ h^{-1} mol$_{Au}$$^{-1}$
			HFCA	FDCA		
1 [50]	Au/ZrO$_2$	100	0	75	75	19 (72)
2	AuZ2.1	100	9	16	25	5 (10)
3	AuZ2.4	100	0	35	35	8 (18)
4	AuZ2.7	100	0	43	43	10 (25)
5	AuZ2.9	100	1	30	31	7 (18)

a: Productivities normalized to surface Au atoms are given in brackets (mol$_{FDCA}$ h^{-1} mol$_{Au,surface}$$^{-1}$).

Comparing the product yields under the same reaction conditions over the catalysts prepared by the colloidal method, although the particle size differences are rather small, a clear trend can be observed. AuZ2.1 gave the lowest carbon balance of 25% (Table 3, entry 2) with a FDCA yield of 16% and 9% yield of the intermediate HFCA. AuZ2.4 gave a higher FDCA yield of 35% and a maximum FDCA yield was observed over AuZ2.7, which produced FDCA in 43% yield. The differences in activity between AuZ2.1 and AuZ2.4 can be attributed to the particle size, since these catalysts have similar phosphorous and chloride content as well as fraction of cationic Au. No HFCA was produced over AuZ2.4 and AuZ2.7, indicating a high activity in the oxidation of the intermediate HFCA. AuZ2.9 with the largest mean particle size among the catalysts prepared by the colloidal method gave FDCA in a lower yield of 30% along with 1% of HFCA. The same trend is also reflected in the productivity, which normalizes the product yield to the reaction time and to the Au content thus eliminating any effect from the slightly different Au loadings. For AuZ2.7, the highest productivity rate of 10 mol$_{FDCA}$ h^{-1} mol$_{Au}$$^{-1}$ was obtained under these reaction conditions. Note, that the productivity was calculated based on the total amount of Au instead of the surface Au. Taking surface atoms into account in the calculation leads to a higher productivity because of the lower fraction of surface Au atoms. The trend persists, but the absolute values and relative differences between the catalysts increase (Table 3). AuZ2.4 and AuZ2.9 perform similarly when normalized to surface Au atoms with AuZ2.7 giving the highest productivity of 25 mol$_{FDCA}$ h^{-1} mol$_{Au,surface}$$^{-1}$.

The influence of the reaction temperature on the oxidation of HMF is shown in Figure 3. Since HMF is unstable in basic aqueous solution and at high temperatures, [18] lowering the reaction temperature results in a higher carbon balance. On the other hand, the oxidation of the hydroxymethyl group is the rate limiting step and often requires higher reaction temperatures [28]. Therefore, an optimal temperature as a compromise between reaction time and high yield has to be determined. As expected, a high carbon balance of above 80% was observed for all catalysts at 50 °C and HFCA was formed as the main product. This shows that all catalysts have a high activity in aldehyde oxidation, although there are slight differences. The highest carbon balance of 98% was observed for AuZ2.7, which yielded 73% of HFCA along with 25% of FDCA. This catalyst was active in alcohol oxidation even at a low temperature of 50 °C. Increasing the reaction temperature led to a reduction of the HFCA yield and the carbon balance for all catalysts. The HFCA yield decreased linearly over AuZ2.1 with increasing the temperature to 75 °C and FDCA yield just increased slightly from 1 to 16%. On the other hand, the selectivity shifted to FDCA at 75 °C for all catalysts except AuZ2.1, for which the selectivity switch occurred only at 100 °C, with a decreasing overall carbon balance. These results show that alcohol oxidation is the rate limiting step in this reaction, in line with literature [28] and is more pronounced for smaller Au particles (<2.4 nm). For the three catalysts with larger mean particle sizes, the highest FDCA yield of 55% was obtained at 75 °C (12 $mol_{FDCA}\ h^{-1}\ mol_{Au}^{-1}$). Also the surface-atom-normalized productivities increase with increasing particle size from 8, 29, 30 to 33 $mol_{FDCA}\ h^{-1}\ mol_{Au,\ surface}^{-1}$ over AuZ2.1, AuZ2.4, AuZ2.7, and AuZ2.9, respectively, showing superior activity of larger particles. Therefore, further catalytic experiments were carried out at this temperature, which was also chosen for AuZ2.1 in order to have comparable reaction conditions.

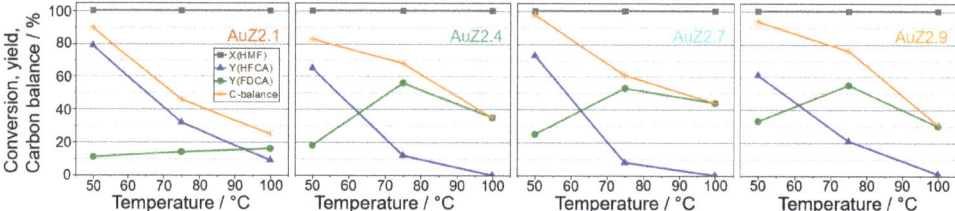

Figure 3. Influence of the reaction temperature on the product distribution in the oxidation of HMF over Au/ZrO$_2$ catalysts with different particle sizes. The AuZ2.1 catalyst with the smallest mean particle size is shown on the left side and the mean particle sizes increase towards the right side of the figure. Reaction conditions: 100 °C, 10 bar air pressure, HMF:NaOH 1:4, 5 h reaction time, 1 mmol HMF in 10 mL H$_2$O, 98.5 mg catalyst.

Reducing the amount of added base minimizes HMF degradation and enhances the sustainability of the reaction. However, the poor solubility of FDCA in water and the formation of a germinal diol as essential intermediate in the claimed reaction mechanism necessitate the use of a homogeneous base in most reactions [51]. Figure 4 shows the influence of NaOH addition on the product yields over all four catalysts. In all reactions that were conducted in the absence of base, a little amount of HMF was converted, showing some catalytic activity in the base-free oxidation of HMF. Under these conditions, the HMF conversion increased from 5 to 11% with increasing mean particle sizes. This indicates that the particle size of Au also affects HMF conversion. In the absence of base, some reactions gave HFCA in high selectivity. For example, HFCA was produced in 100% selectivity over AuZ2.4 at 7% HMF conversion. Increasing the amount of NaOH led to a linear increase in HFCA yield for all catalysts. At 1 equivalent of NaOH referred to HMF, HFCA was exclusively produced indicating that one hydroxide ion is required per oxidized functional group [51]. The highest HFCA yield of 93% was observed over AuZ2.1. Further increasing the amount of added base to two equivalents led to an increase in FDCA production, which was produced in the highest yield of 51% over AuZ2.7. Under these conditions, also AuZ2.1 produced some FDCA in 30% yield. Increasing the amount of added NaOH to 4 equivalents led to a further increase in FDCA production, which was produced in a yield of around 55% over all catalysts other than AuZ2.1, which was the only catalyst that afforded less FDCA in 14% yield upon further increasing the amount of NaOH.

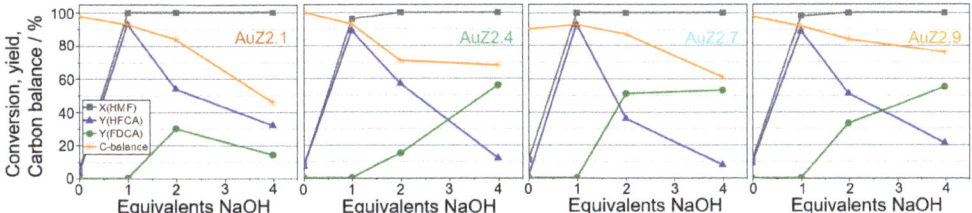

Figure 4. Influence of the amount of added NaOH on the product distribution in the oxidation of HMF over Au/ZrO$_2$ catalysts with different particle sizes. The AuZ2.1 catalyst with the smallest mean particle size is shown on the left side and the mean particle sizes increase toward the right side of the figure. Reaction conditions: 75 °C, 10 bar air pressure, HMF:NaOH 1:0/1/2/4, 5 h reaction time, 1 mmol HMF in 10 mL H$_2$O, 98.5 mg catalyst.

The influence of the air pressure was investigated as a last step and was performed at 75 °C in the presence of 4 equivalents of NaOH (Figure 5). As oxygen is consumed in the oxidation process, presumably by indirect participation in the mechanism, the yield of oxidation products should also increase with increasing pressure. A higher air pressure of 30 bar led to the exclusive production of FDCA for all catalysts, which was additionally produced in higher yields. Under these reaction conditions, FDCA was even produced in a high yield of 80% over AuZ2.1, which did not show a high activity in alcohol oxidation in the previous reactions. However, in line with the aforementioned studies, the FDCA yield over AuZ2.1 was the lowest under these conditions. Also at a higher air pressure of 30 bar, AuZ2.7 showed the highest activity in FDCA synthesis, which was produced in 92% yield at a productivity of 20 mol$_{FDCA}$ h^{-1} mol$_{Au}$. The differences in the productivity normalized to surface Au atoms becomes smaller here with the highest value of 53 mol$_{FDCA}$ h^{-1} mol$_{Au,surface}^{-1}$ for AuZ2.7 further underlining its superior activity. No leaching of Au (<1 ppm) was observed by means of ICP-OES except for AuZ2.9, which corresponded to 0.2% of the total Au content. Thus, the adsorption of pre-formed colloids resulted in strongly bound Au particles.

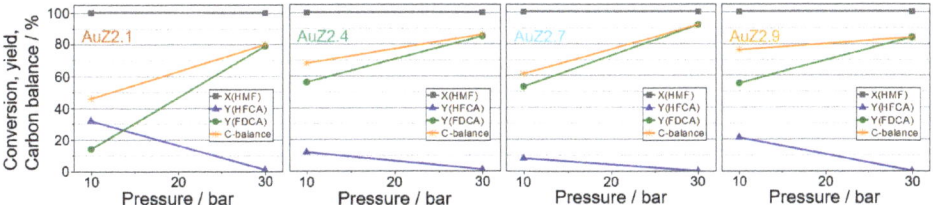

Figure 5. Influence of the air pressure on the product distribution in the oxidation of HMF over Au/ZrO$_2$ catalysts with different particle sizes. The AuZ2.1 catalyst with the smallest mean particle size is shown on the left side and the mean particle sizes increase towards the right side of the figure. Reaction conditions: 75 °C, HMF:NaOH 1: 4, 5 h reaction time, 1 mmol HMF in 10 mL H$_2$O, 98.5 mg catalyst.

3. Discussion

In this study, we investigated the particle size effect of small Au nanoparticles on the catalytic HMF oxidation by fine-tuning the size of Au particles. Mean particle sizes between 2.1 and 2.9 nm (286 atoms for 2.1 nm to 753 atoms for 2.9 nm) showed significant variations in activity and selectivity. The differences in the catalytic activity can most likely be attributed to different particle sizes, since other factors like e.g., the phosphorus or chloride contamination, did not show any clear correlation. In addition, particles in the nanometer range rather than small clusters were most active in the oxidation of alcohols [44]. Recently, Megías-Sayago et al. [52] studied HMF oxidation over Au particles in the range of 4–40 nm supported on carbon. In contrast to our study, both the HMF conversion and HFCA yield were independent from the particle size at the larger particle sizes. Note that HMF conversion is 100% in most reactions due to its instability in alkaline solution. An increasing activity of smaller particles in alcohol oxidation of HMF was attributed to enhanced activity of such particles to reduce oxygen on Au(100), which are present to a larger extent on small particles. Consequently, oxygen is most probably activated on the Au particles in the present study as well, also considering the unlikely activation on the support. However, the used model only applies to particles with a size of more than 4 nm [52], which is larger than the mean sizes of particles used in this study also considering the small fraction of particles in this range for AuZ2.7 and AuZ2.9. Thus, the observed differences in catalytic activity can be attributed to the smaller particle sizes studied here.

The Au particle size is the most decisive factor influencing the catalytic activity of Au-based catalysts in liquid phase reactions, sometimes even more important than the support material [53]. In general, the effect of the support material in HMF oxidation is an active field of research and was attributed to the oxygen storage capacity [54], oxygen vacancies [48], or Brønsted acidity [55]. For ZrO_2, a low oxygen storage capacity was speculated to result in more active catalysts [54]. In addition, the support material affects the particle-support interactions. Carbon-supported Au catalysts might show different particle size dependencies, for example, intermediate particles were more active in the oxidation of ethylene glycol when supported on carbon instead of Al_2O_3, since smaller Au particles were located in micropores thus not accessible for the reactants [56]. The unlikely participation of the used ZrO_2 support material because of its low oxygen storage capacity combined with the fact that the same support material was chosen for all catalysts tested here further underline the difference in activity due to the particle size. This is also supported by no correlation between catalytic activity and Au electronic structure.

AuZ2.1 showed the narrowest particle size distribution and no larger particles were found in TEM analysis, therefore, its catalytic activity can be linked directly to the smallest gold particles. As AuZ2.1 showed the highest activity in HFCA production, i.e., aldehyde oxidation, and just a slight production of FDCA at 2 equivalents of NaOH, the smaller particles seem to be beneficial in the oxidation of the aldehyde but not the alcohol moiety of HMF when supported on ZrO_2. In contrast, AuZ2.9 also had a large fraction of small particles, however, with additional larger particles. Generally, catalysts with larger mean particle sizes were more active in FDCA synthesis, i.e., alcohol oxidation, in line with previous reports [43,57,58]. This might be due to the presence of surface steps, which are active in β-H removal [58], also considering that the reactions were carried out with excess NaOH [52]. This step is also an important part of the reaction mechanism in HMF oxidation, which probably follows a dehydrogenation mechanism [51,59]. During the reaction, the surface of Au particles gets covered by hydrogen [51,60]. Since the interaction between hydrogen and gold is stronger for smaller Au particles [61], active sites of AuZ2.1 may be blocked before being released after the reaction with oxygen, which might be the reason for FDCA production only with increasing air pressure. During the oxidation of the hydroxymethyl group, an aldehyde is formed in the first step [51]. Thus, the alcohol moiety of HMF is transformed into the aldehyde on larger particles, which in turn is converted on the smaller particles.

The instability of HMF in basic aqueous solution leads to humin formation, which lowers the carbon balance. Since, HFCA does not polymerize or decompose under these reaction conditions, it is therefore highly desirable to quickly convert HMF into HFCA before its thermal decomposition. This is particularly relevant with the alcohol oxidation being the rate-limiting step in HMF oxidation under such conditions. Therefore, the presence of a rather large fraction of small particles is beneficial for HMF oxidation, since HFCA is formed quickly. An increasing FDCA yield with the particle size up to AuZ2.7 shows that small but sufficiently large particles then oxidize the alcohol moiety and FDCA is formed.

4. Materials and Methods

4.1. Materials

Analytical grade chemicals were used without further purification: HMF, FDCA, THPC (Sigma-Aldrich, Darmstadt, Germany), HFCA, NaOH (Merck, Darmstadt, Germany), 2,5-diformylfuran, 5-formyl-2-furancarboxylic acid (TCI Chemicals, Eschborn, Germany), ZrO2 1/8″ pellets, $HAuCl_4·3H_2O$ (Alfa Aesar, Karlsruhe, Germany), synthetic air (Air Liquide, Düsseldorf, Germany).

4.2. Catalyst Preparation

The catalysts were prepared by a modified chemical reduction method based on a previously established procedure [43]. In brief, gold colloids of different sizes were prepared by reducing alkaline solutions of $HAuCl_4$ with different molar ratios of the reducing agent THPC, which also acts as a stabilizing agent. After generation of the gold sol, powdered ZrO_2 was added to the acidified (pH = 2) colloidal suspension, stirred for 1 h and the resulting catalyst was separated by centrifugation and washed three times with water.

4.3. Catalyst Characterization

The starting colloids were analyzed, after 1:1 dilution with distilled water, by means of UV-Vis spectroscopy, using a Perkin Elmer Lambda 650 spectrometer.

X-ray diffraction (XRD) measurements of powder catalysts were recorded between diffraction angels 20° and 80° (step size 0.017°, 0.53 s acquisition time) using Cu K_α radiation (1.54600 Å) on a PANalytical (Malvern Panalytical, Kassel, Germany) X'Pert Pro instrument. In addition, rotating sample holders were used.

The gold content of the supported catalysts was determined by X-ray fluorescence (XRF, Bruker S4 Pioneer, Bruker, Billerica, MA, USA) and the concentration of gold in the reaction solutions was investigated using inductively coupled plasma optical emission spectrometry (ICP-OES, Agilent 720/725-ES, Agilent, Waldbronn, Germany).

The presence of organics on the supported catalysts was studied using infrared (IR) spectroscopy. For this purpose, the powder catalysts were pressed into a pellet with KBr and spectra were recorded in transmission mode (Agilent Varian 600-IR, Agilent, Waldbronn, Germany).

Specific surface areas of the catalysts and the corresponding support material were determined by N_2 physisorption according to the Brunauer–Emmet–Teller (BET) method. The catalysts were pretreated by heating to 300 °C for 2 h under reduced pressure, prior to adsorption (Rubotherm BELSORP-mini II, MicrotracBEL, Osaka, Japan).

Particle size distributions were determined based on transmission electron microscopy (TEM). For the measurements, a suspension of the catalysts in ethanol was deposited on holey carbon-coated gold grids. Scanning TEM (STEM) measurements were performed under high annular dark field (HAADF) using a Fischione (Fischione Instruments, Export, PA, USA) model 3000 HAADF-STEM detector on a FEI (Hillsboro, OR, USA) Titan 80–300 aberration corrected TEM instrument (300 kV). The particle size distribution was determined by fitting ellipsoid shapes to the noble metal particles using ImageJ software (National Institute of Health, Rockville, MD, USA). The dispersion of Au was calculated from the mean particle size under the assumption of spherical particles using tabulated values for the volume of Au atoms in bulk and the area of surface Au [62].

The XPS measurements were carried out under ultra-high vacuum with a base pressure in 10^{-10} mbar range. Core-level spectra were recorded under normal emission with a Scienta (Scienta Omicron, Uppsala, Sweden) R4000 electron analyzer using Al-Kα radiation (1486.6 eV). In addition to the survey XP spectra (Figure S3) the detailed C1s, O1s, P2p, Au4f, and Zr3d XP spectra were recorded. For the determination of the atomic percentage the CasaXPS software was used. In case of Au4f line the XP spectra were deconvoluted into two doublets corresponding to Au^0 and Au^{3+} states (Figure S4).

4.4. Selective Oxidation of HMF

Catalytic reactions were conducted in Teflon® inlets placed in home-built 52 mL autoclaves that were magnetically stirred. The appropriate amount of water was added to 5 mL of a 0.2 M HMF and the desired amount of NaOH was added as a 2.5 M solution to give a total volume of 10 mL. After the addition of the catalyst (calculated based on the nominal metal loading of 2 wt.%), the reactors were closed and the desired pressure was adjusted after purging the reactor three times with synthetic air. The stirring speed was set to the maximum value and the reactors were heated with heating sleeves controlled by a thermocouple inside the reaction mixture. The time at which the solution first reached the desired temperature was set as the starting point of the reaction (t = 0). The reactors were cooled to room temperature in an ice bath after the reactions followed by depressurizing and separating the catalyst via decantation. Samples of the liquid phase were taken before and after the reactions, filtered with a 0.45 µm Pall Teflon filter and diluted for HPLC analysis (HitachiPrimaide, Hitachi, Chiyoda, Tokyo, Japan, Bio-Rad Aminex HPX-87H column, solvent 5 mM H_2SO_4, temperature 50 °C, 50 bar).

5. Conclusions

Au particles with different sizes between 2.1 and 2.9 nm supported on ZrO_2 showed varying product distributions in HMF oxidation. While small particles were identified to be highly active in aldehyde oxidation, slightly larger particles favor alcohol oxidation. The differences in the catalytic activity can be attributed to the presence of surface steps on larger particles, which are important in alcohol oxidation as well as strongly bound H on smaller particles hindering further oxidation. These differences could only be revealed by the very fine-tuning of the Au particle size on ZrO_2. Therefore, an ideal HMF oxidation catalysts should have small and larger Au particles to generate a type of bifunctional catalysts of the same metal in order to achieve the highest activity in the oxidation of both functional groups of HMF.

Supplementary Materials: The following are available online at http://www.mdpi.com/2073-4344/10/3/342/s1, Figure S1: UV-Vis spectra of the colloidal suspensions. Figure S2: XRD patterns of the supported as prepared catalysts and the pure ZrO_2 support material. Figure S3: XP spectra of the supported as prepared catalysts. Figure S4: Au4f XP spectra of the supported catalysts. Figure S5: XRD patterns of the spent catalysts after catalytic reactions.

Author Contributions: The topic was developed by J.-D.G. and O.S. Catalysts were prepared by P.D. followed by characterization and testing in the selective oxidation of HMF by O.S. XPS measurements were performed by A.N. and C.W. and data were evaluated by P.D. TEM measurements were performed by X.H. and evaluated by O.S. The concept of the manuscript was discussed by O.S., J.-D.G., P.D., and E.S. and the manuscript was written with contributions from all authors. All authors have read and agreed to the published version of the manuscript.

Funding: This work was funded and supported by KIT and the recently accepted FNR-project (KEFIP, FKZ: 22010718). This work was partly carried out with the support of the Karlsruhe Nano Micro Facility (KNMF, www.knmf.kit.edu), a Helmholtz Research Infrastructure at Karlsruhe Institute of Technology (KIT, www.kit.edu).

Acknowledgments: The authors thank Yakub Fam (ITCP) for performing the TEM measurements and Angela Deutsch for the N_2 physisorption. We acknowledge support by the KIT-Publication Fund of the Karlsruhe Institute of Technology.

Conflicts of Interest: The authors declare no conflict of interest.

References

1. Nikolau, B.J.; Perera, M.A.D.N.; Brachova, L.; Shanks, B. Platform biochemicals for a biorenewable chemical industry. *Plant J.* **2008**, *54*, 536–545. [CrossRef]
2. Henrich, E.; Dahmen, N.; Dinjus, E.; Sauer, J. The role of biomass in a future world without fossil fuels. *Chem. Ing. Tech.* **2015**, *87*, 1667–1685. [CrossRef]
3. Esposito, D.; Antonietti, M. Redefining biorefinery: The search for unconventional building blocks for materials. *Chem. Soc. Rev.* **2015**, *44*, 5821–5835. [CrossRef]

4. van Putten, R.-J.; van der Waal, J.C.; de Jong, E.; Rasrendra, C.B.; Heeres, H.J.; de Vries, J.G. Hydroxymethylfurfural, A Versatile Platform Chemical Made from Renewable Resources. *Chem. Rev.* **2013**, *113*, 1499–1597. [CrossRef]
5. Rosatella, A.A.; Simeonov, S.P.; Frade, R.F.M.; Afonso, C.A.M. 5-Hydroxymethylfurfural (HMF) as a building block platform: Biological properties, synthesis and synthetic applications. *Green Chem.* **2011**, *13*, 754–793. [CrossRef]
6. Kuster, B.F.M. 5-Hydroxymethylfurfural (HMF). A Review Focussing on its Manufacture. *Starch-Stärke* **1990**, *42*, 314–321. [CrossRef]
7. Steinbach, D.; Kruse, A.; Sauer, J. Pretreatment technologies of lignocelulosic biomass in water in view of furfural and 5-hydroxymethylfurfural production- A review. *Biomass Convers. Biorefin.* **2017**, *7*, 247–274. [CrossRef]
8. Chatterjee, M.; Ishizaka, T.; Kawanami, H. Selective hydrogenation of 5-hydroxymethylfurfural to 2,5-bis-(hydroxymethyl)furan using Pt/MCM-41 in an aqueous medium: A simple approach. *Green Chem.* **2014**, *16*, 4734–4739. [CrossRef]
9. Chatterjee, M.; Ishizaka, T.; Chatterjee, A.; Kawanami, H. Dehydrogenation of 5-hydroxymethylfurfural to diformylfuran in compressed carbon dioxide: An oxidant free approach. *Green Chem.* **2017**, *19*, 1315–1326. [CrossRef]
10. Hengst, K.; Schubert, M.; Kleist, W.; Grunwaldt, J.-D. Hydrodeoxygenation of Lignocellulose-Derived Platform Molecules. In *Catalytic Hydrogenation for Biomass Valorization*; Rinaldi, R., Ed.; The Royal Society of Chemistry: Cambridge, UK, 2014; pp. 125–150.
11. Werpy, T.; Petersen, G. *Top Value Added Chemicals from Biomass*; Natural Renewable Energy Laboratory: Golden, CO, USA, 2004.
12. Bozell, J.J.; Petersen, G.R. Technology development for the production of biobased products from biorefinery carbohydrates-the US Department of Energy's "Top 10" revisited. *Green Chem.* **2010**, *12*, 539–554. [CrossRef]
13. Sousa, A.F.; Vilela, C.; Fonseca, A.C.; Matos, M.; Freire, C.S.R.; Gruter, G.-J.M.; Coelho, J.F.J.; Silvestre, A.J.D. Biobased polyesters and other polymers from 2,5-furandicarboxylic acid: A tribute to furan excellency. *Polym. Chem.* **2015**, *6*, 5961–5983. [CrossRef]
14. Zhang, D.; Dumont, M.-J. Advances in polymer precursors and bio-based polymers synthesized from 5-hydroxymethylfurfural. *J. Polym. Sci. Part A Polym. Chem.* **2017**, *55*, 1478–1492. [CrossRef]
15. Miura, T.; Kakinuma, H.; Kawano, T.; Matsuhisa, H. Method for Producing Furan-2,5-dicarboxylic Acid. U.S. Patent 7,411,078, 12 August 2008.
16. Krystof, M.; Pérez-Sánchez, M.; Domínguez de María, P. Lipase-Mediated Selective Oxidation of Furfural and 5-Hydroxymethylfurfural. *ChemSusChem* **2013**, *6*, 826–830. [CrossRef] [PubMed]
17. Dijkman, W.P.; Groothuis, D.E.; Fraaije, M.W. Enzyme-Catalyzed Oxidation of 5-Hydroxymethylfurfural to Furan-2,5-dicarboxylic Acid. *Angew. Chem. Int. Ed.* **2014**, *53*, 6515–6518. [CrossRef] [PubMed]
18. Vuyyuru, K.R.; Strasser, P. Oxidation of biomass derived 5-hydroxymethylfurfural using heterogeneous and electrochemical catalysis. *Catal. Today* **2012**, *195*, 144–154. [CrossRef]
19. Barwe, S.; Weidner, J.; Cychy, S.; Morales, D.M.; Dieckhöfer, S.; Hiltrop, D.; Masa, J.; Muhler, M.; Schuhmann, W. Electrocatalytic Oxidation of 5-(Hydroxymethyl)furfural Using High-Surface-Area Nickel Boride. *Angew. Chem. Int. Ed.* **2018**, *57*, 11460–11464. [CrossRef]
20. Taitt, B.J.; Nam, D.-H.; Choi, K.-S. A Comparative Study of Nickel, Cobalt, and Iron Oxyhydroxide Anodes for the Electrochemical Oxidation of 5-Hydroxymethylfurfural to 2,5-Furandicarboxylic Acid. *ACS Catal.* **2018**, *9*, 660–670. [CrossRef]
21. Partenheimer, W.; Grushin, V.V. Synthesis of 2,5-Diformylfuran and Furan-2,5-Dicarboxylic Acid by Catalytic Air-Oxidation of 5-Hydroxymethylfurfural. Unexpectedly Selective Aerobic Oxidation of Benzyl Alcohol to Benzaldehyde with Metal=Bromide Catalysts. *Adv. Synth. Catal.* **2001**, *343*, 102–111. [CrossRef]
22. Ait Rass, H.; Essayem, N.; Besson, M. Selective aqueous phase oxidation of 5-hydroxymethylfurfural to 2,5-furandicarboxylic acid over Pt/C catalysts: Influence of the base and effect of bismuth promotion. *Green Chem.* **2013**, *15*, 2240–2251. [CrossRef]
23. Davis, S.E.; Houk, L.R.; Tamargo, E.C.; Datye, A.K.; Davis, R.J. Oxidation of 5-hydroxymethylfurfural over supported Pt, Pd and Au catalysts. *Catal. Today* **2011**, *160*, 55–60. [CrossRef]

24. Yu, H.; Kim, K.-A.; Kang, M.J.; Hwang, S.Y.; Cha, H.G. Carbon Support with Tunable Porosity Prepared by Carbonizing Chitosan for Catalytic Oxidation of 5-Hydroxylmethylfurfural. *ACS Sus. Chem. Eng.* **2018**, *7*, 3742–3748. [CrossRef]
25. Siyo, B.; Schneider, M.; Radnik, J.; Pohl, M.-M.; Langer, P.; Steinfeldt, N. Influence of support on the aerobic oxidation of HMF into FDCA over preformed Pd nanoparticle based materials. *Appl. Catal. A* **2014**, *478*, 107–116. [CrossRef]
26. Chen, C.; Li, X.; Wang, L.; Liang, T.; Wang, L.; Zhang, Y.; Zhang, J. Highly Porous Nitrogen- and Phosphorus-Codoped Graphene: An Outstanding Support for Pd Catalysts to Oxidize 5-Hydroxymethylfurfural into 2,5-Furandicarboxylic Acid. *ACS Sus. Chem. Eng.* **2017**, *5*, 1300–11306. [CrossRef]
27. Lei, D.; Yu, K.; Li, M.-R.; Wang, Y.; Wang, Q.; Liu, T.; Liu, P.; Lou, L.-L.; Wang, G.; Liu, S. Facet Effect of Single-Crystalline Pd Nanocrystals for Aerobic Oxidation of 5-Hydroxymethyl-2-furfural. *ACS Catal.* **2017**, *7*, 421–432. [CrossRef]
28. Casanova, O.; Iborra, S.; Corma, A. Biomass into Chemicals: Aerobic Oxidation of 5-Hydroxymethyl-2-furfural into 2,5-Furandicarboxylic Acid with Gold Nanoparticle Catalysts. *ChemSusChem* **2009**, *2*, 1138–1144. [CrossRef] [PubMed]
29. Gorbanev, Y.Y.; Klitgaard, S.K.; Woodley, J.M.; Christensen, C.H.; Riisager, A. Gold-Catalyzed Aerobic Oxidation of 5-Hydroxymethylfurfural in Water at Ambient Temperature. *ChemSusChem* **2009**, *2*, 672–675. [CrossRef] [PubMed]
30. Cai, J.; Ma, H.; Zhang, J.; Song, Q.; Du, Z.; Huang, Y.; Xu, J. Gold Nanoclusters Confined in a Supercage of Y Zeolite for Aerobic Oxidation of HMF under Mild Conditions. *Chem. Eur. J.* **2013**, *19*, 14215–14223. [CrossRef]
31. Miao, Z.; Zhang, Y.; Pan, X.; Wu, T.; Zhang, B.; Li, J.; Yi, T.; Zhang, Z.; Yang, X. Superior catalytic performance of $Ce_{1-x}Bi_xO_{2-\delta}$ solid solution and $Au/Ce_{1-x}Bi_xO_{2-\delta}$ for 5-hydroxymethylfurfural conversion in alkaline aqueous solution. *Catal. Sci. Technol.* **2015**, *5*, 1314–1322. [CrossRef]
32. Gupta, N.K.; Nishimura, S.; Takagaki, A.; Ebitani, K. Hydrotalcite-supported gold-nanoparticle-catalyzed highly efficient base-free aqueous oxidation of 5-hydroxymethylfurfural into 2,5-furandicarboxylic acid under atmospheric oxygen pressure. *Green Chem.* **2011**, *13*, 824–827. [CrossRef]
33. Gao, T.; Gao, T.; Fang, W.; Cao, Q. Base-free aerobic oxidation of 5-hydroxymethylfurfural to 2,5-furandicarboxylic acid in water by hydrotalcite-activated carbon composite supported gold catalyst. *Mol. Catal.* **2017**, *439*, 171–179. [CrossRef]
34. Masoud, N.; Donoeva, B.; de Jongh, P.E. Stability of gold nanocatalysts supported on mesoporous silica for the oxidation of 5-hydroxymethyl furfural to furan-2,5-dicarboxylic acid. *Appl. Catal. A* **2018**, *561*, 150–157. [CrossRef]
35. Schade, O.R.; Dannecker, P.-K.; Kalz, K.F.; Steinbach, D.; Meier, M.A.; Grunwaldt, J.-D. Direct Catalytic Route to Biomass-Derived 2,5-Furandicarboxylic Acid and Its Use as Monomer in a Multicomponent Polymerization. *ACS Omega* **2019**, *4*, 16972–16979. [CrossRef] [PubMed]
36. Haruta, M.; Kobayashi, T.; Sano, H.; Yamada, N. Novel Gold Catalysts for the Oxidation of Carbon Monoxide at a Temperature far Below 0 °C. *Chem. Lett.* **1987**, *16*, 405–408. [CrossRef]
37. Hashmi, A.S.K.; Hutchings, G.J. Gold Catalysis. *Angew. Chem. Int. Ed.* **2006**, *45*, 7896–7936. [CrossRef]
38. Grunwaldt, J.-D.; Kiener, C.; Wögerbauer, C.; Baiker, A. Preparation of Supported Gold Catalysts for Low-Temperature CO Oxidation via "Size-Controlled" Gold Colloids. *J. Catal.* **1999**, *181*, 223–232. [CrossRef]
39. Haruta, M.; Uphade, B.; Tsubota, S.; Miyamoto, A. Selective oxidation of propylene over gold deposited on titanium-based oxides. *Res. Chem. Intermed.* **1998**, *24*, 329–336. [CrossRef]
40. Haruta, M. When gold is not noble: Catalysis by nanoparticles. *Chem. Rec.* **2003**, *3*, 75–87. [CrossRef]
41. Herzing, A.A.; Kiely, C.J.; Carley, A.F.; Landon, P.; Hutchings, G.J. Identification of active gold nanoclusters on iron oxide supports for CO oxidation. *Science* **2008**, *321*, 1331–1335. [CrossRef]
42. Bogdanchikova, N.; Pestryakov, A.; Farias, M.; Diaz, J.A.; Avalos, M.; Navarrete, J. Formation of TEM-and XRD-undetectable gold clusters accompanying big gold particles on TiO_2–SiO_2 supports. *Solid State Sci.* **2008**, *10*, 908–914. [CrossRef]
43. Haider, P.; Kimmerle, B.; Krumeich, F.; Kleist, W.; Grunwaldt, J.-D.; Baiker, A. Gold-catalyzed aerobic oxidation of benzyl alcohol: Effect of gold particle size on activity and selectivity in different solvents. *Catal. Lett.* **2008**, *125*, 169–176. [CrossRef]

44. Adnan, R.H.; Andersson, G.G.; Polson, M.I.; Metha, G.F.; Golovko, V.B. Factors influencing the catalytic oxidation of benzyl alcohol using supported phosphine-capped gold nanoparticles. *Catal. Sci. Technol.* **2015**, *5*, 1323–1333. [CrossRef]
45. Tsunoyama, H.; Ichikuni, N.; Tsukuda, T. Microfluidic synthesis and catalytic application of PVP-stabilized, ~1 nm gold clusters. *Langmuir* **2008**, *24*, 11327–11330. [CrossRef] [PubMed]
46. Tofighi, G.; Lichtenberg, H.; Pesek, J.; Sheppard, T.L.; Wang, W.; Schöttner, L.; Rinke, G.; Dittmeyer, R.; Grunwaldt, J.-D. Continuous microfluidic synthesis of colloidal ultrasmall gold nanoparticles: In Situ study of the early reaction stages and application for catalysis. *React. Chem. Eng.* **2017**, *2*, 876–884. [CrossRef]
47. Sasirekha, N.; Sangeetha, P.; Chen, Y.-W. Bimetallic Au–Ag/CeO$_2$ catalysts for preferential oxidation of CO in hydrogen-rich stream: Effect of calcination temperature. *J. Phys. Chem. C* **2014**, *118*, 15226–15233. [CrossRef]
48. Li, Q.; Wang, H.; Tian, Z.; Weng, Y.; Wang, C.; Ma, J.; Zhu, C.; Li, W.; Liu, Q.; Ma, L. Selective oxidation of 5-hydroxymethylfurfural to 2,5-furandicarboxylic acid over Au/CeO$_2$ catalysts: The morphology effect of CeO$_2$. *Catal. Sci. Technol.* **2019**, *9*, 1570–1580. [CrossRef]
49. Albonetti, S.; Lolli, A.; Morandi, V.; Migliori, A.; Lucarelli, C.; Cavani, F. Conversion of 5-hydroxymethylfurfural to 2,5-furandicarboxylic acid over Au-based catalysts: Optimization of active phase and metal–support interaction. *Appl. Catal. B* **2015**, *163*, 520–530. [CrossRef]
50. Schade, O.R.; Kalz, K.F.; Neukum, D.; Kleist, W.; Grunwaldt, J.-D. Supported gold- and silver-based catalysts for the selective aerobic oxidation of 5-(hydroxymethyl)furfural to 2,5-furandicarboxylic acid and 5-hydroxymethyl-2-furancarboxylic acid. *Green Chem.* **2018**, *20*, 3530–3541. [CrossRef]
51. Davis, S.E.; Zope, B.N.; Davis, R.J. On the mechanism of selective oxidation of 5-hydroxymethylfurfural to 2,5-furandicarboxylic acid over supported Pt and Au catalysts. *Green Chem.* **2012**, *14*, 143–147. [CrossRef]
52. Megías-Sayago, C.; Lolli, A.; Bonincontro, D.; Penkova, A.; Albonetti, S.; Cavani, F.; Odriozola, J.A.; Ivanova, S. Effect of gold particles size over Au/C catalyst selectivity in HMF oxidation reaction. *ChemCatChem* **2020**, *12*, 1177–1183. [CrossRef]
53. Ishida, T.; Kinoshita, N.; Okatsu, H.; Akita, T.; Takei, T.; Haruta, M. Influence of the support and the size of gold clusters on catalytic activity for glucose oxidation. *Angew. Chem. Int. Ed.* **2008**, *47*, 9265–9268. [CrossRef]
54. Sahu, R.; Dhepe, P.L. Synthesis of 2,5-furandicarboxylic acid by the aerobic oxidation of 5-hydroxymethyl furfural over supported metal catalysts. *React. Kinet. Mech. Catal.* **2014**, *112*, 173–187. [CrossRef]
55. Megías-Sayago, C.; Chakarova, K.; Penkova, A.; Lolli, A.; Ivanova, S.; Albonetti, S.; Cavani, F.; Odriozola, J.A. Understanding the role of the acid sites in HMF oxidation to FDCA reaction over gold catalysts: Surface investigation on Ce$_x$Zr$_{1-x}$O$_2$ compounds. *ACS Catal.* **2018**, *8*, 11154–11164. [CrossRef]
56. Porta, F.; Prati, L.; Rossi, M.; Coluccia, S.; Martra, G. Metal sols as a useful tool for heterogeneous gold catalyst preparation: Reinvestigation of a liquid phase oxidation. *Catal. Today* **2000**, *61*, 165–172. [CrossRef]
57. Sun, K.-Q.; Luo, S.-W.; Xu, N.; Xu, B.-Q. Gold nano-size effect in Au/SiO$_2$ for selective ethanol oxidation in aqueous solution. *Catal. Lett.* **2008**, *124*, 238–242. [CrossRef]
58. Guan, Y.; Hensen, E.J. Ethanol dehydrogenation by gold catalysts: The effect of the gold particle size and the presence of oxygen. *Appl. Catal. A* **2009**, *361*, 49–56. [CrossRef]
59. Zope, B.N.; Hibbitts, D.D.; Neurock, M.; Davis, R.J. Reactivity of the Gold/Water Interface during Selective Oxidation Catalysis. *Science* **2010**, *330*, 74–78. [CrossRef]
60. Mondelli, C.; Ferri, D.; Grunwaldt, J.-D.; Krumeich, F.; Mangold, S.; Psaro, R.; Baiker, A. Combined liquid-phase ATR-IR and XAS study of the Bi-promotion in the aerobic oxidation of benzyl alcohol over Pd/Al$_2$O$_3$. *J. Catal.* **2007**, *252*, 77–87. [CrossRef]
61. Kartusch, C.; van Bokhoven, J.A. Hydrogenation over gold catalysts: The interaction of gold with hydrogen. *Gold Bull.* **2009**, *42*, 343–348. [CrossRef]
62. Bergeret, G.; Gallezot, P. Particle size and dispersion measurements. In *Handbook of Heterogeneous Catalysis*; Ertl, G., Knözinger, H., Schüth, F., Weitkamp, J., Eds.; Wiley-VCH: Weinheim, Germany, 2008; pp. 738–765.

© 2020 by the authors. Licensee MDPI, Basel, Switzerland. This article is an open access article distributed under the terms and conditions of the Creative Commons Attribution (CC BY) license (http://creativecommons.org/licenses/by/4.0/).

Article

Autocatalytic Fractionation of Wood Hemicelluloses: Modeling of Multistage Operation

Mar López, Valentín Santos and Juan Carlos Parajó *

Department of Chemical Engineering, Faculty of Science, University of Vigo (Campus Ourense), As Lagoas, 32004 Ourense, Spain; marlopezr@uvigo.es (M.L.); vsantos@uvigo.es (V.S.)
* Correspondence: jcparajo@uvigo.es; Tel.: +34-988-387-033

Received: 21 February 2020; Accepted: 12 March 2020; Published: 17 March 2020

Abstract: *Eucalyptus globulus* wood samples were treated with hot, compressed water (autohydrolysis) in consecutive stages under non-isothermal conditions in order to convert the hemicellulose fraction into soluble compounds through reactions catalyzed by in situ generated acids. The first stage was a conventional autohydrolysis, and liquid phase obtained under conditions leading to an optimal recovery of soluble saccharides was employed in a new reaction (second crossflow stage) using a fresh wood lot, in order to increase the concentrations of soluble saccharides. In the third crossflow stage, the best liquid phase from the second stage was employed to solubilize the hemicelluloses from a fresh wood lot. The concentration profiles determined for the soluble saccharides, acids, and furans present in the liquid phases from the diverse crossflow stages were employed for kinetic modeling, based on pseudohomogeneous reactions and Arrhenius-type dependence of the kinetic coefficients on temperature. Additional characterization of the reaction products by High Pressure Size Exclusion Chromatography, High Performance Anion Exchange Chromatography with Pulsed Amperometric Detection, and Matrix Assisted Laser Desorption/Ionization Time of Flight Mass Spectrometry provided further insight on the properties of the soluble saccharides present in the various reaction media.

Keywords: *Eucalyptus globulus* wood; cross-flow autohydrolysis; kinetic modeling; hemicellulose-derived products

---

1. Introduction

A major concern of sustainability is the consumption of nonrenewable resources, and their replacement by renewable ones. Lignocellulosic materials (LCM) are interesting renewable raw materials for industry, due to their availability, widespread occurrence, carbon-neutral character, and low cost. *Eucalyptus globulus* shows favorable features as a feedstock for industry, including its productivity, high cellulose content, and hemicelluloses largely dominated by acetylated glucuronoxylan [1,2], with minor amounts of other components [3].

The industrial utilization of LCM can be accomplished using the biorefinery approach, which entails the selective separation of the polymeric components of LCM (cellulose, hemicelluloses and lignin) through "fractionation" treatments, and the further transformation of the resulting fractions into commercial products, taking into account the principles of green chemistry and circular economy [4].

The fractionation of woods in the scope of biorefineries can be achieved by diverse methods, depending on the fraction (or fractions) targeted. In particular, hemicelluloses can be selectively separated from cellulose and lignin by performing a mild acidic treatment in aqueous media. Under selected conditions, hemicelluloses can be converted into soluble saccharides (or saccharide-decomposition products), whereas cellulose and lignin are scarcely altered and remain in solid phase. The solids from autohydrolysis can be processed (for example, by delignification, enzymatic hydrolysis or acidic processing under harsh conditions) to yield a wide scope of products, including sugar solutions suitable as fermentation media, furans, cellulose pulp, and organic acids.

The acidic conditions promoting the hemicellulose depolymerization in aqueous media can be achieved by external addition of an acidic catalyst (typically, a mineral acid in prehydrolysis treatments), or just by aqueous processing (autohydrolysis treatments). In the latter case, the breakdown of the glycosidic bonds in hemicelluloses is catalyzed by hydronium ions, which come at the beginning of the reaction from water autoionization and uronic substituents in hemicelluloses. In further reaction stages, the acetic acid progressively released in situ from the acetyl groups becomes the major source of hydronium ions (autocatalytic reaction). As a consequence, the comparatively high acetyl group content of *Eucalyptus globulus* wood is an advantage for this type of reactions.

Besides the ability of autohydrolysis for causing an extensive and selective separation of hemicelluloses from the rest of the polymeric wood components, other advantages (including its "green" character, the limited problems derived from equipment corrosion, and the fact that no neutralization or sludge management stages are necessary) contribute to the interest of this technology for wood fractionation [5].

When autohydrolysis is performed under suitable operational conditions, the acetylated glucuronoxylan in *E. globulus* wood is mainly converted into xylooligosaccharides (XOs), which in turn can be converted into xylose and/or xylose-decomposition products. The kinetic modeling of single-stage *E. globulus* wood autohydrolysis has been considered in literature, including the hydrolysis of xylan under isothermal and non-isothermal conditions [5,6] and the generation of acetic acid from acetyl groups [7].

The hydrolysis of *E. globulus* xylan into soluble products has been interpreted in literature [5,6] using a mechanism based on the following hypotheses:

- The reactions taking place in the reaction medium are irreversible and present a first-order, pseudohomogeneous kinetics
- The kinetic coefficients involved in the mechanism follow the Arrhenius equation
- *Eucalyptus* wood xylan (X_n) is made up of two fractions, one being unreactive under the operational conditions, and the other (susceptible xylan, X_{nS}) can be hydrolyzed to yield high-molecular weight oligosaccharides (XO_H)
- The relative proportion of hydrolyzable xylan is measured by the "susceptible fraction" $\alpha_{X_{nS}}$ (g susceptible xylan/g xylan)
- XO_H are split into low molecular weight oligosaccharides (XO_L)
- XO_L are hydrolyzed into xylose (X), which is dehydrated into furfural (F)

This mechanism can be summarized as follows:

$$X_{nS} \xrightarrow{k_1} XO_H \xrightarrow{k_2} XO_L \xrightarrow{k_3} X \xrightarrow{k_4} F$$

Although xylan can be hydrolyzed at high yield by autohydrolysis treatments, the volumetric concentrations of the soluble saccharides and furfural in liquid phase are limited, owing to both the xylan content of wood and the concentration of solids in the reaction media (typically, 10–12.5 g oven-dried wood/100 g water). As a consequence, the volumetric concentration range is below the desired threshold for a number of applications. Increased concentrations of the target products can be achieved by concentration (for example, by evaporation or membrane technologies) or by coupling reaction stages [8]. This latter strategy entails a number of issues, related to the higher concentrations of catalyst (acetic acid derived from acetyl groups) and to the increased conversion of susceptible substrates by hydrolysis, dehydration and/or condensation reactions.

To our knowledge, no studies have been reported on the kinetic modeling of multistage autohydrolysis. The closest precedents for our study are the articles reported for the single stage autohydrolysis of *Eucalyptus globulus* wood [5,6]. Other studies dealing with wood autohydrolysis have been reported for *Acacia dealbata* (a hardwood with hemicelluloses mainly made up of xylan) [9] and *Pinus pinaster* (a softwood with hemicelluloses mainly made up of glucomannan and xylan) [10]. Additional kinetic studies of single-stage autohydrolysis have been reported for a number or non-wood materials, including corncobs [11], vine shoots [12], *Arundo donax* [13], barley husks [14], bamboo [15], *Cytisus scoparius* [16], rye straw [17], and almond shells [18].

This work deals with the kinetic modeling of *E. globulus* wood processing by multistage autohydrolysis. Wood was processed in three crossflow stages (as indicated in Figure 1), where the first one was a conventional hydrothermal treatment (wood was processed with water), whereas fresh wood lots were treated in the second and third stages with the liquid phases coming from stages 1 and 2, respectively. Operation was carried out under non-isothermal conditions. In experiments performed at temperatures within the range 160–220 °C, samples were withdrawn at selected reaction times, and assayed for the following reaction products: XOs, X, F, glucosyl groups in oligosaccharides (GOs), arabinosyl groups in oligosaccharides (ArOs), glucose (Gl), arabinose (Ar), acetyl groups in oligosaccharides (AcO), hydroxymethylfurfural (HMF), acetic acid (AcH), formic acid (FA), and levulinic acid (LA). Kinetic models giving a close interpretation of the experimental data were developed, and the properties of the soluble products leaving the diverse reaction stages were assessed by High Pressure Size Exclusion Chromatography (HPSEC), High Performance Anion Exchange Chromatography with Pulsed Amperometric Detection (HPAEC-PAD), and Matrix Assisted Laser Desorption/Ionization Time of Flight Mass Spectrometry (MALDI-TOF MS).

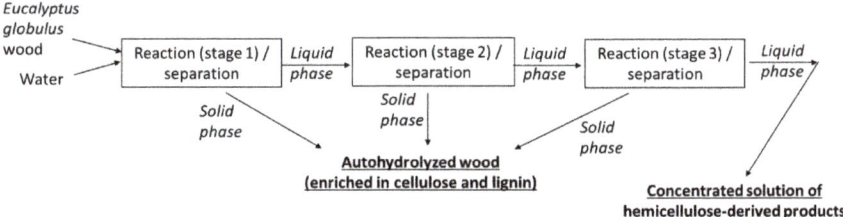

Figure 1. Processing scheme considered in this study.

2. Results and Discussion

2.1. Wood Composition

The composition of lignocellulosic materials depends on the type of substrate and on other specific factors (growth conditions and age of the specimen, sampling method, etc.). The reactions involved in autohydrolysis processing are affected by the composition of the wood sample: for example, the acetyl group content measures the maximum amount of acetic acid that can be released (and so the concentration of hydronium ions with catalytic activity present in the media), whereas the ash fraction may present neutralizing ability. Because of this, the wood lot employed in experiments was characterized in detail, yielding the results listed in Table 1. Cellulose (40.8 wt %) and acid-insoluble lignin (22.5 wt %) were the major wood components; whereas xylan (16.8 wt %), arabinosyl groups (0.63 wt %) and acetyl groups (3.16 wt %) were the target substrates for autohydrolysis.

Table 1. Composition of the *E. globulus* wood lot employed in experiments (data obtained from triplicate determinations).

Fraction	Weight % (Oven-Dry Basis)	Standard Deviations
Glucan (G_n)	40.8	0.71
Xylan (X_n)	16.8	0.15
Arabinosyl groups	0.63	0.02
Acetyl groups	3.16	0.08
Klason lignin	22.5	0.30
Acid soluble lignin	1.11	0.13
Uronic acids	5.35	0.38
Ash	0.29	0.05
Water-soluble extractives	4.36	0.12
Other (average value, by difference)	4.92	-

2.2. Effects Caused by the First Reaction Stage

According to Figure 1, the first reaction stage was a conventional non-isothermal autohydrolysis reaction, similar to the one described in literature dealing with the kinetics of X_n conversion into XOs, X and F [6]. In the present study, the concentrations of additional reaction products (GOs, ArOs, Gl, Ar, AcO, AcH, FA, HMF, and LA) were also determined to allow a deeper understanding of the underlying chemistry. The volumetric concentrations achieved at the considered temperatures are listed in Table 2. The concentration profiles determined for the products derived from xylan (XOs, X and F) were essentially the same as those reported in an earlier study [6], and are not discussed in depth here. In order to facilitate the interpretation of data, the most important experimental trends can be summarized as follows: XOs were generated from X_n by partial hydrolysis, and then converted into X. As a consequence, XOs behaved as reaction intermediates, reaching a maximum concentration (15.9 g/L) under conditions of intermediate severity (195 °C, or reaction time = 35.4 min). The concentration of X increased steadily along the heating process (to reach a maximum concentration of 10.2 g/L under the harshest conditions assayed), but the concentration increases were below the stoichiometric ones calculated from the decrease in XOs concentration, owing to the dehydration of X into F. As observed for X, the concentrations of F (generated by the dehydration of pentoses) increased steadily with the reaction time. The variation pattern observed for AcO was closely related to the one observed for XOs, with a maximum concentration (3.55 g/L) taking place under the same conditions. AcH (resulting from AcO hydrolysis) reached concentrations in the range 0.01–3.04 g/L, providing increased amounts of hydronium ions at increased temperatures. ArOs and Ar reached low concentrations (derived from the scarce amount of arabinosyl groups in the substrate), and were more susceptible to hydrolysis/dehydration than XOs and X, respectively. The concentration of F (produced from pentoses) increased steadily along the experimental domain, up to reach 1.76 g/L under the severest conditions assayed. Glucan (ascribed to cellulose), an abundant fraction in the raw material, was scarcely affected by hydrolysis reactions, resulting in low concentrations of GOs, Gl (generated from GOs) and HMF (coming from Gl rehydration). In the reaction media, FA can be generated from two sources: formyl groups in lignin, and HMF rehydration (which yields equimolar amounts of FA and LA). Since no significant amounts of LA were detected, it can be concluded that FA was produced from formyl groups.

Table 2. Concentrations determined for the target products (in g/L) in the liquid phase from the first reaction stage.

Component	Temperature (°C)/Reaction Time (min)					
	170/26.4	180/30.0	185/31.8	187/32.5	190/33.6	192/34.3
GOs	0.754	0.982	1.02	1.05	1.11	1.14
XOs	1.75	8.13	9.88	13.1	14.1	14.8
ArOs	0.201	0.089	0.079	0.067	0.020	0.041
AcO	0.392	1.77	2.06	2.91	3.17	3.33
Gl	0.160	0.192	0.219	0.249	0.269	0.285
X	0.204	0.701	1.06	1.52	2.09	2.55
Ar	0.160	0.289	0.352	0.408	0.444	0.441
FA	0.025	0.058	0.067	0.079	0.079	0.100
AcH	0.010	0.211	0.275	0.379	0.462	0.554
LA	<0.01	<0.01	<0.01	<0.01	<0.01	<0.01
HMF	0.014	0.017	0.021	0.039	0.045	0.054
F	0.085	0.033	0.044	0.085	0.111	0.150
Component	Temperature (°C)/Reaction Time (min)					
	195/35.4	196/35.7	199/36.8	200/37.2	205/39.0	210/40.8	213/41.9
GOs	1.19	1.18	1.27	1.33	1.28	1.20	1.22
XOs	15.9	15.1	14.9	14.9	11.9	6.04	4.58
ArOs	<0.01	<0.01	<0.01	<0.01	<0.01	<0.01	<0.01
AcO	3.55	3.48	3.51	3.68	3.135	1.74	1.54
Gl	0.309	0.307	0.367	0.374	0.444	0.696	0.734
X	3.52	3.42	5.12	5.41	7.31	9.95	10.2
Ar	0.473	0.483	0.488	0.506	0.536	0.322	0.454
FA	0.111	0.120	0.132	0.154	0.185	0.276	0.291
AcH	0.700	0.705	1.05	1.05	1.56	2.78	3.04
LA	<0.01	<0.01	<0.01	<0.01	<0.01	<0.01	<0.01
HMF	0.065	0.067	0.098	0.106	0.151	0.282	0.310
F	0.207	0.233	0.335	0.377	0.727	1.53	1.76

2.3. Effects Caused by the Second Reaction Stage

Figure 1 shows that the second reaction stage was different from a conventional non-isothermal autohydrolysis reaction, since the liquid phase employed for wood processing contained reactive components (oligosaccharides, sugars, acetyl groups and furans) together with organic acids acting as catalyst sources. Because of this, the second reaction stage is expected to cause similar effects on wood than the first reaction stage, but the reactive components initially present in the liquid phase will undergo hydrolysis, dehydration, rehydration and/or condensation reactions, affecting to both kinetics and product yields. Table 3 shows the compositional data determined for experiments performed at temperatures within the range 160–220 °C using the solution from stage 1 that led to the maximum XOs concentration in the liquid phase. The overall kinetic pattern observed in the second stage was similar to the one observed in stage 1. For example, GOs and XOs behaved as reactions intermediates (reaching their maximum concentrations under conditions of intermediate severity), whereas ArOs reached their maximum concentration under the mildest conditions assayed, owing to their increased susceptibility to hydrolysis. As before, the concentrations of Gl and X increased steadily with the severity of treatments, whereas the concentration of Ar showed a defined maximum within the experimental domain, owing to its higher susceptibility to dehydration into F. The presence of organic acids, oligosaccharides and sugars at the beginning of the reaction resulted in concentration increases in the second stage that were non-proportional respect to the initial amounts of potential substrates available in the reaction medium. Owing to the increased hydrolysis of XOs, their maximum concentration (25.4 g/L, achieved at 190 °C) less than doubled the one achieved in the first stage (15.9 g/L, achieved at 195 C). Oppositely, when the second stage was performed at 190 °C, the concentrations of Gl, X, AcH, FA, F, and HMF more than doubled the respective concentrations achieved in the first stage, owing to the increased concentrations of their precursors in the reactions media.

Table 3. Concentrations determined for the target products (in g/L) in the liquid phase from the second reaction stage.

Component	Temperature (°C)/Reaction Time (min)					
	160/22.8	165/24.6	170/26.4	175/28.2	180/30.0	185/31.8
GOs	1.73	1.83	2.00	2.21	2.15	2.34
XOs	14.5	14.2	15.8	18.6	20.6	24.1
ArOs	0.115	0.198	0.182	0.096	0.044	0.010
AcO	3.37	3.14	3.33	4.15	4.26	5.11
Gl	0.417	0.396	0.456	0.454	0.538	0.506
X	4.10	4.34	4.49	5.29	5.73	7.00
Ar	0.573	0.650	0.672	0.831	0.878	0.943
FA	0.197	0.196	0.222	0.237	0.260	0.312
AcH	0.958	1.06	1.11	1.35	1.46	1.73
LA	< 0.01	< 0.01	< 0.01	< 0.01	< 0.01	< 0.01
HMF	0.057	0.070	0.070	0.090	0.090	0.120
F	0.225	0.278	0.286	0.38	0.446	0.558
Component	Temperature (°C)/Reaction Time (min)						
	190/33.6	195/35.4	200/37.2	205/39.0	210/40.8	215/42.6	220/44.4
GOs	2.29	2.35	2.15	2.49	2.16	1.73	1.77
XOs	25.4	25.0	21.9	18.5	16.6	7.54	1.99
ArOs	< 0.01	< 0.01	< 0.01	< 0.01	< 0.01	< 0.01	< 0.01
AcO	5.86	6.15	5.36	5.06	4.51	2.85	1.27
Gl	0.609	0.622	0.781	0.808	1.01	1.30	1.34
X	8.57	10.9	12.0	15.0	15.3	17.2	16.6
Ar	1.01	1.02	0.972	0.977	0.968	0.648	0.618
FA	0.350	0.386	0.420	0.441	0.456	0.533	0.563
AcH	2.05	2.58	2.90	4.03	4.32	6.37	6.45
LA	< 0.01	< 0.01	< 0.01	< 0.01	< 0.01	< 0.01	< 0.01
HMF	0.150	0.176	0.217	0.306	0.363	0.616	0.652
F	0.752	1.14	1.45	2.34	2.59	4.43	4.61

2.4. Effects Caused by the Third Reaction Stage

According to Figure 1, stage 3 was performed by treating a fresh wood lot with the reaction liquor from stage 2. For this purpose, the solution obtained in the experiment leading to a maximum XOs concentration in the second stage was employed. The experimental trends observed in the experiments performed at 160–215 °C (see concentration profiles in Table 4) were closely related to the ones described above for stage 2. Taking the conditions leading to the maximum XOs generation in the third stage as a basis for comparison, it can be noted that the concentrations of oligomers (GOs and XOs) and acetyl groups bound to oligomers (AcO) increased less than proportionally respect to the concentrations of the potential substrates initially available in the reaction medium. The same behavior was observed for Ar, which was easily converted into F under conditions of intermediate severity. In turn, the concentrations of X, Gl, organic acids, and furans increased more than proportionally, owing to the higher concentrations of their respective precursors.

Table 4. Concentrations determined for the target products (in g/L) in the liquid phase from the third reaction stage.

Component	Temperature (°C)/Reaction Time (min)					
	160/22.8	165/24.6	170/26.4	175/28.2	180/30.0	185/31.8
GOs	2.55	2.24	2.45	2.50	2.62	2.67
XOs	23.0	22.0	23.6	24.5	27.6	28.4
ArOs	0.177	0.095	0.174	0.094	0.031	<0.01
AcO	5.31	4.96	5.19	5.63	6.13	6.56
Gl	0.789	0.746	0.813	0.818	0.934	0.928
X	9.07	9.11	9.44	10.1	11.6	12.3
Ar	0.986	1.00	1.03	1.14	1.23	1.32
FA	0.386	0.401	0.417	0.407	0.450	0.484
AcH	2.34	2.28	2.47	2.6	3.04	3.14
LA	0.025	0.027	0.030	0.028	0.034	0.032
HMF	0.137	0.153	0.143	0.185	0.180	0.234
F	0.722	0.727	0.803	0.900	1.17	1.28
Component	Temperature (°C)/Reaction Time (min)					
	190/33.6	195/35.4	200/37.2	205/39.0	210/40.8	215/42.6
GOs	2.41	2.46	2.39	2.62	2.39	1.89
XOs	29.5	26.8	24.5	20.1	14.4	5.24
ArOs	<0.01	<0.01	<0.01	<0.01	<0.01	<0.01
AcO	6.87	6.88	6.32	5.91	4.41	2.70
Gl	1.10	1.20	1.34	1.46	1.60	1.86
X	15.6	18.7	19.8	22.1	21.4	20.3
Ar	1.38	1.53	1.49	1.27	1.22	0.806
FA	0.534	0.610	0.673	0.721	0.856	1.00
AcH	3.99	4.73	5.42	6.63	7.47	9.59
LA	0.034	0.023	0.033	0.029	0.051	0.099
HMF	0.320	0.406	0.446	0.647	0.722	1.00
F	1.89	2.44	3.09	4.17	5.56	7.67

2.5. Kinetic Modeling of Multistage Operation

The concentrations listed in Tables 2–4 were employed for kinetic modeling, in order to provide a deeper understanding of the major phenomena involved in the solubilization of the hemicellulosic polymers, using a unified model for the multistage processing suitable for quantitative assessments, design and economic evaluation.

Preliminary modeling attempts of the reactions governing the consumption of X_n, arabinosyl groups in solid phase (Ar_n), acetyl groups in solid phase (Ac_n), and the products derived from them were performed using models reported for single-stage autohydrolysis of diverse lignocellulosic materials, [4–6,9–17].

The preliminary validation of the various models confirmed that some common assumptions (kinetic modeling, individual reactions with first order kinetics involving coefficients with Arrhenius-type dependence on temperature) were suitable for modeling the multistage autohydrolysis of *Eucalyptus* wood. The best results for X_n solubilization were achieved assuming that the xylan in the native wood was made up of a "resistant xylan" (X_{nR}) fraction, and a "susceptible xylan" (X_{nS}) fraction for which the relative abundance was measured by the "susceptible fraction" (α_{XnS}, expressed as g hydrolyzable fraction/g xylan in wood). Concerning the generation of XOs, the existence of high molecular weight and low molecular weight fractions improved the fitting of data in the first stage, but this hypothesis did not improve significantly the interpretation of data regarding the second and third stages. This finding was ascribed to the higher participation of hydrolysis reactions of high molecular weight oligosaccharides, which resulted in products of lower average molecular weight (see Section 2.6). In turn, the assumption that a single type of XOs was generated from X_n led to satisfactory interpretation of data for each of the three processing stages. Similarly, the decomposition of ArOs into

Ar and the generation of AcH from AcO were well interpreted by single reactions. The material balances showed that AcH was not consumed by further reactions. Oppositely, the literature models did not provide a suitable interpretation of data regarding the generation of pentoses (X and Ar) and their further dehydration into furfural. Models involving the direct generation of XOs and ArOs from their precursors followed by the formation of the monomers and their dehydration into F (and, eventually, the partial conversion of this latter compound into decomposition products, DecP) overestimated the concentrations of X and Ar. Oppositely, modified models involving both the dehydration of pentoses into F and their consumption by parasitic reactions (as shown in Figure 2) led to satisfactory results. The consumption of xylose in acidic media by parasitic reactions different from rehydration has been postulated in literature. For example, xylose can be consumed by retroaldol fragmentation [19,20], or through the formation of reactive cyclic or acyclic intermediates [21–24] able to yield furfural or to participate in non-productive reactions with other reactive species present in the reaction media. On the other hand, satisfactory interpretation of the AcO concentration profiles was achieved when the presence of resistant and susceptible acetyl groups in wood (denoted Ac_{nR} and Ac_{nS}) was assumed, in relative proportions measured by the mass fraction of susceptible fraction respect to the total acetyl groups in wood (α_{AcnS}).

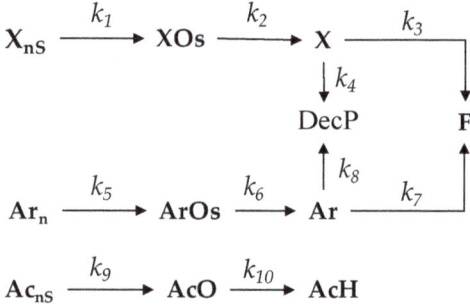

Figure 2. Kinetic model considered in this study.

Based on these ideas, the model presented in Figure 2 was employed for kinetic modeling of the multistage autohydrolysis.

Table 5 lists the equations governing the overall reaction scheme, and Table 6a presents the results determined for the susceptible fractions (α_{XnS} and α_{AcnS}) and for the pre-exponential factors and activation energies corresponding to the kinetic coefficients k_1 to k_{10} in Figure 2.; whereas Table 6b presents the values of the statistical parameter R^2 measuring the correlation between experimental and calculated data. R^2 of 0.921, 0.937 and 0.960 were calculated for the XOs production in the first, second and third stages, respectively. The increased R^2 obtained when fitting the XOs concentrations determined for the second and third stages were ascribed to the higher contents of low molecular weight oligomers. $R^2 > 0.95$ were calculated for X, F and AcH, whereas the poorer coefficients of determination observed for Ar, ArOs and AcO were ascribed to the increased relative errors resulting from their low concentrations.

Table 5. Equations employed for kinetic modeling of the reactions involving hemicellulosic polysaccharides and acetyl groups in wood.

$\alpha_{XnS} = \left[\dfrac{X_{nS}}{X_n}\right]$	(1)
$\dfrac{d[X_{nS}]}{dt} = -k_1 \cdot [X_{nS}]$	(2)
$\dfrac{d[XOs]}{dt} = k_1 \cdot [X_{nS}] - k_2 \cdot [XOs]$	(3)
$\dfrac{d[X]}{dt} = k_2 \cdot [XOs] - k_3 \cdot [X] - k_4 \cdot [X]$	(4)
$\dfrac{d[Ar_n]}{dt} = -k_5 \cdot [Ar_n]$	(5)
$\dfrac{d[ArOs]}{dt} = k_5 \cdot [Ar_n] - k_6 \cdot [ArOs]$	(6)
$\dfrac{d[Ar]}{dt} = k_6 \cdot [ArOs] - k_7 \cdot [Ar] - k_8 \cdot [Ar]$	(7)
$\dfrac{d[F]}{dt} = k_3 \cdot [X] + k_7 \cdot [Ar]$	(8)
$\alpha_{AcnS} = \left[\dfrac{Ac_{nS}}{Ac_n}\right]$	(9)
$\dfrac{d[Ac_{ns}]}{dt} = -k_9 \cdot [Ac_{nS}]$	(10)
$\dfrac{d[AcO]}{dt} = k_9 \cdot [Ac_{nS}] - k_{10} \cdot [AcO]$	(11)
$\dfrac{d[AcH]}{dt} = k_{10} \cdot [AcO]$	(12)

Table 6. (a) Results achieved for the pre-exponential factors (k_{0i}) and activation energies (Ea_i) of the coefficients involved in the kinetic models, and for the susceptible fractions of xylan and acetyl groups. (b) Coefficients of determination (R^2) calculated for the various coefficients and stages.

(a)

Reaction	Coefficient	Stage 1		Stage 2		Stage 3	
		Ln k_{0i} (k_{0i}, min^{-1})	Ea_i (kJ·mol^{-1})	Ln k_{0i} (k_{0i}, min^{-1})	Ea_i (kJ·mol^{-1})	Ln k_{0i} (k_{0i}, min^{-1})	Ea_i (kJ·mol^{-1})
$X_{nS} \to XOs$	k_1	46.70	182.9	46.84	183.1	46.91	183.3
$XOs \to X$	k_2	27.75	118.5	27.85	118.8	28.30	119.9
$X \to F$	k_3	16.40	78.2	16.65	79.1	16.90	79.9
$X \to DecP$	k_4	17.42	83.1	21.82	99.8	23.60	105.6
$Ar_n \to ArOs$	k_5	25.80	103.9	26.04	104.2	26.32	104.5
$ArOs \to Ar$	k_6	32.69	124.7	33.12	124.7	20.52	79.1
$Ar \to F$	k_7	22.02	99.8	22.40	100.3	22.46	100.4
$Ar \to DecP$	k_8	16.44	83.1	17.03	83.3	17.21	83.5
$Ac_{nS} \to AcO$	k_9	33.23	133.0	33.50	133.2	33.54	133.2
$AcO \to AcH$	k_{10}	29.14	124.7	28.9	124.8	29.23	124.7
-	α_{XnS}	0.88		0.90		0.90	
-	α_{AcnS}	0.90		0.91		0.93	

(b)

Compound	Coefficients of Determination, R^2		
	Stage 1	Stage 2	Stage 3
XOs	0.921	0.937	0.960
X	0.990	0.972	0.964
ArOs	0.888	0.686	0.772
Ar	0.764	0.884	0.844
F	0.989	0.969	0.995
AcO	0.823	0.813	0.830
AcH	0.963	0.952	0.983

Additionally, Figure 3a–c show the experimental concentrations and the concentration profiles calculated for the compounds involved in models. The ability of the models for giving a quantitative interpretation of results was confirmed by the close interrelationship between experimental and calculated data.

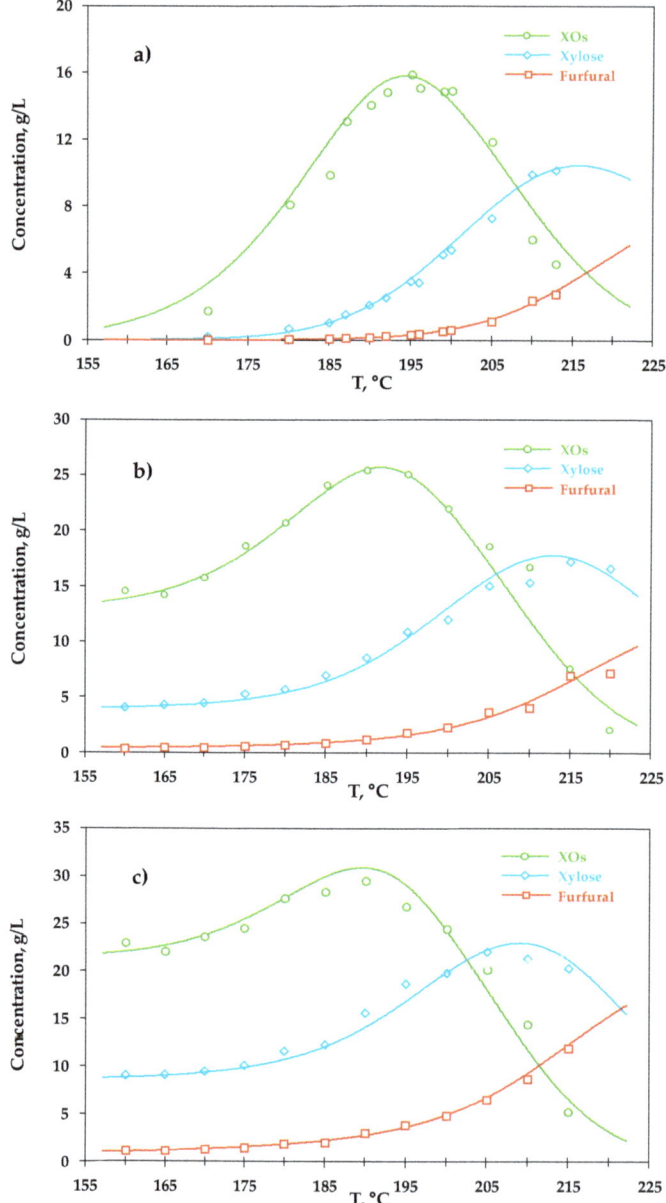

Figure 3. Experimental and calculated concentrations of the soluble products involved in the kinetic model. (**a**) Data determined for the first stage; (**b**) data determined for the second stage; (**c**) data determined for the third stage.

2.6. Characterization of the Reaction Products

In order to get further insight on the properties of the products from the diverse reaction stages, samples from the reaction media were analyzed by HPSEC, HPAEC-PAD and MALDI-TOF MS.

The HPSEC chromatogram shown in Figure 4 confirmed that both the amount of monosaccharides and the mass ratio monosaccharides/oligosaccharides increased from stage 1 to stage 3. In all cases, the major products were monosaccharides and higher saccharides with degrees of polymerization (DP) ≤ 7, among which xylobiose was predominant.

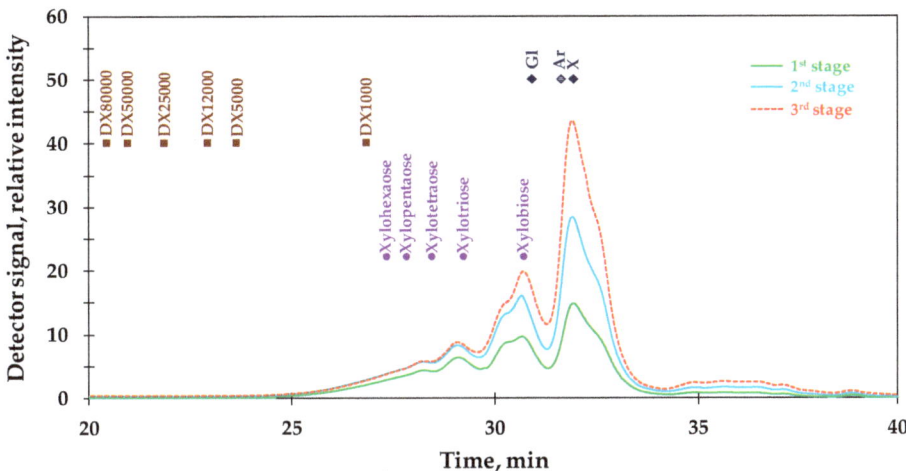

Figure 4. High Pressure Size Exclusion Chromatography (HPSEC) data obtained for the liquid phases from the 1st, 2nd and 3rd reaction stages, operating under conditions leading to the maximum concentration of XOs.

The HPAEC-PAD elution profiles shown in Figure 5 supported the above findings. Small amounts of oligosaccharides with DP > 6 were present in the liquid phases from stages 1 to 3, whereas monosaccharides and DP2–DP4 were the most abundant products.

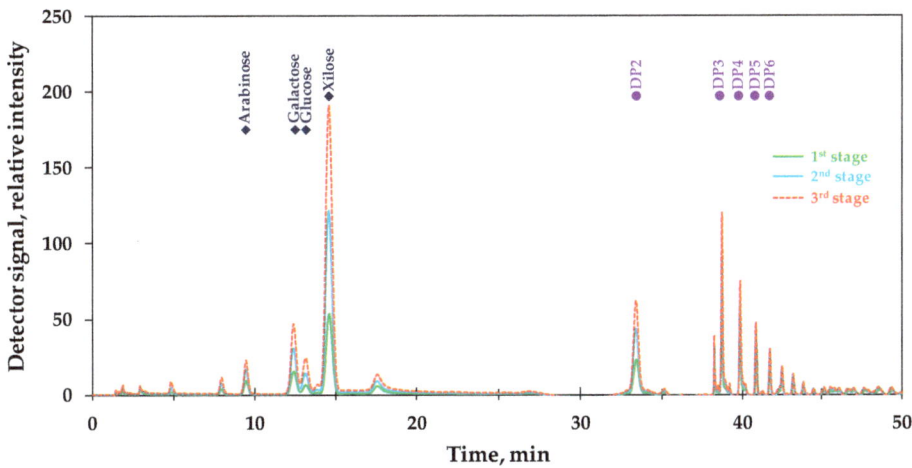

Figure 5. High Performance Anion Exchange Chromatography with Pulsed Amperometric Detection (HPAEC-PAD) data obtained for the liquid phases from the 1st, 2nd and 3rd reaction stages, operating under conditions leading to the maximum concentration of XOs.

Additional information on the structure of the major reaction compounds was obtained by MALDI-TOF analysis. The samples from the reaction media 1 to 3 showed a variety of structures substituted with acetyl and O-methyluronic groups. As an example, Table 7 lists the structures identified in the liquid stream from the third reaction stage, which corresponded mainly to heavily substituted oligosaccharides made up of 2–11 pentose units, containing up to 6 acetyl groups and up to 2 O-methyluronic groups. This rich substitution pattern is in agreement with literature [3,25], and are known to show biological properties, including prebiotic activity [26].

Table 7. Matrix Assisted Laser Desorption/Ionization Time of Flight Mass Spectrometry (MALDI-TOF) data determined for the liquid stream from the third autohydrolysis results (the m/z values corresponded to the sodium adducts of the identified compounds).

m/z	Pentose Units/ Uronic Subst./ Acetyl Subst.	m/z	Pentose Units/ Uronic Subst./ Acetyl Subst.	m/z	Pentose Units/ Uronic Subst./ Acetyl Subst.
537.54	2/1/1	1034.34	4/2/2	1323.85	7/1/4
611.58	4/0/1	1049.71	7/0/2	1339.77	6/2/3
653.50	4/0/2	1059.57	5/1/4	1365.84	7/1/5
669.27	3/1/1	1065.72	6/1/1	1371.96	8/1/2
695.50	4/0/3	1091.75	7/0/3	1397.98	9/0/4
711.45	3/1/2	1107.74	6/1/2	1413.96	8/1/3
785.54	5/0/2	1133.75	7/0/4	1455.95	8/1/4
801.53	4/1/1	1149.73	6/1/3	1471.90	7/2/3
827.56	5/0/3	1165.69	5/2/2	1497.91	8/1/5
843.50	4/1/2	1176.43	7/0/5	1546.06	9/1/3
869.51	5/0/4	1191.67	6/1/4	1588.10	9/1/4
885.49	4/1/3	1207.65	5/2/3	1630.06	9/1/5
917.67	6/0/2	1223.90	8/0/3	1672.03	9/1/6
933.61	5/1/1	1233.72	6/1/5	1720.13	10/1/4
959.66	6/0/3	1239.87	7/1/2	1762.04	10/1/5
975.61	5/1/2	1265.94	8/0/4	1804.05	10/1/6
1001.65	6/0/4	1281.85	7/1/3	1894.19	11/1/5
1017.63	5/1/3	1307.78	8/0/5	1936.20	11/1/6

3. Materials and Methods

3.1. Raw Material and Chemical Processing

Eucalyptus globulus wood samples were collected locally, milled in a Wiley instrument fitted with an 8 mm screen, air-dried, mixed to ensure a constant composition, and stored until use. Wood samples were processed either with water (in first autohydrolysis stage) or with liquid streams from previous autohydrolysis treatments (in the second and third stages) using a solid charge of 1 kg oven-dry wood/8 kg liquid phase. Reaction was performed in a 3.75 L stainless steel reactor (Parr Instruments Company, Moline, IL, USA). Operation was carried out in non-isothermal mode (the media was heated up to reach the target temperature, and then cooled immediately). The solid phase was recovered after press-filtration, and subjected to displacement washing. Process water, reaction media, and washing effluents were processed according to the integrated scheme reported in an earlier work [8].

3.2. Analysis of Wood and Samples from Hydrothermal Treatments

Native wood and the exhausted solids from the diverse hydrothermal stages were washed with deionized water and assayed for moisture, extractives and ash using NREL standard methods. Klason lignin, acetyl groups and hemicellulose-derived sugars were measured by quantitative acid hydrolysis (NREL/TP-510- 42618 method) followed by HPLC determination, operating as reported elsewhere [8]. The liquid phases from the various reaction media were assayed by the same HPLC method (directly and after quantitative posthydrolysis according to the NREL/TP-510-42623 method), for sugars, furfural, HMF and organic acids. The concentrations of oligosaccharides and bound acetyl groups were calculated from the differences in the respective concentrations determined in the analysis of direct and posthydrolyzed samples. Uronic substituents were determined as per Blumenkrantz and Asboe-Hansen [27]. Non-volatile compounds in liquors were measured as the residue remaining after oven drying at 105 °C until constant weight.

3.3. Additional Characterization of the Reaction Products

HPSEC analysis was employed to assess the molecular weight distribution of oligosaccharides. Operation was performed at 30 °C using two TSKGel G3000PWXL columns in series in combination with a PWX-guard column (all of them from Tosoh Bioscience, Stuttgart, Germany), using Milli Q water as a mobile phase. XOs (DP 2-6, from Megazyme, Ireland) and dextrans (1000–80000 g·mol^{-1}, from Fluka) were employed as standards low- and high- molecular weight fractions, respectively.

Liquid samples from the diverse reaction media were analyzed by HPAEC-PAD using an ICS3000 instrument (Dionex, Sunnyvale, CA, USA) equipped with a CarboPac PA-1 and a CarboPac PA guard column [28].

MALDI-TOF MS analysis of soluble saccharides was performed using an Autoflex III Smartbeam instrument (from Bruker Daltonics, Bremen, Germany). Flex Control 3.0 and Flex Analysis 3.0 were employed for data acquisition.

3.4. Error Assessment

Wood analysis was performed in triplicate, and the standard deviations are reported in Table 1. Analysis of the liquid phases from the reaction media were performed in duplicate, and the average results are reported. The relative error depended on the concentration range of each type of compounds. For example, considering the results listed in Tables 2–4, the average relative errors (measured as absolute values respect to the mean for each duplicate determination, ε) were 0.92 and 1.18%, for GOs and XOs respectively; in comparison with 8.64% for ArOs (which appeared in little concentrations, and were of little importance for the purposes of this study). The values of ε determined for the rest of target compounds were as follows: Gl, 039%; X, 0.30%; Ar, 0.35%; FA, 0.72%; AcH, 0.61%; LA, 0.75%; HMF, 0.63%; and F, 1.09%.

3.5. Fitting of Data

The set of differential equations describing the considered mechanism were solved using the 4th order Runge–Kutta method implemented in an Excel spreadsheet. The values of the parameters involved in the diverse model equations were calculated by regression of the experimental data. The optimal values of the parameters were obtained by minimization of the sum of the squares of the deviations between the experimental and calculated data, using an optimization routine (Solver) built in the Excel spreadsheet.

4. Conclusions

Crossflow coupling of autohydrolysis stages enables the manufacture of solutions containing a number of valuable products at increased concentrations. When *Eucalyptus globulus* was used as a feedstock for autohydrolysis, most of the hemicellulose fraction is converted into soluble products (including oligosaccharides, monosaccharides, furans, organic acids). The kinetic principles of conventional autohydrolysis processing has been established for a number of lignocellulosic materials (including wood), but (to our knowledge), no literature has been reported on the kinetic modeling of multistage autohydrolysis. This problem shows specific features, derived from the presence (from the beginning) of reactive intermediates and catalysts at increased concentrations. In this study, the concentration profiles determined for the target products present in the liquid phases from the diverse crossflow stages were employed for kinetic modeling. Several mechanisms based on pseudohomogeneous reactions and Arrhenius-type dependence of the kinetic coefficients on temperature were assessed. Overestimation of the predicted concentrations of pentoses was observed for kinetic models involving the generation of oligosaccharides from their precursors, with further formation of monosaccharides, and the generation of furans from sugars (with possible consumption of furfural to yield decomposition products). Oppositely, satisfactory results were achieved when the models were modified to include both the dehydration of pentoses into F and their consumption by parasitic reactions. Additional characterization of the reaction products by HPSEC, HPAEC-PAD, and MALDI-TOF MS confirmed that the major reaction products were monosaccharides and higher saccharides with degrees DP ≤ 7, which presented a rich substitution pattern by O-methyluronic and acetyl groups.

Author Contributions: Conceptualization, V.S. and J.C.P.; methodology, M.L., V.S. and J.C.P.; validation, M.L., V.S. and J.C.P.; investigation, M.L., V.S. and J.C.P.; resources, J.C.P.; data curation, M.L.; writing—original draft preparation, M.L., V.S. and J.C.P.; writing—review and editing, M.L., V.S. and J.C.P.; supervision, V.S. and J.C.P.; project administration, J.C.P.; funding acquisition, V.S. and J.C.P. All authors have read and agreed to the published version of the manuscript.

Funding: This research was funded by the "Ministry of Economy and Competitiveness" of Spain (research project "Modified aqueous media for wood biorefineries", reference CTQ2017-82962-R).

Acknowledgments: Mar López thanks "Xunta de Galicia" and the European Union (European Social Fund – ESF) for her predoctoral grant (reference ED481A-2017/316).

Conflicts of Interest: The authors declare no conflict of interest.

References

1. Ebringerová, A.; Hromádková, Z.; Heinze, T. Hemicellulose. *Adv. Polym. Sci.* **2005**, *186*, 1–67.
2. Vázquez, M.J.; Alonso, J.L.; Domínguez, H.; Parajó, J.C. Production and refining of soluble products from *Eucalyptus globulus* glucuronoxylan. *Collect. Czech. Chem. Commun.* **2007**, *72*, 307–320. [CrossRef]
3. Kabel, M.A.; Carvalheiro, F.; Garrote, G.; Avgerinos, E.; Koukios, E.; Parajó, J.C.; Gírio, F.M.; Schols, H.A.; Voragen, A.G.J. Hydrothermally treated xylan rich by-products yield different classes of xylo-oligosaccharides. *Carbohydr. Polym.* **2002**, *50*, 47–52. [CrossRef]
4. Rivas, S.; Vila, C.; Santos, V.; Parajó, J.C. Furfural production from birch hemicelluloses by two-step processing: A potential technology for biorefineries. *Holzforschung* **2016**, *70*, 901–910. [CrossRef]
5. Garrote, G.; Domínguez, H.; Parajó, J.C. Mild autohydrolysis: An environmentally friendly technology for xylooligosaccharide production from wood. *J. Chem. Technol. Biotechnol.* **1999**, *74*, 1101–1109. [CrossRef]
6. Garrote, G.; Parajó, J.C. Non-isothermal autohydrolysis of *Eucalyptus* wood. *Wood Sci. Technol.* **2002**, *36*, 111–123. [CrossRef]
7. Garrote, G.; Domínguez, H.; Parajó, J.C. Study on the deacetylation of hemicelluloses during the hydrothermal processing of *Eucalyptus* wood. *Holz Roh Werkst.* **2001**, *59*, 53–59. [CrossRef]
8. López, M.; Penín, L.; Vila, C.; Santos, V.; Parajó, J.C. Multi-stage hydrothermal processing of *Eucalyptus globulus* wood: An experimental assessment. *J. Wood Chem. Technol.* **2019**, *39*, 329–342. [CrossRef]
9. Yáñez, R.; Romaní, A.; Garrote, G.; Alonso, J.L.; Parajó, J.C. Processing of *Acacia dealbata* in aqueous media: First step of a wood biorefinery. *Ind. Eng. Chem. Res.* **2009**, *48*, 6618–6626. [CrossRef]
10. González-Muñoz, M.J.; Rivas, S.; Santos, V.; Parajó, J.C. Aqueous processing of *Pinus pinaster* wood: Kinetics of polysaccharide breakdown. *Chem. Eng. J.* **2013**, *231*, 380–387. [CrossRef]
11. Garrote, G.; Domínguez, H.; Parajó, J.C. Kinetic modelling of corncob autohydrolysis. *Process. Biochem.* **2001**, *36*, 571–578. [CrossRef]
12. Gullón, B.; Eibes, G.; Dávila, I.; Vila, C.; Labidi, J.; Gullón, P. Valorization of vine shoots based on the autohydrolysis fractionation optimized by a kinetic approach. *Ind. Eng. Chem. Res.* **2017**, *56*, 14164–14171. [CrossRef]
13. Caparrós, S.; Garrote, G.; Ariza, J.; López, F. Autohydrolysis of *Arundo donax* L., a kinetic assessment. *Ind. Eng. Chem. Res.* **2006**, *45*, 8909–8920. [CrossRef]
14. Garrote, G.; Domínguez, H.; Parajó, J.C. Production of substituted oligosaccharides by hydrolytic processing of barley husks. *Ind. Eng. Chem. Res.* **2004**, *43*, 1608–1614. [CrossRef]
15. González, D.; Santos, V.; Parajó, J.C. Manufacture of fibrous reinforcements for biocomposites and hemicellulosic oligomers from bamboo. *Chem. Eng. J.* **2011**, *167*, 278–287. [CrossRef]
16. González, D.; Campos, A.R.; Cunha, A.M.; Santos, V.; Parajó, J.C. Manufacture of fibrous reinforcements for biodegradable biocomposites from *Cytisus scoparius*. *J. Soc. Chem. Ind.* **2011**, *86*, 575–583.

17. Gullón, B.; Yáñez, R.; Alonso, J.L.; Parajó, J.C. Production of oligosaccharides and sugars from rye straw: A kinetic approach. *Bioresour. Technol.* **2010**, *101*, 6676–6684. [CrossRef]
18. Nabarlatz, D.; Farriol, X.; Montané, D. Autohydrolysis of almond shells for the production of xylo-oligosaccharides: Product characteristics and reaction kinetics. *Ind. Eng. Chem. Res.* **2005**, *44*, 7746–7755. [CrossRef]
19. Aida, T.M.; Shiraishi, N.; Kubo, M.; Watanabe, M.; Smith, R.L., Jr. Reaction kinetics of D-xylose in sub- and supercritical water. *J. Supercrit. Fluids* **2010**, *55*, 208–216. [CrossRef]
20. Lange, J.P.; van der Heide, E.; van Buijtenen, J.; Price, R. Furfural—A promising platform for lignocellulosic biofuels. *ChemSusChem* **2012**, *5*, 150–166. [CrossRef]
21. Antal, M.J., Jr.; Leesomboon, T.; Mok, W.S.; Richards, G.N. Mechanism of formation of 2-furaldehyde from D-xylose. *Carbohydr. Res.* **1991**, *217*, 71–85. [CrossRef]
22. Enslow, K.R.; Bell, A.T. The kinetics of Brönsted acid-catalyzed hydrolysis of hemicellulose dissolved in 1-ethyl-3-methylimidazolium chloride. *RSC Adv.* **2012**, *2*, 10028–10036. [CrossRef]
23. Nimlos, M.R.; Qian, X.; Davis, M.; Himmel, M.E.; Johnson, D.K. Energetics of xylose decomposition as determined using quantum mechanics modeling. *J. Phys. Chem. A* **2006**, *110*, 11824–11838. [CrossRef] [PubMed]
24. Rasmussen, H.; Sørensen, H.R.; Meyer, A.S. Formation of degradation compounds from lignocellulosic biomass in the biorefinery: Sugar reaction mechanisms. *Carbohydr. Res.* **2014**, *385*, 45–57. [CrossRef] [PubMed]
25. Kabel, M.A.; Schols, H.A.; Voragen, A.G.J. Complex xylo-oligosaccharides identified from hydrothermally treated *Eucalyptus* wood and brewery's spent grain. *Carbohydr. Polym.* **2002**, *50*, 191–200. [CrossRef]
26. Gullón, P.; González-Muñoz, M.J.; van Gool, M.P.; Schols, H.A.; Hirsch, J.; Ebringerová, A.; Parajó, J.C. Structural features and properties of soluble products derived from *Eucalyptus globulus* hemicelluloses. *Food Chem.* **2011**, *127*, 1798–1807. [CrossRef]
27. Blumenkrantz, N.; Asboe-Hansen, G. New method for quantitative determination of uronic acids. *Anal. Biochem.* **1973**, *54*, 484–489. [CrossRef]
28. Peleteiro, S.; Santos, V.; Garrote, G.; Parajó, J.C. Furfural production from *Eucalyptus* wood using an acidic ionic liquid. *Carbohydr. Polym.* **2016**, *146*, 20–25. [CrossRef]

© 2020 by the authors. Licensee MDPI, Basel, Switzerland. This article is an open access article distributed under the terms and conditions of the Creative Commons Attribution (CC BY) license (http://creativecommons.org/licenses/by/4.0/).

Article

5-Hydroxymethylfurfural and Furfural Base-Free Oxidation over AuPd Embedded Bimetallic Nanoparticles

Camila P. Ferraz [1,2], Natalia J. S. Costa [1], Erico Teixeira-Neto [3], Ângela A. Teixeira-Neto [3], Cleber W. Liria [4], Joëlle Thuriot-Roukos [2], M. Teresa Machini [4], Rénato Froidevaux [5], Franck Dumeignil [2], Liane M. Rossi [1] and Robert Wojcieszak [2,*]

1. Departamento de Química Fundamental, Instituto de Química, Universidade de São Paulo, São Paulo 05508-000, Brazil; camila.ferraz@univ-lille.fr (C.P.F.); nataliajscosta@gmail.com (N.J.S.C.); lrossi@iq.usp.br (L.M.R.)
2. Univ.Lille, CNRS, Centrale Lille, ENSCL, Univ.Artois, UMR 8181—UCCS—Unité de Catalyse et Chimie du Solide, F-59000 Lille, France; joelle.thuriot@univ-lille.fr (J.T.-R.); franck.dumeignil@univ-lille.fr (F.D.)
3. Laboratório de Microscopia Eletrônica, LNNano-CNPEM, C.P. 6192, Campinas 13083-970, Brazil; erico.teixeira.neto@gmail.com (E.T.-N.); angelaalb@gmail.com (Â.A.T.-N.)
4. Departamento de Bioquímica, Instituto de Química, Universidade de São Paulo, São Paulo 05508-000, Brazil; cwliria@iq.usp.br (C.W.L.); mtmachini@iq.usp.br (M.T.M.)
5. Univ.Lille, INRA, ISA, University Artois, Univ.Littoral Côte d'Opale, EA 7394—ICV—Institute Charles Viollette, F-59000 Lille, France; renato.froidevaux@univ-lille.fr
* Correspondence: robert.wojcieszak@univ-lille1.fr

Received: 4 December 2019; Accepted: 2 January 2020; Published: 4 January 2020

Abstract: The heterogeneous catalytic partial oxidation of alcohols and aldehydes in the liquid phase usually needs the addition of a homogeneous base, which in turn makes the products' recovery cumbersome, and can further induce undesired side reactions. In the present work, we propose the use of novel catalysts based on metallic Au, Pd and bimetallic AuPd nanoparticles embedded in a titanosilicate matrix. The as-prepared catalysts showed good efficiency in the base-free partial oxidation of furfural and 5-hydroxymethylfurfural. Au_4Pd_1@SiTi catalyst showed high selectivity (78%) to monoacids (namely, 5-formyl-2-furancarboxylic acid and 5-hydroxymethyl-2-furancarboxylic acid) at 50% 5-hydroxymethylfurfural (HMF) conversion. The selectivity even reached 83% in the case of furfural oxidation to furoic acid (at 50% furfural conversion). The performances of the catalysts strongly depended on the Au–Pd ratio, with an optimal value of 4:1. The pH of the solution was always below 3.5 and no leaching of metals was observed, confirming the stabilization of the metal nanoparticles within the titanosilicate host matrix.

Keywords: bimetallic nanoparticles; base-free; green oxidation; embedded catalysts; biomass

1. Introduction

5-hydroxymethylfurfural (HMF) is one of the top 12 highly valued compounds identified early as promising platform molecules within the frame of the development of biorefineries [1–4]. HMF is readily obtained by the dehydration of glucose and through the intermediate isomerization of fructose, the glucose being potentially obtained by cellulose hydrolysis [2,4]. The shared enthusiasm around HMF lies in its great versatility in enabling the production of a wide variety of fuels and high value-added chemicals [2,4,5]. In particular, the products obtained via partial oxidation processes are of high interest for the chemical and polymer industries. Such derivatives have the potential to replace petrochemical-based monomers [2,6]. Indeed, 2-,5-furandicarboxylic acid (FDCA), for example, is used for the synthesis of polyethylene furoate (PEF), which is a promising alternative to polyethylene terephthalate (PET). In addition to being a green alternative, PEF does not bio-accumulate and is biodegradable [6].

Like HMF, furfural is also attracting much attention. It is also part as of the aforementioned highly valued compounds list, with a very high potential for biorefineries development [1–4,7]. It is generally obtained from hemicellulose, mainly via the hydrolysis of xylose polymers and subsequent dehydration. The partial oxidation of furfural gives furoic acid (furan-2-carboxylic acid), which has various applications in the agrochemical, flavor, pharmaceutical, and fine chemistry industries.

HMF and furfural oxidation reactions can be carried out using different methods, including chemical oxidation, biochemical transformation and homogeneous or heterogeneous catalytic conversions [8–10]. However, there are still several drawbacks related to the difficulties in the control of undesired side product formation and the final separation step. Moreover, very often, the oxidation is carried out in the presence of an inorganic base, such as NaOH or KOH. Basic media can be responsible for HMF and furfural degradation. Indeed, placing HMF or furfural under high pH conditions in the absence of a catalyst yields high substrate conversion, but without the formation of the desired oxidation products. The formation of a black precipitate-like compound is frequently observed, which is attributed to the formation of humins [11–14]. A high pH of the medium also favors the formation of low molar weight acids such as levulinic acid and formic acid, consecutively to C–C bond cleavage. Previous studies [12–14] show that the challenge in developing catalysts for furfural and HMF partial oxidation lies in the design of heterogeneous catalytic systems capable of maintaining high activity and selectivity while getting rid of the use of a homogeneous base. Regarding Au catalysts, strategies to prevent the use of a base consist of adopting a basic and/or nanometric support and/or adding a second metal (bimetallic). In the latter case, Au with high selectivity forms bimetallic nanomaterials that can combine the advantages of different components at the atomic level. It can also increase catalytic activity and stability in oxidation reactions of organic compounds in water [15,16]. Therefore, the development of a titanosilicate support in which Au, Pd and bimetallic AuPd nanoparticles are encapsulated should increase the stability of the catalysts while preventing leaching of the metal.

Herein, we report a study of embedded catalysts for base-free furfural and HMF partial oxidation. Novel Au_xPd_y@SiTi nanomaterials were developed, where the AuPd nanoparticles were embedded in a titanosilicate matrix (Scheme 1). Different Au–Pd ratios were used in order to optimize base-free furfural and HMF partial oxidation to the desired compounds.

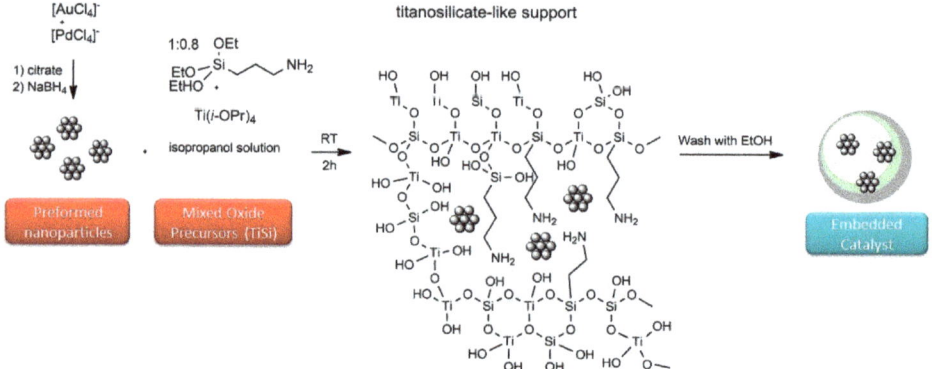

Scheme 1. Schematic representation of the synthesis procedure.

2. Results

In this work, two different oxidation reactions were studied. The Au_xPd_y@SiTi systems were first used in HMF oxidation to identify/optimize the main factors governing the activity of bimetallic

AuPd nanoparticles during the base-free oxidation reaction in water using O_2 as oxidant. Au@SiTi and Pd@SiTi monometallic catalysts were also developed to understand the effect of AuPd's alloying activity by comparison with their bimetallic counterparts (Figure 1). The Au–Pd ratio was varied in bimetallic catalysts from Au_4Pd_1@SiTi to Au_1Pd_4@SiTi, and the effect of Au or Pd enrichment in the alloy on catalysis activity and selectivity was evaluated. In addition, the corresponding supported versions of the embedded catalysts, denoted as Au_4Pd_1/SiTi, Au_1Pd_1/SiTi and Au_1Pd_4/SiTi, were also synthesized for comparison (Figure 1).

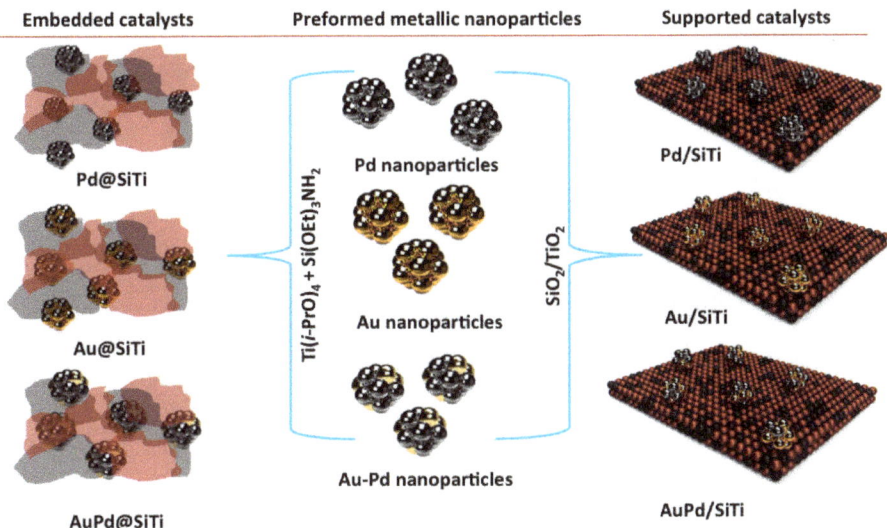

Figure 1. Schematic representation of the compared catalytic systems. Embedded catalysts are denoted as Au_xPd_y@SiTi and supported catalysts as Au_xPd_y/SiTi (red—TiO_2, gray—SiO_2).

Initially, AuPd equimolar catalyst was synthesized, characterized and evaluated. Au_1Pd_1@SiTi with Au–Pd molar ratio of 1 was prepared using equivalent quantities of APTES and TIP (500 mg) for comparison to its monometallic counterparts (Au@SiTi and Pd@SiTi). The catalysts were prepared using preformed NPs, which were stabilized by sodium citrate and reduced by $NaBH_4$ in a well-known and reproducible method [17] to provide Au, Pd and AuPd nanoparticles of about 4 nm. The as-prepared nanoparticles were then soaked in a mixed titanosilicate oxide and had their size preserved after immobilization in the support (as represented in Scheme 1). Embedded Au_1Pd_1 nanoparticles were monodispersed with a mean size of 3.8 ± 0.8 nm (Figure 2). Following this, two other catalysts with Au–Pd molar ratios of 4:1 and 1:4 (Au_4Pd_1@SiTi and Au_1Pd_4@SiTi, respectively) were synthetized and characterized. In both cases, the mean particle size was again about 4 nm (Figure 2).

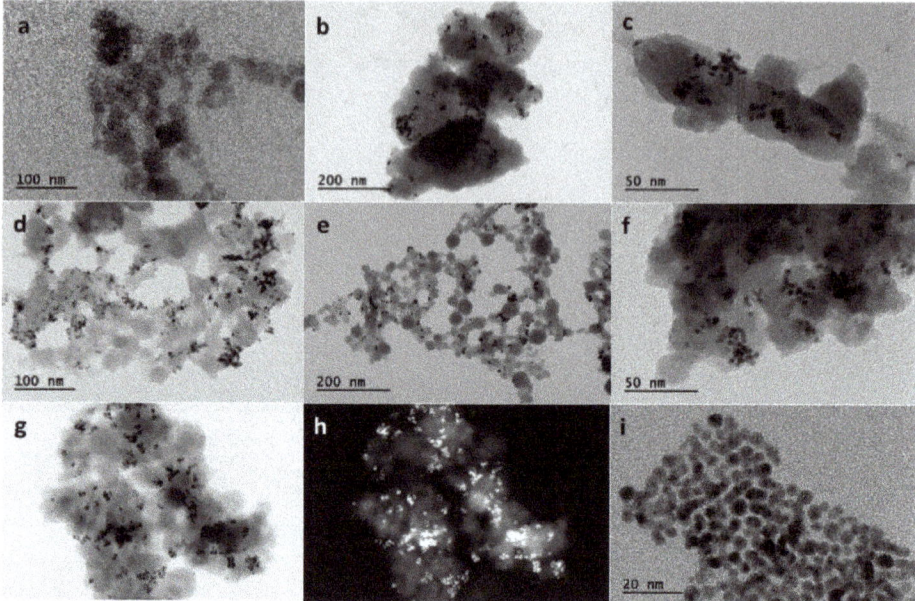

Figure 2. Transmission electron microscopy (TEM) images of (**a**) bare SiTi; (**b**) Au@SiTi; (**c**) Pd@SiTi; (**d**) Au_4Pd_1@SiTi; (**e**) Au_1Pd_1@SiTi; (**f**) Au_1Pd_4@SiTi; (**g**) Au_1Pd_1@SiTi; (**h**) Au_1Pd_1@SiTi in dark field measurement and (**i**) preformed Au_1Pd_1 nanoparticles stabilized by sodium citrate and used for the embedded catalysts synthesis.

Considering the same metal loading (ca. 2.2 wt.%, Table 1) and particle size, the catalytic performances are expected to be governed mainly by the interaction of the metal with the support and by the chemical composition of the bimetallic nanoparticles. The formation of Pd-rich or Au-rich nano alloys can particularly strongly affect the catalytic activity. The Brunauer-Emmett-Teller (BET) analysis showed that different porosities were obtained for the embedded samples. The BET surface area of the Au_4Pd_1@SiTi sample was twice as big as that of Au_1Pd_1 sample (78 and 39 m^2/g, respectively). No result was observed for the Pd-rich sample (BET surface area of <1 m^2g^{-1}) after being performed twice. This could be due to the collapse of the structure during the degassing step (150 °C) or the complete filling of the pores by the organic precursors.

The high-resolution transmission electron microscopy (TEM) images and Energy Dispersive X-Ray Spectrometry (EDS) analysis of the Au_1Pd_1@SiTi catalyst are shown in Figure 3. This sample was chosen because of the equimolar Au:Pd ratio. It was confirmed that the encapsulation of the bimetallic nanoparticles in SiTi preserved the mean particle size. Moreover, chemical mapping of the images suggested some segregation of Au and Pd in the NPs. A core shell structure with a Pd-rich shell was formed. Regarding the titanosilicate structure, Si and Ti were uniformly distributed in the solid, but Ti seemed to surround the bimetallic nanoparticles (Figure 3).

Table 1. Chemical composition of Au$_x$Pd$_y$@SiTi and Au$_x$Pd$_y$/SiTi catalysts as determined by inductively coupled plasma optical emission spectrometry (ICP—OES) and X-ray fluorescence spectrometry (XRF) for the former.

Catalyst	Au (%)	Pd (%)	Metal % (Au + Pd)	Si (%)	Ti (%)	Au:Pd[a]	Si:Ti[b]
Au$_4$Pd$_1$@SiTi	1.68	0.42	2.11	5.02	21.69	79:21 (77:23) *80:20*	19:81 (15:85)
Au$_1$Pd$_1$@SiTi	1.37	0.97	2.34	3.75	20.41	58:42 (50:50) *50:50*	16:84 (16:84)
Au$_1$Pd$_4$@SiTi	0.69	1.60	2.29	4.14	22.08	30:70 (22:78) *20:80*	16:84 (17:83)
Au$_4$Pd$_1$/SiTi	5.41	0.80	6.21	8.83	48.61	78:22 *80:20*	15:85
Au$_1$Pd$_1$/SiTi	2.30	1.31	3.61	8.83	48.61	50:50 *50:50*	15:85
Au$_1$Pd$_4$/SiTi	1.21	2.21	3.43	8.83	48.61	23:77 *20:80*	15:85
Au@SiTi	6.40	-	6.40	-	-	-	-
Pd@SiTi	-	3.01	3.01	-	-	-	-
SiTi	-	-	-	8.83	48.61	-	15:85

[a] XRF results given in brackets, theoretical ratio in italic, [b] theoretical value of 60:40.

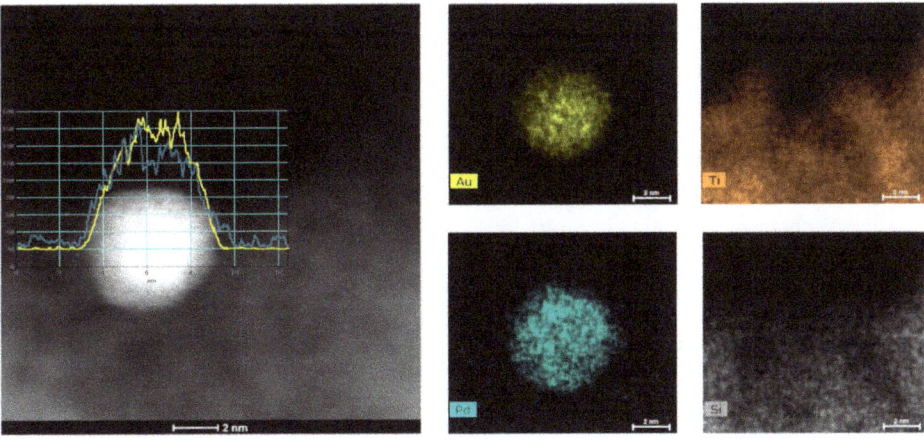

Figure 3. HRTEM HAADF image of the Au$_1$Pd$_1$@SiTi catalyst and X-ray mapping of the sample.

All catalysts were characterized by inductively coupled plasma optical emission spectrometry (ICP-OES) to determine their Au, Pd, Si and Ti contents (Table 1). The Au–Pd ratios for the catalysts were reasonably consistent with the expected values, resulting in catalysts with about 2.2 wt.% of metal loading (for embedded catalysts). Higher amounts of metals were obtained for the supported Au$_1$Pd$_1$/SiTi and Au$_1$Pd$_4$/SiTi samples (about 3.3 wt.%). In the case of the Au$_4$Pd$_1$/SiTi sample, the metal loading reached 6 wt.% (Table 1). In all cases, the Ti–Si ratio was close to 4.

X-ray photoelectron spectroscopy (XPS) analysis confirmed that AuPd nanoparticles were embedded in the Si–Ti matrix. The results can be seen in the Table 2 below. It can be clearly observed the strong decrease in the Au and Pd content for the embedded samples as compared to the corresponding supported catalysts. This can be explained by the covering of the metal nanoparticles with the oxide phases.

Table 2. X-ray photoelectron spectroscopy (XPS) analysis of the supported and embedded catalysts.

	Au at%	Pd at%	Au/Pd mol	Pd0/PdII
Au_1Pd_1/SiTi	47 ± 1	24.0 ± 0.7	1.05 ± 0.01	4 ± 1
Au_1Pd_1@TiSi	6.4 ± 0.2	3.5 ± 0.2	1.00 ± 0.09	2.2 ± 0.2
Au_1Pd_4/SiTi	24 ± 1	42.4 ± 0.8	0.31 ± 0.02	2.4 ± 0.2
Au_1Pd_4@SiTi	2.2 ± 0.1	5.1 ± 0.1	0.24 ± 0.01	2.7 ± 0.5
Au_4Pd_1/SiTi	61 ± 1	9.7 ± 0.1	3.40 ± 0.03	3.8 ± 0.4
Au_4Pd_1@SiTi	9.4 ± 0.3	1.3 ± 0.03	3.95 ± 0.06	3.5 ± 0.8

A good accordance between the surface (XPS) and bulk analyses (XRF, ICP) was observed in term of chemical composition. As can be seen from Tables 1 and 2, the molar ratio between both metals is similar for all techniques used. This could be explained by the homogeneous composition of the bimetallic nanoparticles.

The solids had Ti contents much higher than expected (theoretical Si–Ti value of 60:40), thus with relatively low amounts of Si (actual average Si–Ti ratio of 15:85). Under these conditions, the hydrolysis of the precursors may not be complete; in particular, APTES (silica precursor) could remain in the solution, which explains the higher Ti content as compared to the expected ones and also the fact that the Si–Ti ratio is the same for all the synthetized embedded samples. The catalysts were also characterized by XRF analysis (Table 1), and the Au–Pd and Si–Ti ratios were found the same as those obtained by ICP-OES. The XRF values were closer to the theoretical ones. The Au–Pd molar ratios of 4:1, 1:1 and 1:4 were confirmed by ICP and XRF analyses.

2.1. HMF Oxidation

Effect of Au:Pd Ratio on Catalytic Activity

The results of HMF oxidation are presented in Figures 4 and 5, and, as expected, all the catalysts were active. Variation in the AuPd composition affected both conversions and selectivities, with better catalysts being obtained when the Au proportion was increased in the bimetallic systems (Figure 4a). A volcano profile was observed when plotting the conversion observed after 24 h as a function of the catalysts' composition (Figure 4b), where a maximum activity was reached for the Au_4Pd_1 gold-rich composition. The distribution of the products was also greatly affected by the composition variation (Figure 5). Pure Au favors the formation and accumulation of 5-hydroxymethyl-2-furancarboxylic acid (HFCA), as already observed in a previous study focused on Au catalysts in different supports [18]. The addition of Pd makes possible the formation of 2,5-diformylfuran (FDC) by the chemisorption of HMF via the aldehyde group [19], which was detected in all AuPd (Figure 5b) and pure Pd compositions (Figure 5c). In the Au-rich Au_4Pd_1 composition, the formed FDC was quickly consumed, leading to higher yields of furandicarboxylic acid (FDCA) (Figure 5).

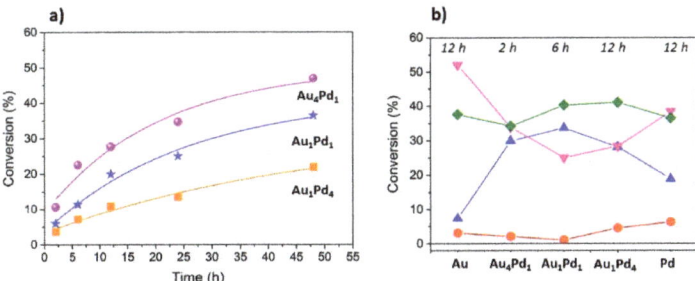

Figure 4. 5-hydroxymethylfurfural (HMF) oxidation in the aqueous phase in the absence of base using Au_xPd_y@SiTi catalysts: conversion and selectivity as a function of time (**a**), and as a function of Au_xPd_y composition at *iso*-conversion of 10% (**b**): selectivity to: ● 2-, 5-furandicarboxylic acid (FDCA), ▲ 2, 5-diformylfuran (FDC), ▼ 5-hydroxymethyl-2-furancarboxylic acid HFCA, ♦ 5-formyl-2-furancaboxylic acid (FFCA). Aqueous solution of HMF (30 µmol), substrate/metal = 18 (mol/mol), O_2 (1 bar), 100 °C, 600 rpm.

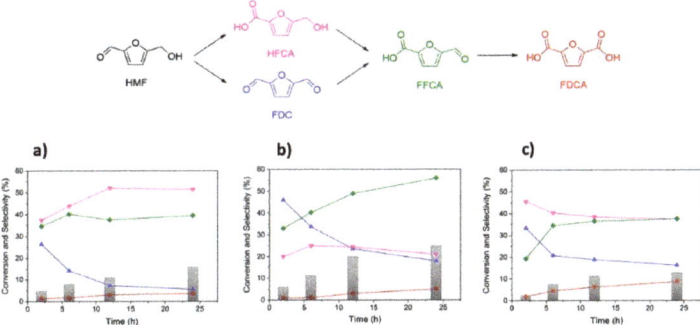

Figure 5. HMF oxidation in the aqueous phase in the absence of base using Au_xPd_y@SiTi catalysts: conversion and selectivity as a function of time: (**a**) Au@SiTi, (**b**) Au_1Pd_1@SiTi, (**c**) Pd@SiTi; selectivity to: ● FDCA, ▲ FDC, ▼ HFCA, ♦ FFCA. Aqueous solution of HMF (30 µmol), substrate/metal = 18 (mol/mol), O_2 (1 bar), 100 °C, 600 rpm.

The activity of the embedded catalysts was compared to that of corresponding supported conventional materials (Table 3). Au/TiO_2 and Au/ZrO_2 catalysts were also studied for comparison (Table 3, Entries 8 and 9).

Table 3. Oxidation of HMF in the aqueous phase in the absence of base for 24 h using embedded Au_xPd_y@SiTi and supported Au_xPd_y/SiTi catalysts. Aqueous solution of HMF (30 µmol), substrate/metal = 18 (mol/mol), O_2 (1 bar), 100 °C, 600 rpm, [a] sample calcined at 500 °C.

Entry	Catalyst	C Balance	S_{FDCA}%	X%	Y_{FDCA}%	TON
1	Au_4Pd_1@SiTi	74	22	35	8	67
2	Au_1Pd_1@SiTi	86	5	25	2	65
3	Au_1Pd_4@SiTi	88	4	14	1	38
4	Au_4Pd_1/SiTi	74	1	13	<1	43
5	Au_1Pd_1/SiTi	88	1	6	<1	19
6	Au_1Pd_4/SiTi	94	6	5	<1	11
7	Au_1Pd_{1PVA}@SiTi	78	3	27	1	38
8	Au@SiTi	12	4	16	<1	35
9	Pd@SiTi	16	8	13	<1	32
10[a]	Au_4Pd_1@SiTi$_{cal}$	80	1	11	1	21

The conversion of HMF observed for the embedded samples after 24 h (Table 3, Entries 1–3) was strongly influenced by the composition of the bimetallic nanoparticles. The maximum level of conversion was obtained for the rich gold sample (Au_4Pd_1@SiTi) and the lowest for the Pd-rich sample (Au_1Pd_4@SiTi). Interestingly the performances of the monometallic Au and Pd samples (Table 3, Entries 8 and 9 respectively) were low and close to that of the Pd-rich Au_1Pd_4@SiTi sample. Very low activity was observed for the supported catalysts (Table 3, Entries 4–6). Moreover, important differences in FDCA selectivity were also observed. Indeed, the Au_4Pd_1@SiTi sample (Figure 6) was much more selective to FDCA after 24 h than the two other samples (Figure 6a). A conversely higher selectivity to HFCA was observed for Au_1Pd_1@SiTi (Figure 6a). Almost no FDCA formation was observed for the supported materials (Figure 6b). In contrast to the embedded samples, the FDC was the main product formed in the case of the supported catalysts.

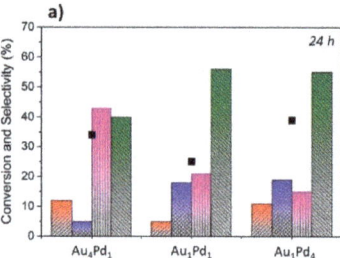 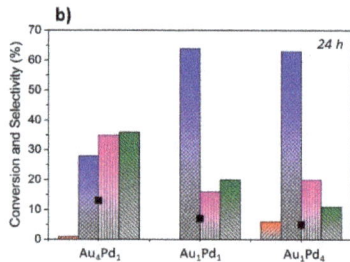

Figure 6. HMF oxidation in the aqueous phase in the absence of a base using (**a**) embedded and (**b**) supported catalysts; black—conversion, selectivity to: red—FDCA, blue—FDC, pink—HFCA, green—FFCA. Aqueous solution of HMF (30 μmol), substrate/metal = 18 (mol/mol), O_2 (1 bar), 100 °C, 600 rpm.

As the total metal content was similar for all samples, the activity in the case of embedded catalysts was strongly affected by the composition of the alloyed nanoparticles, which is in good agreement with data earlier reported [20].

2.2. Furfural Oxidation

Catalytic tests for furfural oxidation using Au, Pd and AuPd–titanosilicate catalysts were performed using air as an oxidant at high pressure (26 bar). The results showed that the catalysts were also able to oxidize furfural to furoic acid (FA) in the absence of a base (Table 4). However, in all cases, the carbon balance was quite low (Table 4). This could be explained by the degradation of furfural on the acid supports and/or the adsorption of furfural on the SiTi matrix.

The results presented in Table 4 highlight the synergistic effect of Au_xPd_y bimetallic catalysts compared to their monometallic Au and Pd counterparts. Au_4Pd_1 and Au_1Pd_1 were the most active samples, although the former led to a higher yield of FA. The Au_4Pd_1@SiTi catalyst presented a slightly higher performance (expressed in TON) than other compositions, which is also in good agreement with the HMF oxidation results presented above.

The effect of the stabilizing agent on catalytic activity was also evaluated when comparing Au, Pd and AuPd NPs stabilized with PVA and citrate (Table 4, Entry 9). The tests revealed no difference in activity, which means that both stabilizers were efficient in inducing an active catalytic formulation. No significant differences in mean particle sizes were observed for both stabilizers (*ca.* 3.8 nm with citrate and 4.1 nm with PVA). As observed above, a higher catalytic efficiency was achieved with Au_xPd_y@SiTi catalysts when compared to Au_xPd_y/SiTi supported catalysts (Table 4) for producing FA at higher yields. The exception was the Au_4Pd_1/SiTi catalyst, which has led to 50% of the FA yield; however, when considering the TON rates, embedded catalysts were more efficient (Table 4). It is important to highlight the efficiency of these titanosilicate-based catalysts when compared to titanium and zirconium catalysts (Table 4, Entries 10 and 11), as was already the case in HMF oxidation (Table 3). The latter achieved complete conversion in 10 h of reaction but showed low selectivity to FA. A blank test for the 2 h reaction showed a furfural conversion of 23%, selectivity to FA of 0%, carbon balance = 77%, and a conversion of 52% after 10 h of reaction, selectivity to FA of 0% and a carbon balance of 48% (Table 4, Entry 12). In the absence of a catalyst, it is not possible to obtain furoic acid, nor when using pure titanosilicate, although some conversion was observed. Apparently, some degradation through the formation of humins and humic acids occurred, considering that no other probable product was detected. This could also explain the low carbon balance values observed for the catalysts, since adsorption tests were performed and excluded the possibility of adsorption of FA or furfural in the pores and/or surface of the catalyst.

Table 4. Oxidation of furfural in the absence of a base for 10 h using embedded Au_xPd_y@SiTi and supported Au_xPd_y/SiTi, Au/TiO$_2$, Au/ZrO$_2$ catalysts. (Conditions: Aqueous solution of furfural (49.4 μmol), substrate/metal = 50 (mol/mol), air (26 bar), 110 °C, 600 rpm, [a] samples calcined at 500 °C).

Entry	Catalyst	C Balance %	S_{FA}%	X%	Y_{FA}%	TON
1	Au_4Pd_1@SiTi	54	45	83	37	91
2	Au_1Pd_1@SiTi	47	30	77	23	80
3	Au_1Pd_4@SiTi	58	23	54	12	57
6	Au_4Pd_1/SiTi	49	48	99	48	50
7	Au_1Pd_1/SiTi	56	18	54	10	28
8	Au_1Pd_4/SiTi	66	11	38	4	19
9	Au_1Pd_{1PVA}@SiTi	48	41	87	36	44
10	Au/TiO$_2$	8	8	100	8	43
11	Au/ZrO$_2$	37	32	92	29	45
12	blank	48	0	52	0	-
13	Au@SiTi	80	32	30	10	15
14	Pd@SiTi	73	20	34	7	17
15[a]	Au_4Pd_1@SiTi$_{cal}$	38	2	63	1	-
16[a]	Au_1Pd_1@SiTi$_{cal}$	86	12	16	2	-
17[a]	Au_1Pd_4@SiTi$_{cal}$	12	1	88	1	-

3. Discussion

Embedded systems exhibit a higher level of intermolecular interaction between embedded metallic nanoparticles and the oxide from the titanosilicate matrix as compared to classical oxide-supported metal nanoparticles. The synthesis method presented herein to produce embedded catalysts allows us to maintain high metal dispersion during utilization, even for relatively high metal contents. The degree of metal support interactions could be controlled, resulting in different stabilization mechanisms. Embedded metal nanoparticle catalysts present a covalent link between preformed AuPd nanoparticles (citrate or PVP method of preparation) and the growing support. This allows better thermal and chemical stability and also enables us to tune the selectivity in catalytic oxidation reactions as well as obtaining interesting yields even without base addition. Indeed, monometallic Au and Pd catalysts showed very low activity in the oxidation of furfural and HMF. This is in good agreement with the literature data. The activity of monometallic catalysts in base free oxidation is generally low due to the adsorption of products on the metal surface, impeding the conversion of the substrates into the final products. Up to now only Au supported on basic oxides such as MgO, CuO or hydrotalcites reached high reaction yields. This is due to the partial leaching of the support and in situ base formation, as discussed in [21].

Several oxidation steps occur during the aerobic oxidation of HMF. Moreover, all these steps could occur simultaneously (Figure 5). Depending on the active metal used for the reaction, the oxidation to FDCA could proceed through the formation of HFCA or FDC. The mechanism of the HMF oxidation follows two pathways: (1) in the first step, the aldehyde group is oxidized to carboxylic group to form 5-hydroxymethyl-2-furancarboxylic acid (HFCA). In the second step, the 5-formyl-2-furancaboxylic acid (FFCA) is formed due to the alcohol group oxidation to aldehyde; or (2) the hydroxyl group is oxidized to an aldehyde to form 2, 5-diformylfuran (FDC), and, in the second step, the oxidation of carbonyl group gives a monoacid: 5-formyl-2-furancaboxylic acid (FFCA). The FDCA is formed through oxidation of the aldehyde group of FFCA. It is well known that the reaction medium and catalyst composition govern the reaction pathway. Under high pH conditions (<7) the first pathway is favored, especially when gold-based catalysts are used [22]. As the oxidation of the alcohol group is the rate-determining step in the case of gold-based catalysts, the HFCA formation is generally favored. Casanova et al. [23] reported the formation of HFCA as the only intermediate detected on Au/TiO_2 and Au/CeO_2 catalysts in the HMF oxidation in basic medium. According to these authors, this might be due to the fast conversion of FFCA to FDCA. The formation of HFCA in basic medium is favored because of the aldehyde group, which can rapidly undergo hydration to a geminal diol. The further step, β-hydride elimination of the geminal diol to form a carboxylic acid, is also favored at a high pH by the OH^- adsorbed on the metal surface. FDC and FFCA intermediate formation in base-free oxidation of HMF using AuPd/CNT catalyst was observed by Wang et al. [24]. The authors observed that the hydroxyl group oxidation was much faster than the aldehyde group. For the Au/CNT, the aldehyde group was oxidized, forming HFCA, as was also observed for the Au-based catalysts in basic medium (Au/CeO_2 and Au/TiO_2). However, in this case, the ring-opening reaction of HFCA occurred, forming side products. In the case of monometallic Pd and AuPd bimetallic catalysts the reaction pathway changed. The formation of FDC was observed, as Pd seems to facilitate the oxidation of the alcohol group. In addition, the rapid oxidation of FFCA to FDCA was favored in the case of the Pd samples. This is one of the limiting steps over gold-based catalysts under base-free conditions.

In order to confirm the base-free conditions during the furfural and HMF oxidation, pH measurements were performed during the reaction, as shown in Table 5. The furfural and HMF solutions are already acidic, showing a pH of 3.96 and 4.25, respectively.

Table 5. Final pH values of furfural and HMF oxidation reactions using embedded Au_xPd_y@SiTi and supported Au_xPd_y/SiTi catalysts.

Catalyst	Furfural		HMF	
	2 h	10 h	2 h	10 h
blank test	-	2.93	-	3.04
Au_4Pd_1@SiTi	3.81	3.36	3.93	2.55
Au_1Pd_1@SiTi	3.89	3.22	3.28	2.60
Au_1Pd_4@SiTi	3.74	3.69	3.30	2.93
Au_4Pd_1/SiTi	-	2.78	-	2.53
Au_1Pd_1/SiTi	-	3.73	-	3.20
Au_1Pd_4/SiTi	-	3.55	-	3.49
Au@SiTi	-	3.21	-	3.30
Pd@SiTi	-	3.34	-	3.41

The low pH observed due to acid formation after the oxidation reactions did not increase the leaching of the metals to the solution (which was confirmed by ICP analysis). Considering that the reactions occurred in an acidic medium, lower reaction rates are justified when compared to the reactions performed in a neutral or alkaline medium [20,25]. This means that the concept for the development of catalysts for reactions in aqueous medium in the absence of a base was successful, although the search for an increase in catalytic efficiency for these systems remains topical.

To benchmark the results obtained under base-free conditions, tests in the presence of a base were also conducted. The results are given in Table 6.

Table 6. Catalytic results obtained in the oxidation of HMF in the presence of the base. Conditions: Aqueous solution of HMF (30 μmol), substrate/metal = 18 (mol/mol), O_2 (1 bar), 100 °C, 600 rpm, 2 molar equivalents of K_2CO_3 were added to the reactant solution before catalytic test.

Entry	Catalyst	Base	X%	Selectivity (%)			
				FDCA	FDC	HFCA	FFCA
1	Au_4Pd_1@SiTi	2 eq	37	3	14	34	49
2	Au_1Pd_1@SiTi	2 eq	18	4	30	37	29
3	Au_1Pd_4@SiTi	2 eq	4	3	34	53	10
4	Au_4Pd_1/SiTi	2 eq	34	2	15	32	51
5	Au_1Pd_1/SiTi	2 eq	5	8	15	60	17
6	Au_1Pd_4/SiTi	2 eq	5	9	14	59	18

The HMF conversion was much lower in the case of Au_4Pd_1@SiTi sample in the presence of a base (37%) as compared to base-free conditions (Table 3, Entry 1). However, the selectivity to FDCA was much higher (17%) in the reaction carried out without the base as compared to that in basic conditions (3%). In both cases, the activity of the catalysts was affected by the Au–Pd ratio. The monometallic catalysts showed very low oxidation activity of both molecules.

The activity of the catalysts was strongly affected by the calcination at 500 °C, as can be seen from Tables 3 and 4. For the Au_4Pd_1@SiTi$_{cal}$ calcined sample, the selectivity to furoic acid decreased from 45% to 2%. The selectivity to FDCA reached only 1% instead of 22% for the non-calcined material (Table 3). The decrease in the activity is due to the collapse of the structure and suppression of the porosity of the material (decrease of the BET surface after calcination from 78 m^2/g to 2 m^2/g in the case of Au_4Pd_1 sample). These results are comparable to that observed for the bare support. As the metal particles are embedded into the matrix and the pores are blocked, there is no contact between the substrate and the metal. The conversion is due to the furfural and HMF degradation.

In these reactions, the highest activity was observed for Au_4Pd_1 compositions (Section 2.1, Figure 2b), which is in good agreement with the previous data. Indeed, the arrangement between gold and palladium atoms plays a crucial role in catalytic performances of the bimetallic catalytic systems [26,27]. Therefore, the optimization and control of the fine bimetallic nanoparticles structure constitute the major objective to obtain high catalytic performances (the so-called "catalysis by design"), which finally allows the design of the most effective catalytic materials. The synthesis protocol used for bimetallic nanoparticle preparation strongly affects the type of the catalyst morphology. The interaction between two metals strongly influences the catalytic performance, as already reported [28–30]. The optimal metal-to-metal ratio must be determined for a given reaction. In the case of bimetallic catalysts based on AuPd systems, the remarkable increase in the catalytic oxidation activity was already observed [31–36]. In liquid phase oxidation of aldehydes and alcohols, the AuPd nanoparticles have already proved to be twice as efficient as the Au or Pd alone. It has been proposed that Au plays the role of an electronic promoter of Pd [35,36], which is in good agreement with the observed core-shell structure. A compositional effect has also been reported, where a gradual increase in catalytic activity was observed with the increase in Pd content. An optimal Au–Pd molar ratio of about 1:1.86 was reported for gas phase CO oxidation [37]. In our case, the catalytic activity also passes through the maximum for Au_4Pd_1 composition. This is also in good agreement with the literature for liquid phase oxidation of glucose and benzyl alcohol [20,25]. It could be stated that an optimum ratio between Pd^0 and Pd^{2+} species is necessary to maintain the high activity of these catalysts, as well to obtain high selectivity to desired products [20,38].

4. Materials and Methods

4.1. Materials

Chloroauric acid trihydrate (HAuCl$_4$·3H$_2$O, 99.9%, Sigma-Aldrich (St. Louis, MO, USA)), sodium borohydride (NaBH$_4$, 98%, Sigma-Aldrich (St. Louis, MO, USA)), poly(vinyl alcohol) (PVA, Sigma-Aldrich (St. Louis, MO, USA), MW 55,000 g/mol), sodium citrate (Na$_3$C$_6$H$_5$O$_7$, 99%, Sigma-Aldrich (St. Louis, MO, USA)), furfural (C$_5$H$_4$O$_2$, >99%, Sigma-Aldrich (St. Louis, MO, USA)), 5-hydroxymethylfurfural (HMF, 99%, Sigma-Aldrich (St. Louis, MO, USA)), furoic acid (FA, 99% Sigma-Aldrich (St. Louis, MO, USA)), 5-hydroxymethyl-2-furancarboxylic acid (HMFCA, >99%, Sigma-Aldrich (St. Louis, MO, USA)), 2, 5-diformylfuran (DFF, >99%, Sigma-Aldrich (St. Louis, MO, USA)), 5-formyl-2-furoic acid (FFCA, >99%, Sigma-Aldrich (St. Louis, MO, USA)), 2, 5-furandicarboxylic acid (FDCA, 99%, Sigma-Aldrich (St. Louis, MO, USA)) were of analytical grade and were used as received. Deionized water (18.2 MΩ) was used for the preparation of all the needed solutions.

4.2. Synthesis of Au, Pd and AuPd Titanosilicate Catalysts

In 200 mL of water with vigorous stirring, Au and Pd precursors (HAuCl$_4$ solution 30 wt.% and PdCl$_2$, 10.11 × 10^{-5} mol of metal) and 2 mL of a sodium citrate solution (0.17 M, metal:citrate = 1:3.4) were added. Then, a freshly prepared solution of NaBH$_4$ (50 mM, 10 mL, NaBH$_4$/Au (mol/mol) = 5) was then added drop by drop to form a metallic sol; the color of the sol was dark brown (Au$_1$Pd$_1$, Au$_1$Pd$_4$) and light-red (Au, Au$_4$Pd$_1$). After 20 min of sol generation, 400 mL of water were added. Then, 10 mL of a solution containing titanium isopropoxide (TIP) and (3-aminopropyl)triethoxysilane (APTES) (Si:Ti (mol/mol) = 3:2) were filtered using a 0.22 μm filter membrane and slowly added. After the hydrolysis of the oxide precursors (2 h), the slurry was filtered, the solid washed with ethanol (2 × 25 mL) and dried at room temperature (RT) for 24 h.

4.3. Synthesis of Au, Pd and AuPd Supported Catalysts

ZrO$_2$- and TiO$_2$- supported catalysts were prepared by the sol-immobilization method described in [21]. SiTi support used for the immobilization of bimetallic AuPd nanoparticles was synthetized similarly to the method described above. In the standard procedure, the preformed AuPd nanoparticles were put into contact with the SiTi support in water for 2 h under stirring. After this, the solid was removed by centrifugation, washed with ethanol and dried in air for 24 h.

4.4. Catalytic Reactions

Base-free oxidation of furfural: catalytic reactions under air atmosphere were performed using the Freeslate MultiReactor available on EQUIPEX platform REALCAT. It consisted of 24 parallel batch reactors for high throughput screening, in which each reactor was loaded with an aqueous furfural solution (2 mL, 24.7 mmol L^{-1}) and the Au-based catalyst (ca. 10 mg, 0.9 μmol Au). The reaction was carried out at 20 bar of air (26 bar final pressure), 110 °C, 600 rpm, 2 h (SPR reactions). Base free oxidation of HMF: catalytic reactions under O$_2$ atmosphere were performed using a 50 mL Fischer-Porter loaded with an aqueous HMF solution (2 mL, 30 μmol) and the Au-based catalyst (ca. 100 mg). The reaction was carried out at 1 bar of O$_2$, 100 °C, 600 rpm, with different reaction times, from 0.5 h to 24 h. In both cases, after the reaction the catalyst was removed by filtration, the products were diluted in water and analyzed. The analysis of the products was performed by UHPLC chromatography using the Aminex HPX-87H Ion Exclusion (Agilent (Santa Clara, CA, USA)) column for the analysis of furfural oxidation. Formic acid (0.5% v/v) was used as a mobile phase at a flowrate of 0.30 mL/min and product formation was observed on a UV-VIS detector/photodiode array detector at 253 nm. For the analysis of HMF oxidation products, a Phenomenex (Torrance, CA, USA) column (ROA, organic acid H$^+$; 300 × 7.8 mm) was used. Sulphuric acid (5 mmol/L) was used as a mobile phase at a flowrate of 0.60 mL/min and the products were detected on a Shodex RI-101 detector at 265 nm.

4.5. Characterization

TEM electron microscopy images were recorded by placing a drop of the particle's dispersion in ethanol over a carbon film supported on a cooper grid. The samples were studied using a FEI Titan Themis 60–300 microscope (FEI, Hillsboro, OR, USA). The average Au nanoparticle size was determined, taking into account at least 300 particles. X-ray photoelectron spectroscopy (XPS) experiments were performed with K-alpha surface analysis (Thermo Scientific (Waltham, MA, USA)) equipment with an Al-Kα X-ray source (1486.6 eV) and a flood gun. The investigated area was approximately circular (approx. 300 µm in diameter) and three different areas of each sample were examined. The binding energy (BE) of the spectra was corrected with that of adventitious carbon C1s (C–C, C–H) at 284.8 eV. Nitrogen adsorption and desorption analysis was performed using a TriStar II Plus analyzer (Micromeritics (Norcross, GA, USA)). The samples were subjected to a pretreatment before the analyses to eliminate impurities that were adsorbed onto the surface. Namely, the samples were heated up to 150 °C with a temperature ramp of 10 °C min^{-1} and then maintained at this temperature for 60 min under vacuum. To determine the total surface area of the analysed catalysts the BET (Brunauer–Emmett–Teller) model was used. The pore volume was also calculated using the BJH (Barrett–Joyner–Halenda) method. ICP-OES (inductively coupled plasma optical emission spectrometry) analysis was performed using Agilent 720-ES ICP-OES (Santa Clara, CA, USA) equipment combined with the Vulcan 42S automated digestion system. The digestion procedure of the samples prior to analysis was as follows; first, 10 mg of catalysts were weighted and then 500 µL HF and 1.5 mL of aqua regia were distributed in sample holder (Polypropylene tube) by the syringe robot. Then, the tubes were heated up to 65 °C overnight. All samples were neutralized and diluted up to 20 mL with ultrapure water prior to analysis. XRF (M4 Tornado) analysis was performed using an Energy Dispersive X Ray Fluorescence (EDXRF) spectrometer provided from Bruker (Billerica, MA, USA). This spectrometer is equipped with X-ray Rhodium tube and the beam is micro-focused using a polycapillary lens enabling excitation of an area of 200 µm. The measurement was done under vacuum (20 mbar). The X-ray generator was operated at 50 kV and 200 µA and different filters were used to reduce the background (100 µm Al/50 µm Ti/25 µm Cu). Quantitative analysis was done using the fundamental parameter (FP) (standardless).

5. Conclusions

Bimetallic AuPd embedded systems were developed and applied in the furfural and HMF base-free oxidation reactions. The catalysts consisted of AuPd nanoparticles embedded in a titanosilicate support. In this embedded catalyst, as shown above, the effect of AuPd bimetallic nanoparticles (obtained by citrate stabilization) coupled with the specific properties of the support resulted in an efficient catalyst for the oxidation of both molecules in an aqueous medium and in the absence of a homogenous base. The results shown in this article highlight the advantages of Au and AuPd catalysts for the selective oxidation of biomass-derived substrates, such as HMF and furfural, in the absence of a base. The Au_xPd_y@SiTi catalysts were effective for the oxidation of furfural and HMF to their respective FA and FDCA acids in the absence of a base, reaching 99% and 89% of conversion, respectively, after 10 h of reaction. The catalyst based on bimetallic AuPd NPs surrounded by a titanosilicate matrix allowed the production of the corresponding acids more efficiently than the monometallic versions of Au and Pd (Au@SiTi and Pd@SiTi) or, even, their supported versions (Au/SiTi, Pd/SiTi and AuPd/SiTi). No catalyst leaching was detected (Au, Ti, Si); in fact, the entire reaction occurred in acid medium (initial and final pH <4), allowing the isolation of FA and FDCA directly from the reaction medium after reaction took place. This type of catalyst could also be applied for the oxidation of other alcohols and aldehydes in base-free conditions.

Author Contributions: Conceptualization, C.P.F., L.M.R. and R.W.; methodology, J.T.-R.; N.J.S.C. and C.W.L.; writing—original draft preparation, C.P.F.; R.F. and R.W.; writing—review and editing, C.P.F.; R.W.; E.T.-N.; Â.A.T.-N.; L.M.R. and F.D.; visualization, N.J.S.C.; M.T.M. and R.F.; supervision, R.W. and L.M.R.; project administration, C.P.F. and L.M.R.; funding acquisition, L.M.R. All authors have read and agreed to the published version of the manuscript.

Funding: This research was funded by Fundação de Amparo à Pesquisa do Estado de São Paulo (FAPESP), grant number 2014/10824-3 2017/03235-0, 2014/15159-8.

Acknowledgments: The authors acknowledge Gustavo Saraiva Silveira for the help with figures and drawings and CNPq for the support. The REALCAT platform benefits from a state subsidy administrated by the French National Research Agency (ANR) within the frame of the 'Investments for the Future' program (PIA), with the contractual reference 'ANR-11-EQPX-0037'. The European Union, through the FEDER funding administered by the Hauts-de-France Region, has co-financed the platform. Centrale Lille, CNRS, and the University of Lille as well as the Centrale Initiatives Foundation, are thanked for their financial contributions to the acquisition and implementation of the equipment of the REALCAT platform. Chevreul Institute (FR 2638), Ministère de l'Enseignement Supérieur, de la Recherche et de l'Innovation, Hauts-de-France Region and FEDER are acknowledged for supporting and funding partially this work. We also acknowledge CNPEM for the FEI Titan Themis 60–300 microscope use and LNNano for XPS measurments.

Conflicts of Interest: The authors declare no conflict of interest.

References

1. Chheda, J.N.; Huber, G.W.; Dumesic, J.A. Liquid-Phase Catalytic Processing of Biomass-Derived Oxygenated Hydrocarbons to Fuels and Chemicals. *Angew. Chem. Int. Ed.* **2007**, *46*, 7164–7183. [CrossRef]
2. Corma, A.; Iborra, S.; Velty, A. Chemical Routes for the Transformation of Biomass into Chemicals. *Chem. Rev.* **2007**, *107*, 2411–2502. [CrossRef]
3. Davis, S.E.; Ide, M.S.; Davis, R.J. Selective oxidation of alcohols and aldehydes over supported metal nanoparticles. *Green Chem.* **2013**, *15*, 17–45. [CrossRef]
4. Dumeignil, F.; Capron, M.; Katryniok, B.; Wojcieszak, R.; Löfberg, A.; Girardon, J.-S.; Desset, S.; Araque-Marin, M.; Jalowiecki-Duhamel, L.; Paul, S. Biomass-derived Platform Molecules Upgrading through Catalytic Processes: Yielding Chemicals and Fuels. *J. Jpn. Pet. Inst.* **2015**, *58*, 257–273. [CrossRef]
5. Li, X.; Jia, P.; Wang, T. Furfural: A Promising Platform Compound for Sustainable Production of C4 and C5 Chemicals. *ACS Catal.* **2016**, *6*, 7621–7640. [CrossRef]
6. Aresta, M.; Dibenedetto, A.; Dumeignil, F. *Biorefineries an Introduction*; Walter de Gruyter GmbH & Co. KG: Berlin, Germany, 2015.
7. Guimarães, L.H.S. Carbohydrates from Biomass: Sources and Transformation by Microbial Enzymes. In *Carbohydrates-Comprehensive Studies on Glycobiology and Glycotechnology*; Intech: London, UK, 2012; Volume 20, pp. 441–456.
8. Zhang, Z.; Deng, K. Recent Advances in the Catalytic Synthesis of 2,5-Furandicarboxylic Acid and Its Derivatives. *ACS Catal.* **2015**, *5*, 6529–6544. [CrossRef]
9. Zhu, Y.; Shen, M.; Xia, Y.; Lu, M. Au/MnO_2 nanostructured catalysts and their catalytic performance for the oxidation of 5-(hydroxymethyl)furfural. *Catal. Commun.* **2015**, *64*, 37–43. [CrossRef]
10. Yang, B.; Dai, Z.; Ding, S.-Y.; Wyman, C.E. Enzymatic hydrolysis of cellulosic biomass. *Biofuels* **2011**, *2*, 421–449. [CrossRef]
11. Besson, M.; Gallezot, P.; Pinel, C. Conversion of Biomass into Chemicals over Metal Catalysts. *Chem. Rev.* **2014**, *114*, 1827–1870. [CrossRef]
12. Wojcieszak, R.; Cuccovia, I.M.; Silva, M.A.; Rossi, L.M. Selective oxidation of glucose to glucuronic acid by cesium-promoted gold nanoparticle catalyst. *J. Mol. Catal. A Chem.* **2016**, *422*, 35–42. [CrossRef]
13. Biella, S.; Castiglioni, G.L.; Fumagalli, C.; Prati, L.; Rossi, M. Application of gold catalysts to selective liquid phase oxidation. *Catal. Today* **2002**, *72*, 43–49. [CrossRef]
14. Delidovich, I.V.; Taran, O.P.; Matvienko, L.G.; Simonov, A.N.; Simakova, I.L.; Bobrovskaya, A.N.; Parmon, V.N. Selective Oxidation of Glucose over Carbon-supported Pd and Pt Catalysts. *Catal. Lett.* **2010**, *140*, 14–21. [CrossRef]
15. Liu, C.; Zhang, J.; Huang, J.; Zhang, C.; Hong, F.; Zhou, Y.; Li, G.; Haruta, M. Efficient Aerobic Oxidation of Glucose to Gluconic Acid over Activated Carbon-Supported Gold Clusters. *ChemSusChem* **2017**, *10*, 1976–1980. [CrossRef] [PubMed]

16. Miedziak, P.J.; Alshammari, H.; Kondrat, S.A.; Clarke, T.J.; Davies, T.E.; Morad, M.; Morgan, D.J.; Willock, D.J.; Knight, D.W.; Taylor, S.H.; et al. Base-free glucose oxidation using air with supported gold catalysts. *Green Chem.* **2014**, *16*, 3132–3141. [CrossRef]
17. Suchomel, P.; Kvitek, L.; Prucek, R.; Panacek, A.; Halder, A.; Vajda, S.; Zboril, R. Simple size-controlled synthesis of Au nanoparticles and their size-dependent catalytic activity. *Sci. Rep.* **2018**, *8*, 4589. [CrossRef] [PubMed]
18. Lolli, A.; Albonetti, S.; Utili, L.; Amadori, R.; Ospitali, F.; Lucarelli, C.; Cavani, F. Insights into the reaction mechanism for 5-hydroxymethylfurfural oxidation to FDCA on bimetallic Pd-Au nanoparticles. *Appl. Catal. A Gen.* **2015**, *504*, 408–419. [CrossRef]
19. Lilga, M.A.; Hallen, R.T.; Gray, M. Production of oxidized derivatives of 5-hydroxymethylfurfural (HMF). *Top. Catal.* **2010**, *53*, 1264–1269.
20. Wojcieszak, R.; Ferraz, C.P.; Jin Sha, J.; Houda, S.; Rossi, L.M.; Paul, S. Advances in Base-Free Oxidation of Bio-Based Compounds on Supported Gold Catalysts. *Catalysts* **2017**, *7*, 352. [CrossRef]
21. Ferraz, P.C.; Zieliński, M.; Pietrowski, M.; Heyte, S.; Dumeignil, F.; Rossi, L.M.; Wojcieszak, R. Influence of Support Basic Sites in Green Oxidation of Biobased Substrates Using Au-Promoted Catalysts. *ACS Sustain. Chem. Eng.* **2018**, *6*, 16332–16340. [CrossRef]
22. Ishida, T.; Kinoshita, N.; Okatsu, H.; Akita, T.; Takei, T.; Haruta, M. Influence of the Support and the Size of Gold Clusters on Catalytic Activity for Glucose Oxidation. *Angew. Chem. Int. Ed.* **2008**, *120*, 9405–9408. [CrossRef]
23. Casanova, O.; Iborra, S.; Corma, A. Biomass into Chemicals: Aerobic Oxidation of 5-Hydroxymethyl-2-furfural into 2,5-Furandicarboxylic Acid with Gold Nanoparticle Catalysts. *ChemSusChem* **2009**, *2*, 1138–1144. [CrossRef] [PubMed]
24. Wan, X.; Zhou, C.; Chen, J.; Deng, W.; Zhang, Q.; Yang, Y.; Wang, Y. Base-Free Aerobic Oxidation of 5-Hydroxymethyl-furfural to 2,5-Furandicarboxylic Acid in Water Catalyzed by Functionalized Carbon Nanotube-Supported Au-Pd Alloy Nanoparticles. *ACS Catal.* **2014**, *4*, 2175–2185. [CrossRef]
25. Sha, J.; Paul, S.; Dumeignil, F.; Wojcieszak, R. Au-based bimetallic catalysts: How the synergy between two metals affects their catalytic activity. *RSC Adv.* **2019**, *9*, 29888–29901. [CrossRef]
26. Ferrer, D.; Torres-Castro, A.; Gao, X.; Sepúlveda-Guzmán, S.; Ortiz-Méndez, U.; José-Yacamán, M. Three-layer core/shell structure in Au-Pd bimetallic nanoparticles. *Nano Lett.* **2007**, *7*, 1701–1705. [CrossRef] [PubMed]
27. Mu, R.; Fu, Q.; Xu, H.; Zhang, H.; Huang, Y.; Jiang, Z.; Zhang, S.; Tan, D.; Bao, X. Synergetic effect of surface and subsurface Ni species at Pt-Ni bimetallic catalysts for CO oxidation. *JACS* **2011**, *133*, 1978–1986. [CrossRef] [PubMed]
28. Benkó, T.; Beck, A.; Frey, K.; Srankó, D.F.; Geszti, O.; Sáfrán, G.; Maróti, B.; Schay, Z. Bimetallic Ag-Au/SiO$_2$ catalysts: Formation, structure and synergistic activity in glucose oxidation. *Appl. Catal. A Gen.* **2014**, *479*, 103–111. [CrossRef]
29. Heidkamp, K.; Aytemir, M.; Vorlop, K.-D.; Prüße, U. Ceria supported gold-platinum catalysts for the selective oxidation of alkyl ethoxylates. *Catal. Sci. Technol.* **2013**, *3*, 2984–2992. [CrossRef]
30. Comotti, M.; Pina, C.D.; Rossi, M. Mono- and bimetallic catalysts for glucose oxidation. *J. Mol. Catal. A Chem.* **2006**, *251*, 89–92. [CrossRef]
31. Silva, T.A.G.; Teixeira-Neto, E.; López, N.; Rossi, L.M. Volcano-like behavior of Au-Pd core-shell nanoparticles in the selective oxidation of alcohols. *Sci. Rep.* **2014**, *4*, 5766. [CrossRef]
32. Enache, D.I.; Edwards, J.K.; Landon, P.; Solsona-Espriu, B.; Carley, A.F.; Herzing, A.A.; Watanabe, M.; Kiely, C.J.; Knight, D.W.; Hutchings, G.J. Solvent-Free oxidation of primary alcohols to aldehydes using Au-Pd/TiO$_2$ Catalysts. *Science* **2006**, *311*, 362–365. [CrossRef]
33. Dimitratos, N.; Villa, A.; Wang, D.; Porta, F.; Su, D.; Prati, L. Pd and Pt catalysts modified by alloying with Au in the selective oxidation of alcohols. *J. Catal.* **2006**, *244*, 113–121. [CrossRef]
34. Ketchie, W.C.; Murayama, M.; Davis, R.J. Selective oxidation of glycerol over carbon-supported AuPd catalysts. *J. Catal.* **2007**, *250*, 264–273. [CrossRef]
35. Kesavan, L.; Tiruvalam, R.; Rahim, M.H.A.; bin Saiman, M.I.; Enache, D.I.; Jenkins, R.L.; Dimitratos, N.; Lopez-Sanchez, J.A.; Taylor, S.H.; Knight, D.W.; et al. Solvent-free oxidation of primary carbon-hydrogen bonds in toluene using Au-Pd alloy nanoparticles. *Science* **2011**, *331*, 195–199. [CrossRef] [PubMed]

36. Silva, T.A.G.; Landers, R.; Rossi, L.M. Magnetically recoverable AuPd nanoparticles prepared by a coordination capture method as a reusable catalyst for green oxidation of benzyl alcohol. *Catal. Sci. Technol.* **2013**, *3*, 2993–2999. [CrossRef]
37. Pritchard, J.; Kesavan, L.; Piccinini, M.; He, Q.; Tiruvalam, R.; Dimitratos, N.; Lopez-Sanchez, J.A.; Carley, A.F.; Edwards, J.K.; Kiely, C.J.; et al. Direct synthesis of hydrogen peroxide and benzyl alcohol oxidation using Au–Pd catalysts prepared by sol immobilization. *Langmuir* **2010**, *26*, 16568–16577. [CrossRef]
38. Silva, T.A.G.; Ferraz, C.; Gonçalves, R.V.; Teixeira-Neto, E.; Wojcieszak, R.; Rossi, L.M. Restructuring of gold-palladium alloyed nanoparticles: A step towards more active catalysts for oxidation of alcohols. *ChemCatChem* **2019**, *11*, 4021–4027. [CrossRef]

© 2020 by the authors. Licensee MDPI, Basel, Switzerland. This article is an open access article distributed under the terms and conditions of the Creative Commons Attribution (CC BY) license (http://creativecommons.org/licenses/by/4.0/).

Article

Liquid Phase Furfural Oxidation under Uncontrolled pH in Batch and Flow Conditions: The Role of In Situ Formed Base

Alessandra Roselli [1], Yuri Carvalho [2], Franck Dumeignil [2], Fabrizio Cavani [1], Sébastien Paul [2] and Robert Wojcieszak [2,*]

1. Dipartimento di Chimica Industriale "Toso Montanari", Università di Bologna, Viale Risorgimento 4, 40136 Bologna, Italy; alessandra.roselli@studio.unibo.it (A.R.); fabrizio.cavani@unibo.it (F.C.)
2. Univ. Lille, CNRS, Centrale Lille, ENSCL, Univ. Artois, UMR 8181, UCCS, Unité de Catalyse et Chimie du Solide, F-59000 Lille, France; yuri.carvalho@centralelille.fr (Y.C.); Franck.dumeignil@univ-lille.fr (F.D.); sebastien.paul@centralelille.fr (S.P.)
* Correspondence: robert.wojcieszak@univ-lille.fr

Received: 7 December 2019; Accepted: 2 January 2020; Published: 3 January 2020

Abstract: Selective oxidation of furfural to furoic acid was performed with pure oxygen in aqueous phase under mild conditions and uncontrolled pH using hydrotalcite-supported gold nanoparticles as catalyst. Hydrotalcites with different Mg: Al ratios were tested as support. The effects of reaction time, temperature and furfural/catalyst ratio were evaluated. The catalyst Au/HT 4:1 showed the highest activity and selectivity to the desired product, achieving a complete conversion of furfural to furoic acid after 2 h at 110 °C. Further, stability tests were carried out in a continuous stirred-tank reactor and a progressive deactivation of the catalyst due to the leaching of Mg^{2+} cations from the support inducing changes in the pH of the reaction medium was observed.

Keywords: furfural; furoic acid; gold; hydrotalcite; oxidation

1. Introduction

The increase of the CO_2 emissions associated with the still increasing global production based on fossil resources is a source of motivation for researchers to find renewable feedstocks to produce chemicals and fuels. Within this frame, the conversion of biomass into fuels and top-valued chemicals is considered as one of the most attractive alternatives to the use of fossil resources [1]. Furfural (FF), the other name for 2-furaldehyde ($C_5H_4O_2$), is a hetero-aromatic furan ring with an aldehyde functional group. It is a natural precursor of furan-based chemicals and has high potential to become a major renewable platform molecule to produce biosourced chemicals and biofuels. Furfural is a natural product that can be easily obtained by dehydration of xylose, a monosaccharide found in large quantities in lignocellulosic biomass [2]. Industrial production of furfural started in 1922, when Quaker Oats Company gave destiny to oat hulls by processing them in a digester with sulphuric acid. Since then, around 50,000 tons per year of oat hulls were being transformed to furfural and sold for 25 cents a pound [3]. It was mainly used as a raw material for the production of other furanic compounds, such as insecticides or herbicides, and in lubricating oils and resins production. Furfural is one of the most promising platform molecules, not only because of its very reactive chemical structure but also because it can be further transformed into higher value-added molecules that find applications mainly as fuels or monomers for the polymers industry [4]. Through the catalytic selective oxidation of the furfural aldehyde moiety, it is possible to produce 2-furoic acid, which is the first down-line oxidation derivative of furfural having many applications in the pharmaceutical (esters forms [5]), agrochemical, flavour and fragrance industries. In the domain of resins, new routes for producing 5-hydroxyvaleric acid from furoic acid can also be highlighted [6]. However, other products can also be obtained from furfural oxidation, such as maleic acid, succinic acid, 2(5H)-furanone and CO_2 [4,7]. Industrially, furoic acid (FA) is nowadays produced through Cannizzaro disproportionation reaction in aqueous NaOH [4]. However, this process presents some drawbacks since the disproportionation reaction also coproduces furfuryl alcohol, leading to maximal selectivity to furoic acid of 50%. Moreover, the final solution has to be neutralized with sulfuric acid, forming sodium bisulfate that must be separated and then treated or disposed. The Cannizzaro reaction is also highly exothermic, so that temperature control of the reactor must be implemented [4]. For these reasons, many studies were done to combine heterogeneous catalysis in presence of NaOH for improving the furfural conversion to furoic acid [8–11]. Within this frame, direct catalytic furfural oxidation using pure oxygen or, even better, air would open the gate to higher selectivity to furoic acid. Douthwaite et al. demonstrated that furfural selective oxidation to furoic acid could take place under base-free conditions when $AuPd/Mg(OH)_2$ was employed as a catalyst [12]. However, they also proved that $AuPd/Mg(OH)_2$ was not stable at low pH since Mg^{2+} leaching was detected when no NaOH was added [12]. Therefore, the high conversion observed under initially supposed base-free conditions was finally attributed to the in situ formation of a homogeneous $Mg(OH)_2$ base. However, no stability test was performed to verify any catalyst deactivation due to its dissolution in aqueous medium. Apart from furfural oxidation, many works in the literature describe the use of basic solid supports, such as MgO, CaO and hydrotalcites, to perform catalytic reactions in aqueous medium but not taking in consideration the connection between catalyst activity and instability related to soluble base formation [13–15]. Considering the use of MgO, the authors already observed the support leaching phenomena on furfural oxidation [16] and further studied the use of more stable oxides such as MnO_2 [17]. In the present work, a heterogeneously-catalyzed process to produce furoic acid by selective oxidation of furfural using different hydrotalcites-supported gold nanoparticles is proposed as an alternative to the Cannizzaro reaction. The reaction scheme is shown in Scheme 1. Further, the study of the stability of such catalysts was carried out in a continuous stirred-tank reactor (CSTR), and a relation between activity and catalyst stability was proposed.

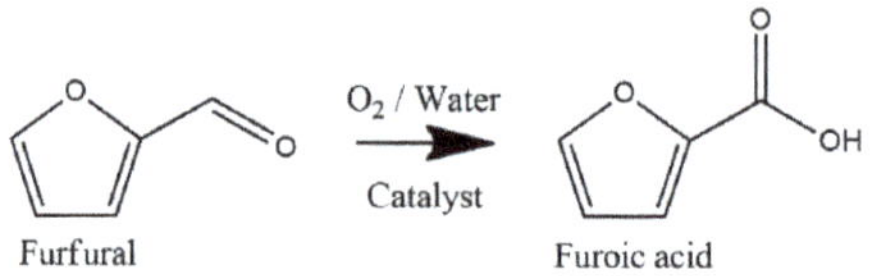

Scheme 1. Reaction scheme of the catalytic oxidation of furfural to furoic acid.

2. Results

2.1. Catalysts Characterization

Hydrotalcite supports were prepared as described in the experimental section using a co-precipitation method. After calcination under static air at 500 °C during 3 h, the transformation of the hydrotalcite structure into $MgO-Al_2O_3$ mixed oxides occurred as shown by XRD (Figure 1). The XRD patterns exhibit the typical features of a mixed oxide of the Mg(Al)O type. The characteristic diffraction peaks clearly observed at $2\theta \approx 43°, 63°$ and also $36°$ correspond to the periclase phase (MgO, powder diffraction file 45-0946 from the International Centre for Diffraction Data, ICDD). In addition, the formation of spinel structures is not observed. Furthermore, the samples were found clean of any impurity such as residual nitrate salts.

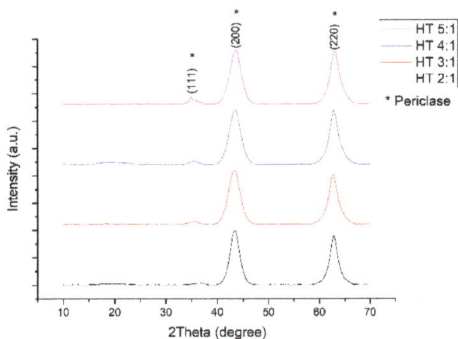

Figure 1. Hydrotalcites X-ray diffractograms after calcination of the samples.

After gold nanoparticles immobilization, the prepared samples were characterized again using XRD technique (Figure 2). The XRD patterns presented typical diffraction peaks from hydrotalcite (powder diffraction file 89-460 from ICDD). Due to hydrotalcite memory effect, it was able to recover its original structure upon rehydration of the mixed oxides when subjected to the gold nanoparticles immobilization protocol. However, it is still possible to observe the presence of periclase, by the appearance of the peaks at $2\theta \approx 43°$ and $63°$, which is more remarkable for the samples with higher Mg/Al ratio.

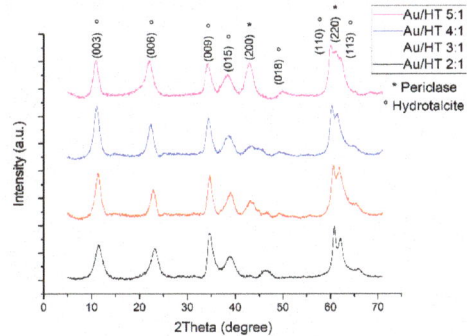

Figure 2. Au/HTs catalysts X-ray diffractograms.

ICP-OES analyses were performed for determining the actual gold content in each catalyst (theoretical value 2 wt.%) as well as the Mg/Al ratio for all the hydrotalcite supports (theoretical values 2, 3, 4 and 5). The Al, Au and Mg contents are displayed in Table 1. Au content in the catalysts was a bit lower than expected with a value of 1.4 ± 0.3 wt.%, while the samples contained a lower proportion of Mg that expected, with a deviation of ca. 20% compared to the theoretical value.

Table 1. ICP-OES analyses of the prepared catalysts.

Catalyst	Au (w/w %)	Al (w/w %)	Mg (w/w %)	Mg/Al Ratio
Au/HT 2:1	1.3%	13.8%	23.0%	1.7
Au/HT 3:1	1.7%	11.1%	27.8%	2.5
Au/HT 4:1	1.2%	8.7%	27.5%	3.2
Au/HT 5:1	1.4%	7.6%	30.0%	4

The Au/HT 4:1 catalyst was selected for characterization by TEM to determine the metal dispersion on the surface due to its performance (see below). TEM images and particle size distribution can be seen in Figure 3. The images showed that the gold nanoparticles are highly dispersed on the support surface. The particles sizes are distributed between 2.5 and 5.5 nm with an average particle size of 3.7 nm.

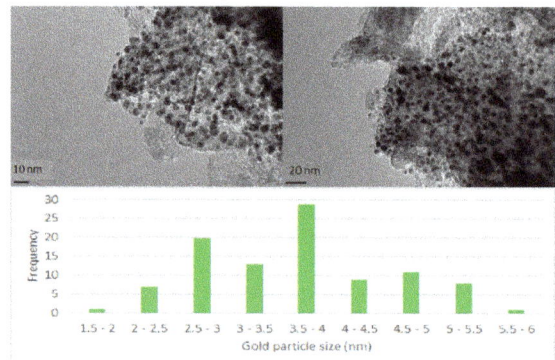

Figure 3. TEM images and particle size distribution from Au/HT 4:1.

The surface areas of the catalysts were determined using the BET model whereas the pore volumes were calculated using the BJH method. The results are given in Table 2. The surface area and pore volume clearly drop from the lowest Mg/Al ratio catalyst, Au/HT 2:1, to higher Mg/Al ratio catalysts. In addition, surface area and pore volume also drop when comparing the bare hydrotalcite support with the corresponding catalyst. For instance, surface area and pore volume for the bare HT 2:1 support were 91 m^2g^{-1} and 0.33 cm^3g^{-1}, respectively, in comparison with 73 m^2g^{-1} and 0.17 cm^3g^{-1}, respectively, after impregnation. It clearly shows the influence of the gold nanoparticles deposition method on the textural properties of the support.

Table 2. Textural properties.

Catalyst	BET Surface Area (m^2g^{-1})	Pore Volume (cm^3g^{-1})
Au/HT 2:1	73	0.17
Au/HT 3:1	25	0.07
Au/HT 4:1	30	0.08
Au/HT 5:1	25	0.07

Finally, to characterize the acid-base properties of the supports, TPD experiments were performed using either 10 mol.% NH_3 or CO_2 in He gas mixtures. The adsorption of the gas molecules were carried out at 50 °C and 40 °C, respectively, for 1 h. For desorption, the temperature ramp used was 10 °C min^{-1} up to 500 °C, which is the calcination temperature used to prepare hydrotalcite supports [18]. The desorption curves of NH_3 for the different supports are shown in Figure 4.

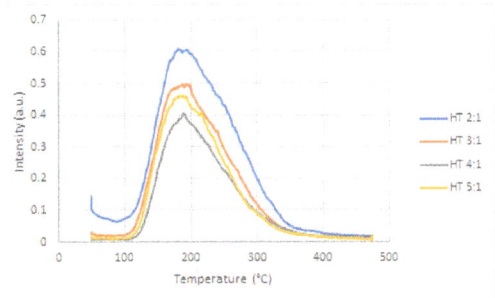

Figure 4. NH$_3$ desorption curves for the different HT supports.

As expected, the results obtained from NH$_3$ desorption are correlated with the Mg: Al molar ratios. The desorption curve in the case of HT Mg: Al = 2:1 catalyst is much more intense than for the other supports. This is probably due to the high surface area of this sample (90 m^2/g) as compared to other samples. The HT 3:1 and 5:1 led to very close desorption curves while the HT 4:1 showed the least intense curve. It is also worth to note that all the HT supports seem to have only one kind of acid sites corresponding to desorption at low temperature (around 190 °C), suggesting weak or weak to medium acid sites. Quite similar but reversed trend was observed for 2:1, 3:1 and 5:1 samples with CO$_2$ desorption data as shown in Figure S1. As expected, the HT Mg: Al = 5:1 support adsorbed more CO$_2$ than the other 2:1 and 3:1 HT samples, being the most basic support used in this study. The hydrotalcite samples seem to have two kinds of basic sites represented by the CO$_2$ desorption at different temperatures, the strongest at higher temperature (around 275 °C) and the second one at lower temperature (around 120 °C). Table 3 displays the amounts, in arbitrary units per gram of material, of NH$_3$ and CO$_2$ desorbed in TPD.

Table 3. Desorbed NH$_3$ and CO$_2$ amounts in TPD for HTs.

Support	NH$_3$ (a.u.g^{-1}) [1]	CO$_2$ (a.u.g^{-1}) [1]	Mg/Al Ratio
HT 2:1	90	33	1.7
HT 3:1	63	68	2.5
HT 4:1	48	–	3.2
HT 5:1	56	105	4

[1] Arbitrary units per gram of support.

2.2. Blank Test with Hydrotalcite Supports Only

The reaction was first performed with the bare hydrotalcites (without any gold) to check if the supports themselves were active in furfural oxidation (Figure 5). Up to 45% conversion was observed, however, the yield to furoic acid was not higher than 11%. The conversion could be attributed to the degradation of furfural into undetected byproducts, as no other oxidation products such as furfuryl alcohol could be detected, leading to a low carbon balance (always less than 80%). Without support the blank test done in the same conditions led to only 5% furfural conversion.

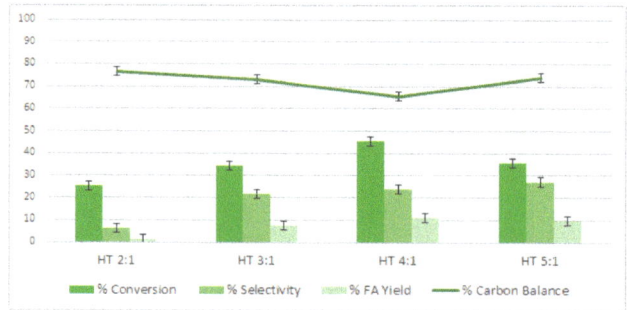

Figure 5. Comparison of the performance in furfural oxidation of HT supports with different molar Mg: Al ratios. Batch reactor, p = 6 bar (O_2), t = 110 °C, 2 h, stirring rate 600 rpm, [FF] = 26 mM, 100 mg of HT.

2.3. Reactivity of Gold Nanoparticles Supported on Hydrotalcites

Strong increases both in the FF conversion and the yield to furoic acid were observed using supported gold nanoparticles catalysts. Indeed, the gold-mediated oxidation is known to avoid radical pathways [19], thus, the degradation of furfural is less favoured as confirmed by the higher carbon balance (always close to or higher than 90%). At the same time, different Mg:Al molar ratios were evaluated to study whether the increase in the Mg content and correlated basic properties would result in an enhancement of the catalytic performances. Actually, increases in furfural conversion and furoic acid yield were observed as the Mg content in Au/HT increased as shown in Figure 6. Full conversion and selectivity to furoic acid were achieved with the Au/HT 4:1 and Au/HT 5:1 catalysts. This was probably due to an increase in the overall basicity of the support, which would promote the activity of the catalysts. Indeed, the catalytic results were well correlated with the basic properties of the HT supports. Obviously, the 100% yield to furoic acid (and hence the 100% carbon balance) means no degradation or formation of undesired products at all.

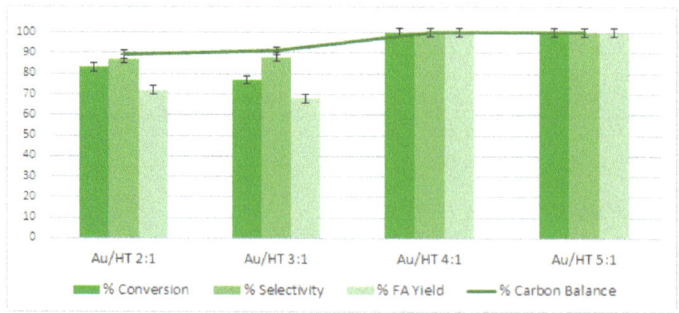

Figure 6. Comparison of the performance in furfural oxidation of Au/HT catalysts with supports having different molar Mg: Al ratios. Batch reactor, p = 6 bar (O_2), t = 110 °C, 2 h, 600 rpm, FF/Au molar ratio = 200:1, [FF] = 26 mM, 25 mg of catalyst.

2.4. Study on the Effect of the Reaction Time

The effect of the reaction time on catalytic activity was further studied using the Au/HT 4:1 catalyst. Complete furfural conversion was observed after 120 min of reaction as shown in Figure 7. The selectivity to furoic acid was above 90% in all cases. In addition, the carbon balance was above 90%, indicating minimum degradation of furfural in these conditions. It shows that the catalyst is very active even at short reaction times.

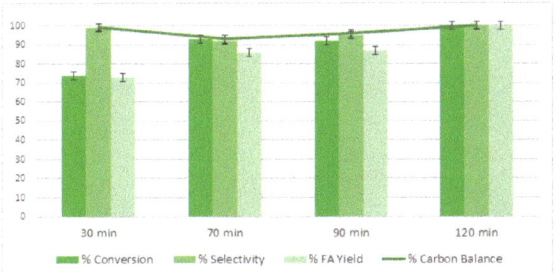

Figure 7. Effect of the reaction time using Au/HT 4:1. Furfural oxidation, batch reactor, $p = 6$ bar (O_2), $t = 110\ °C$, 600 rpm, FF/Au = 80:1, [FF] = 26 mM, 63 mg of catalyst.

In addition, ICP-OES analyses of samples of the liquid phase were performed for monitoring if any leaching was taking place during the reaction. Table 4 shows the concentration of Au and Mg in the reaction medium for different reaction times (analysed twice at each time), as well as the pH value.

Table 4. ICP-OES analyses and pH of the reaction samples.

Time	[Au] mgL^{-1}	[Mg] mgL^{-1}	[Mg]$_{leach}$/[Mg]$_{total}$ w/w %	pH
0	–	–	0	3
30	–	13.4	1.55%	7
70	–	17.7	2.04%	7
90	–	19.8	2.29%	7
120	–	23.0	2.66%	7

According to ICP-OES no gold leaching was detected even after 2 h of batch reaction. However, the Mg concentration in the reaction medium constantly increased with the reaction time, indicating that some leaching occurs. In addition, the pH of all reaction samples was 7, which was not expected since both furfural and furoic acid solutions have a pH around 3 in a base-free medium. This is a strong evidence that a soluble base was formed in situ even if the Mg leaching was limited, as a 10 mgL^{-1} increase of concentration was observed between 30 and 120 min of the reaction time.

2.5. Study on the Effect of the Temperature

The reaction was also studied regarding the influence of the temperature. The results are presented in Figure 8. At an FF/Au molar ratio of 80, the catalyst displayed full conversion, both at 110 and 130 °C, with high values of selectivity and carbon balance. However, at higher temperature (150 °C) some degradation of furfural occurred as illustrated by the lower carbon balance values. At 80 °C the conversion was not complete after 2 h of reaction.

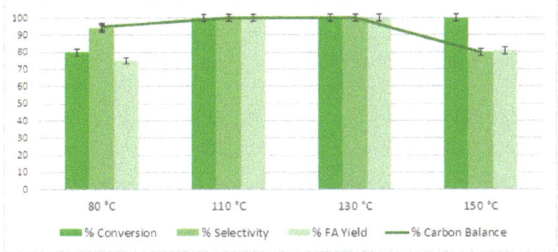

Figure 8. Effect of the reaction temperature using Au/HT 4:1. Furfural oxidation, batch reactor, $p = 6$ bar (O_2), 600 rpm, 2 h, FF/Au = 80, [FF] = 26 mM, 63 mg of catalyst.

2.6. Study of the Effect of the FF/Au Molar Ratios

Au/HT 4:1 was tested using three different theoretical molar ratios between furfural and gold: 80, 200 and 400. As can be seen in Figure 9, the catalyst showed full conversion and high furoic acid yield at low FF/Au molar ratios.

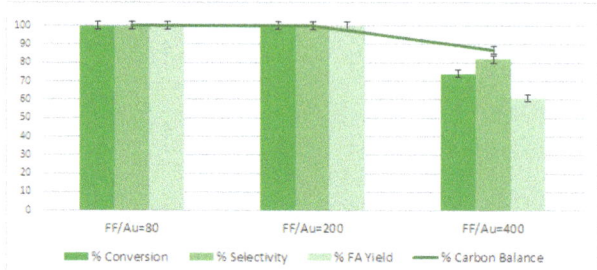

Figure 9. Effect of the FF/Au molar ratios using Au/HT 4:1 catalyst. Furfural oxidation, batch reactor, $p = 6$ bar (O_2), $t = 110\ °C$, 600 rpm, 2h, [FF] = 26 mM.

2.7. Stability of Au/HT 4:1 in a Continuous Reactor

Following the results obtained from the ICP-OES analyses of the liquid sampled during the batch reactions and suspecting that some support leaching takes place, Au/HT 4:1 was tested in a continuous reactor (CSTR-type) working at a constant feed flow rate of 0.5 mL min^{-1}. The continuous reaction system allows a better monitoring of catalyst stability and suitability for potential industrial use. In this system, the reactant was pumped into the vessel and the solution under reaction was collected at the same flow rate, while the catalyst remained trapped inside the reactor by using a filter preventing any catalyst entrainment in the outflow. Figure 10 depicts the reaction behavior over 6 h of time on stream.

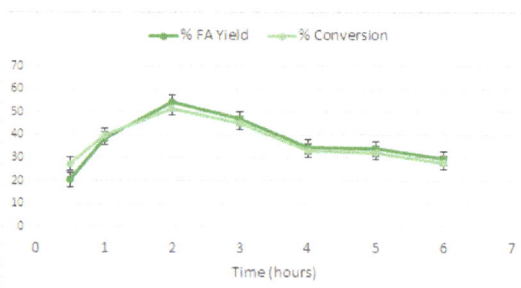

Figure 10. Furfural oxidation in CSTR employing Au/HT 4:1 as catalyst. $p = 5$ bar air; $t = 110\ °C$; FF/Au = 50; feed: 0.5 mL min^{-1} [FF] = 26 mM; 100 mg of catalyst.

Initially, the catalyst seemed to have a relatively low activity, which increased progressively up to 2 h under stream and then constantly decreased. It can be seen that the selectivity to furoic acid and carbon balance were high during the whole reaction since the conversion, and yield values were very close. The steady state was not achieved during the 6 h on stream, indicating that deactivation was taking place in parallel, even during the transient state in the first hours. The pH of the outflow was measured for each reaction sample and was found constant at 7 all along the reaction, as it was observed for batch reaction. This means that leaching of the support was taking place constantly. It is proposed that homogeneous Mg(OH)$_2$ [12] is progressively released in the reaction medium and eliminated by the flow, then causing the activity fall down. One should also consider that for maintaining the pH at 7, the leaching phenomena should be stronger than in batch reaction, since a fresh furfural solution (pH = 3) is being introduced continuously into the CSTR. As described before in the literature [12–15], it was believed that catalysts based on solid basic supports such as $Mg(OH)_2$, MgO, CaO, hydrotalcites, etc., would promote base-free catalytic transformations in aqueous medium. However, this work, presents an evidence that catalysts based on hydrotalcite supports can be active in apparently base-free conditions but in reality, it is not the case because of their partial dissolution in the reaction medium, then generating in situ a soluble base.

3. Discussion

This work was conducted using water as solvent, pure oxygen or air as oxidants and gold nanoparticles immobilized on hydrotalcite (HT) supports. The aim was to develop a greener and sustainable process to selectively oxidize furfural to furoic acid. Using appropriate method, gold nanoparticles were immobilized on different HT supports. A study on the acid-base properties of the supports was done for different HT materials prepared with different Mg: Al molar ratios. Under mild conditions, the synergistic effect between gold nanoparticles and properties of the HT support, allowed a fast and complete transformation of furfural to furoic acid (100% yield). However, by performing catalytic tests in a continuous reaction system, it was possible to observe the instability of the catalyst and to correlate it with Mg leaching from the support in aqueous medium. Based on ICP analysis and pH measurements, it was evidenced that an in situ homogeneous base formation was taking place because of support dissolution under hydrothermal conditions. If, on the one hand, the basic environment would contribute to the catalytic activity in a batch reactor, on the other hand, the catalyst deactivates faster in a CSTR regime. Ultimately, this work provides evidences that the use of catalysts based on HT supports should not be considered as base-free catalysis promoters under the aforementioned hydrothermal conditions.

4. Materials and Methods

4.1. Hydrotalcite Supports Preparation

Hydrotalcite samples (HTs) were prepared by co-precipitating aqueous solutions of Mg and Al salts as precursors with a highly basic solution [18]. Different molar ratios Mg: Al (2:1, 3:1, 4:1 and 5:1) were used to prepare the supports noted HT 2:1, HT 3:1, HT 4:1 and HT 5:1, respectively. The precursors solution ($Mg(NO_3)_2 \bullet 6\ H_2O$ and $Al(NO_3)_3 \bullet 9\ H_2O$ dissolved in deionized H_2O) was added dropwise to 1 M NaOH and the pH was maintained in the range of 10.5 ± 0.1. The temperature during the synthesis was 55 °C ± 0.2. The solution was stirred for one hour, keeping the temperature constant. Then, the suspension was filtered and the solid was washed with warm distilled water (50 °C). The final solid was dried overnight at 100 °C in an oven and then ground using a mortar. To transform HTs to mixed oxides, the samples were calcined at 500 °C for 3 h under static air with a temperature ramp of 5 °C min^{-1}. Finally, XRD analyses were carried out to verify the crystalline structure of the samples.

4.2. Gold NPs on Hydrotalcite Supports

The 2 wt.% Au/HTs catalysts were prepared by a sol immobilization method using $NaBH_4$ and polyvinyl alcohol (PVA) as reducing and dispersing agents, respectively. The gold nanoparticles were prepared following the method described by N. Dimitratos et al. [20,21]. First, a 2% wt.% solution of PVA in distilled water was prepared (PVA/Au (*w/w*) = 1.2). When PVA was completely dissolved, this solution was added to an aqueous solution of $HAuCl_4 \bullet 3H_2O$ (5.08×10^{-4} mol L^{-1}) under vigorous stirring. A fresh $NaBH_4$ solution (0.1 mol L^{-1}) was prepared to yield a molar $NaBH_4$ to Au ratio of 5 and then added to the previous solution in order to form the metallic sol. The color of the sol was deep purple. After 30 min of sol generation, the gold nanoparticles were immobilized by adding different supports under vigorous stirring. The amount of support was calculated to give a final theoretical gold loading of 2 wt.%. After 2 h the slurry was filtered, the solid washed 3 times with 50 mL of warm water (50 °C) and 50 mL of ethanol and further dried in oven at 100 °C for 1 h. The solids obtained were finally grinded to get the catalysts noted Au/HT 2:1, Au/HT 3:1, Au/HT 4:1 and Au/HT 5:1. With this preparation method, it was expected to obtain very small gold nanoparticles, with a diameter around 2–4 nm.

4.3. Catalysts Characterization

The X-ray diffraction (XRD) measurements of the solids were performed using a Bruker D8-Advance Powder X-ray diffractometer (Billerica, MA, USA). The patterns were obtained using CuKα radiation with an accelerating voltage of 40 kV and an emission current of 40 mA. The samples were scanned over a 2θ range of 10° to 70°, with a step size of 0.014° and a time of 19.2 s per step. Gold amount on each catalyst was determined by ICP-OES from Agilent Technologies (Santa Clara, CA, USA). A Vulcan 42 S—Questron Technologies/Horiba automated digester was associated to ICP to digest the powder catalysts precisely. For digestion, 2 mL of aqua regia was added to 5 mg of solid and heated to 65 °C for overnight. For Transmission Electron Microscopy (TEM) analysis, the catalysts were dispersed in ethanol and left for 10 min in the ultrasonic bath. The instrument used was a TEM/STEM FEI TECNAI F20 microscope combined with an Energy Dispersive X-ray Spectrometer (EDS) (Hillsboro, OR, USA) at 200 kV. To calculate the average gold nanoparticles size, the diameters of 100 particles were measured from TEM images and used for statistics. Nitrogen adsorption and desorption analysis on the different catalysts were performed using a TriStar II Plus analyzer (Norcross, GA, USA) (Micromeritics). The samples were subjected to a pretreatment before the analyses to eliminate impurities that were adsorbed on the surface. Namely, the samples were heated to 250 °C with a temperature ramp of 10 °C min^{-1} and then maintained at this temperature for 60 min under vacuum. To determine the total surface area of the analyzed catalysts, the BET model was used. The pore volume was also calculated using the BJH method. The total acidity or basicity of the catalysts was determined using a TPD/TPR/TPO Micromeritics instrument equipped with an MKS MS Spectrometer (Norcross, GA, USA). Generally, 15–30 mg of catalyst were pretreated up to 500 °C (calcination temperature of the samples) under helium flow. For the NH_3-TPD and CO_2-TPD experiments, the samples were cooled to 50 °C and 40 °C, respectively, and the adsorption of each considered probe molecule was performed for 1 h flowing a mixture of 10% NH_3 or 10% CO_2 in helium. Then, helium was flowed for 30 min at the adsorption temperature to remove the excess of physisorbed molecules. Finally, the temperature was increased to 500 °C (10 °C min^{-1}) and maintained at this temperature for 1 h.

4.4. Catalytic Tests

The catalytic tests were performed in a Top Industry autoclave. The reactant solutions were prepared by diluting a certain amount of furfural in 21 mL of H_2O, giving the desired concentration, and stirring the solution to dissolve furfural before adding it into the vessel. Starting solution (1 mL) was taken off for HPLC analysis, and the amount of catalyst, or bare support, (around 100 mg) was added in the autoclave. The reactor was purged three times with pure oxygen before reaching the pressure. After reaching 6 bar, the heating system was turned on. The first 10 min, to reach the desired temperature, were not considered in the reaction time and the final solutions were discharged only at the end of the reactions. All tests were carried out under uncontrolled pH conditions. At the beginning and at the end of the catalytic tests, the pH was measured. At the end, the reactor was cooled by an external air cooler system. The reaction mixtures were filtered, and 1 mL of the final solutions were diluted for HPLC analysis in a Phenomenex column (ROA, organic acid H$^+$; 300 × 7.8 mm). Sulphuric acid (5 mmol L^{-1}) was used as a mobile phase with a flow rate of 0.60 mL min^{-1}, and the products were detected on a Shodex RI-101 and UV-Vis detectors at 253 nm.

For the stability test, a continuous reaction system composed of a Parr reactor coupled to an Eldex reciprocating piston pump was employed, composing a CSTR-like system. First, 20 mL of a 2.5 gL^{-1} of furfural solution (FF/Au = 50) was added to the vessel together with 100 mg of Au/HT 4:1. The reactor was pressurized with 5 bar of synthetic air before turning the heating on. Finally, when the desired temperature (110 °C) was reached, the stirring was turned on and a freshly prepared 2.5 gL^{-1} furfural solution started to be pumped into the vessel at a flow rate of 0.5 mL min^{-1}. A collector equipped with a filter was dipped into the reaction medium to draw the liquid inside the vessel avoiding any catalyst removal. The outflow is controlled by a valve at around 0.5 mL min^{-1} to keep a constant volume of liquid in the reactor (steady state). The final pressure in steady state was around 6 bar. Furfural and furoic acid were quantified by HPLC as described before.

Supplementary Materials: The following are available online at http://www.mdpi.com/2073-4344/10/1/73/s1, Figure S1: TPD-CO_2 of hydrotalcite samples.

Author Contributions: Conceptualization, F.C., S.P. and R.W.; methodology, F.D.; validation, F.D.; formal analysis, A.R. and Y.C.; investigation, A.R. and Y.C.; resources, R.W.; writing—original draft preparation, A.R.; writing—review and editing, Y.C., S.P. and R.W.; supervision, F.C., S.P. and R.W. All authors have read and agreed to the published version of the manuscript.

Funding: This research received no external funding.

Acknowledgments: The REALCAT platform is benefiting from a state subsidy administrated by the French National Research Agency (ANR) within the frame of the 'Future Investments' program (PIA), with the contractual reference 'ANR-11-EQPX-0037'. The European Union, through the ERDF funding administered by the Hauts-de-France Region, has co-financed the platform. Centrale Lille, the CNRS, and Lille University, as well as the Centrale Initiatives Foundation, are thanked for their financial contributions to the acquisition and implementation of the equipment of the REALCAT platform. Chevreul Institute (FR 2638), and Ministère de l'Enseignement Supérieur, de la Recherche et de l'Innovation are also acknowledged for supporting and funding partially this work.

Conflicts of Interest: The authors declare no conflict of interest.

References

1. Sousa-Aguiar, E.F.; Appel, L.G.; Zonetti, P.C.; do Couto Fraga, A.; Bicudo, A.A.; Fonseca, I. Some important catalytic challenges in the bioethanol integrated biorefinery. *Catal. Today* **2014**, *234*, 13–23. [CrossRef]
2. Dunlop, A.P. Furfural formation and behavior. *Ind. Eng. Chem.* **1948**, *40*, 204–209. [CrossRef]
3. Brownlee, H.J.; Miner, C.S. Industrial development of furfural. *Ind. Eng. Chem.* **1948**, *40*, 201–204. [CrossRef]
4. Mariscal, R.; Maireles-Torres, P.; Ojeda, M.; Sádaba, I.; Granados, M.L. Furfural: A renewable and versatile platform molecule for the synthesis of chemicals and fuels. *Energy Environ. Sci.* **2016**, *9*, 1144–1189. [CrossRef]
5. Asano, T.; Tamura, M.; Nakagawa, Y.; Tomishige, K. Selective Hydrodeoxygenation of 2-Furancarboxylic Acid to Valeric Acid over Molybdenum-Oxide-Modified Platinum Catalyst. *ACS Sustain. Chem. Eng.* **2016**, *4*, 6253–6257. [CrossRef]
6. Asano, T.; Takagi, H.; Nakagawa, Y.; Tamura, M.; Tomishige, K. Selective hydrogenolysis of 2-furancarboxylic acid to 5-hydroxyvaleric acid derivatives over supported platinum catalysts. *Green Chem.* **2019**, *21*, 6133–6145. [CrossRef]
7. Li, X.; Lan, X.; Wang, T. Selective oxidation of furfural in a bi-phasic system with homogeneous acid catalyst. *Catal. Today* **2016**, *276*, 97–104. [CrossRef]
8. Dunlop, A.P. Process for Manufacturing Furoic Acid and Furoic Acid Salts. U.S. Patent No. 2/407/066, 3 September 1946.
9. Tian, Q.; Shi, D.; Sha, Y. CuO and Ag2O/CuO catalyzed oxidation of aldehydes to the corresponding carboxylic acids by molecular oxygen. *Molecules* **2008**, *13*, 948–957. [CrossRef] [PubMed]
10. Verdeguer, P.; Merat, N.; Gaset, A. Lead/platinum on charcoal as catalyst for oxidation of furfural. Effect of main parameters. *Appl. Catal. A Gen.* **1994**, *112*, 1–11. [CrossRef]
11. Verdeguer, P.; Merat, N.; Rigal, L.; Gaset, A. Optimization of experimental conditions for the catalytic oxidation of furfural to furoic acid. *J. Chem. Technol. Biotechnol. Int. Res. Process Environ. Clean Technol.* **1994**, *61*, 97–102. [CrossRef]
12. Douthwaite, M.; Huang, X.; Iqbal, S.; Miedziak, P.J.; Brett, G.L.; Kondrat, S.A.; Edwards, J.K.; Sankar, M.; Knight, D.W.; Bethell, D.; et al. The controlled catalytic oxidation of furfural to furoic acid using AuPd/Mg (OH) 2. *Catal. Sci. Technol.* **2017**, *7*, 5284–5293. [CrossRef]
13. Yuan, Z.; Wu, P.; Gao, J.; Lu, X.; Hou, Z.; Zheng, X. Pt/solid-base: A predominant catalyst for glycerol hydrogenolysis in a base-free aqueous solution. *Catal. Lett.* **2009**, *130*, 261–265. [CrossRef]

14. Gupta, N.K.; Nishimura, S.; Takagaki, A.; Ebitani, K. Hydrotalcite-supported gold-nanoparticle-catalyzed highly efficient base-free aqueous oxidation of 5-hydroxymethylfurfural into 2, 5-furandicarboxylic acid under atmospheric oxygen pressure. *Green Chem.* **2011**, *13*, 824–827. [CrossRef]
15. Tongsakul, D.; Nishimura, S.; Ebitani, K. Platinum/gold alloy nanoparticles-supported hydrotalcite catalyst for selective aerobic oxidation of polyols in base-free aqueous solution at room temperature. *ACS Catal.* **2013**, *3*, 2199–2207. [CrossRef]
16. Ferraz, C.P.; Zieliński, M.; Pietrowski, M.; Heyte, S.; Dumeignil, F.; Rossi, L.M.; Wojcieszak, R. Influence of Support Basic Sites in Green Oxidation of Biobased Substrates Using Au-Promoted Catalysts. *ACS Sustain. Chem. Eng.* **2018**, *6*, 16332–16340. [CrossRef]
17. Ferraz, C.P.; Da Silva, A.G.M.; Rodrigues, T.S.; Camargo, P.H.C.; Paul, S.; Wojcieszak, R. Furfural Oxidation on Gold Supported on MnO2: Influence of the Support Structure on the Catalytic Performances. *Appl. Sci.* **2018**, *8*, 1246. [CrossRef]
18. Zhao, R.; Yin, C.; Zhao, H.; Liu, C. Synthesis, characterization, and application of hydotalcites in hydrodesulfurization of FCC gasoline. *Fuel Process. Technol.* **2003**, *81*, 201–209. [CrossRef]
19. Wojcieszak, R.; Ferraz, C.P.; Sha, J.; Houda, S.; Rossi, L.M.; Paul, S. Advances in base-free oxidation of bio-based compounds on supported gold catalysts. *Catalysts* **2017**, *7*, 352. [CrossRef]
20. Dimitratos, N.; Lopez-Sanchez, J.A.; Morgan, D.; Carley, A.; Prati, L.; Hutchings, G.J. Solvent free liquid phase oxidation of benzyl alcohol using Au supported catalysts prepared using a sol immobilization technique. *Catal. Today* **2007**, *122*, 317–324. [CrossRef]
21. Ferraz, C.P.; Garcia, M.A.S.; Teixeira-Neto, É.; Rossi, L.M. Oxidation of benzyl alcohol catalyzed by gold nanoparticles under alkaline conditions: Weak vs. strong bases. *RSC Adv.* **2016**, *6*, 25279–25285. [CrossRef]

© 2020 by the authors. Licensee MDPI, Basel, Switzerland. This article is an open access article distributed under the terms and conditions of the Creative Commons Attribution (CC BY) license (http://creativecommons.org/licenses/by/4.0/).

Article

Capping Agent Effect on Pd-Supported Nanoparticles in the Hydrogenation of Furfural

Shahram Alijani [1], Sofia Capelli [1], Stefano Cattaneo [1], Marco Schiavoni [1], Claudio Evangelisti [2], Khaled M. H. Mohammed [3], Peter P. Wells [3,4,5], Francesca Tessore [1] and Alberto Villa [1,*]

1. Department of Chemistry, University of Milan, via Golgi 19, 20133 Milano, Italy; shahram.alijani@unimi.it (S.A.); sofia.capelli@unimi.it (S.C.); stefano.cattaneo2@unimi.it (S.C.); marco.schiavoni@unimi.it (M.S.); francesca.tessore@unimi.it (F.T.)
2. ISTM-Institute of Molecular Sciences and Technologies, CNR—National Research Council, 20138 Milan, Italy; claudio.evangelisti@cnr.it
3. School of chemistry, University of Southampton, Southampton SO17 1BJ, UK; K.Mohammed@soton.ac.uk (K.M.H.M.); peter.wells@diamond.ac.uk (P.P.W.)
4. UK Catalysis Hub, Research Complex at Harwell, Rutherford Appleton Laboratory, Harwell Oxon, Didcot OX11 0FA, UK
5. Diamond Light Source Ltd., Harwell Science and Innovation Campus, Chilton, Didcot OX11 0DE, UK
* Correspondence: Alberto.villa@unimi.it; Tel.: +39-0250314361

Received: 6 December 2019; Accepted: 18 December 2019; Published: 19 December 2019

Abstract: The catalytic performance of a series of 1 wt % Pd/C catalysts prepared by the sol-immobilization method has been studied in the liquid-phase hydrogenation of furfural. The temperature range studied was 25–75 °C, keeping the H_2 pressure constant at 5 bar. The effect of the catalyst preparation using different capping agents containing oxygen or nitrogen groups was assessed. Polyvinyl alcohol (PVA), polyvinylpyrrolidone (PVP), and poly (diallyldimethylammonium chloride) (PDDA) were chosen. The catalysts were characterized by ultraviolet-visible spectroscopy (UV-Vis), Fourier transform infrared spectroscopy (FTIR), transmission electron microscopy (TEM), and X-ray photoelectron spectroscopy (XPS). The characterization data suggest that the different capping agents affected the initial activity of the catalysts by adjusting the available Pd surface sites, without producing a significant change in the Pd particle size. The different activity of the three catalysts followed the trend: $Pd_{PVA}/C > Pd_{PDDA}/C > Pd_{PVP}/C$. In terms of selectivity to furfuryl alcohol, the opposite trend has been observed: $Pd_{PVP}/C > Pd_{PDDA}/C > Pd_{PVA}/C$. The different reactivity has been ascribed to the different shielding effect of the three ligands used; they influence the adsorption of the reactant on Pd active sites.

Keywords: furfural; hydrogenation; palladium; nanoparticles; capping agent; sol-immobilization

---

1. Introduction

Diminishing fossil fuel resources, a growing energy demand, and the increased environmental concerns caused by CO_2 emissions have led to the search for new sustainable energy resources [1,2]. In this regard, waste lignocellulosic biomass, which mainly contains cellulose, hemicellulose, and lignin, is considered a promising sustainable resource. It is an abundant alternative carbon resource that can be used to produce chemicals and biofuels [3,4].

Among the value-added molecules, furfural is an important precursor in the generation of biofuels and chemical intermediates. It can be readily obtained from hemicellulose by acid-catalyzed cascade hydrolysis and dehydration [4,5]. Owing to its high unsaturation, the selective hydrogenation of furfural has attracted significant attention to produce a range of valuable C_4 and C_5 molecules including furfuryl alcohol (FA), tetrahydrofurfuryl alcohol (THFA), 2-methylfuran (2-MF), 2-methyltetrahydrofuran

(2-MTHF), and others (Figure 1). FA is widely used in the chemical industry, primarily for the production of foundry resins, polymers, synthetic fibers, and a chemical intermediate for the production of perfume and vitamin [6,7]. THFA is considered a green solvent and is usually used in printer ink, agricultural applications, and electronics cleaners [7,8]. 2-MF has applications in biorefineries as a consequence of its high octane number, and in chemical industries, e.g., the production of toluene [9].

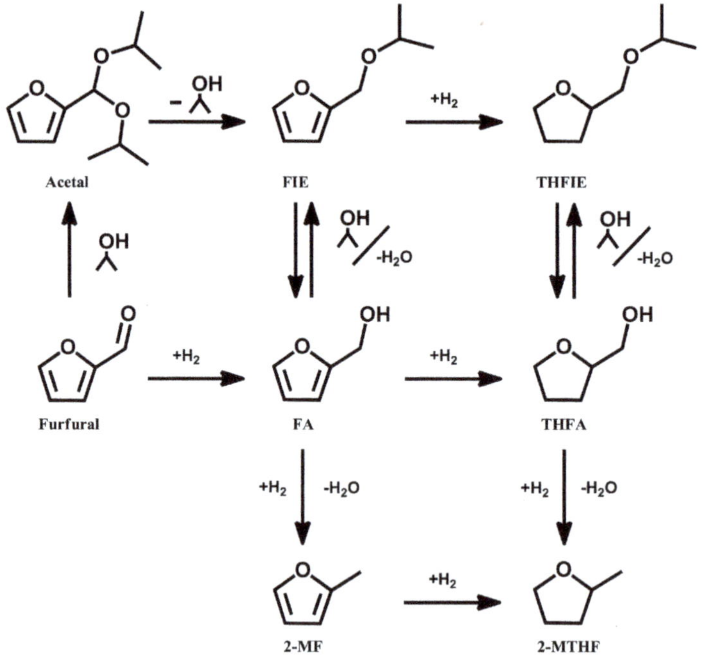

Figure 1. Possible reaction pathway of furfural hydrogenation. Reaction products: furfuryl alcohol (FA), tetrahydrofurfuryl alcohol (THFA), 2-methylfuran (2-MF), 2-methyltetrahydrofuran (2-MTHF), furfuryl isopropyl ether (FIE), and tetrahydrofurfuryl isopropyl ether (THFIE).

The main issue with furfural hydrogenation is controlling its reaction route and hydrogenation degree through a selective catalyst [10]. Depending on the process conditions [11], the nature of the catalyst [12], and the type of the solvent [13], furfural hydrogenation proceeds through various pathways and with the formation of a significant number of products. A variety of heterogeneous catalysts mainly based on Pd [14,15], Pt [16], Ni [17], Cu [18], and Ru [19]. supported on carbon [20], titanium oxide [21], alumina [22], silica. or zeolite [22], have been reported for the hydrogenation of furfural.

In terms of supported metal nanoparticles, the capping agent can affect the particle size, size distribution, morphology, and stability [23,24]. Recently, a number of researchers revealed an additional benefit of capping agents, which act as promoters and/or selectivity modifiers by blocking specific surface sites in various liquid-phase reactions [25]. Medlin et al. used self-assembled monolayers (SAMs) to create a more favorable surface environment for specific product formation [26]. They used alkanethiolate SAM-modified Pd catalysts for the selective hydrogenation of furfural, increasing the selectivity to value-added compounds, i.e., FA and 2-MF, by selectively blocking facets, leaving only particle edges/corners exposed [26]. It was also observed that supported Pd nanoparticles (NPs) protected by PVA led to higher selectivity to benzaldehyde in the liquid phase oxidation of benzyl alcohol, despite the conversion being lower. This was mainly because of the preferential blocking of Pd (111) facets, which have been recognized to facilitate the decarbonylation process [27]. Indeed,

Rogers et al. proposed that the interaction between the solvent and PVA alters the PVA binding on the metal surface. Consequently, this interaction contributes to the different Pd corner and edge sites available [21].

To further understand the role of capping agents in controlling the selectivity of furfural hydrogenation, we prepared carbon-supported Pd nanoparticles through controlled sol-immobilization routes [28], using different capping agents including polyvinyl alcohol (PVA), polyvinylpyrrolidone (PVP), and poly(diallyldimethylammonium) chloride (PDDA). The aim of this work was to perform a more systematic study of the role of stabilizers under mild reaction conditions (25–75 °C, and 5 bar) and to enhance the understanding of this relatively under-reported effect.

2. Results

Pd colloids have been prepared in the presence of PVA, PVP, and PDDA as the stabilizer and NaBH$_4$ as the reducing agent. PVA contains –OH groups, whereas PVP has a pyrrole-type N species in close proximity to a carbonyl group and PDDA contains dimethyl-ammonium groups (Figure 2). The metal loading was confirmed by atomic absorption spectroscopy (AAS).

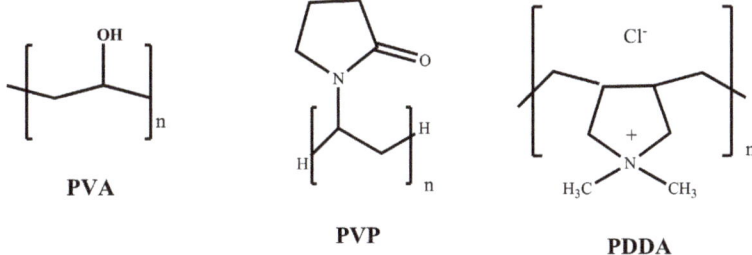

Figure 2. Structure of the capping agents used.

2.1. UV-Visible Characterization

UV-Vis spectroscopy was used to demonstrate the complete reduction of the Pd^{2+} precursor to Pd^0 after the addition of NaBH$_4$, before immobilization on the support (Figure S1).

The peak at 236 nm, corresponds to the Pd–Cl metal charge transfer in [PdCl$_4$]$^{2-}$, whereas the peak at 420 nm is related to d–d transitions [29,30]. The capping agent–H$_2$O sols are transparent in the visible range; however, PVA has a band at 280 nm, which can be attributed to the $\pi \rightarrow \pi^*$ transition of the carbonyl groups (C=O) associated with ethylene unsaturation [31]. The spectra of the Pd precursor+capping agent and Pd sols demonstrate the interaction of the metal with PVA and PVP. A shift at 286 nm (blue curve, Figure S1c) after adding PDDA to the Pd precursor, suggests a possible ligand-to-metal charge transfer interaction between PDDA and [PdCl$_4$]$^{2-}$, as already evidenced in the case of Au [31]. The disappearance of the peaks related to Pd^{2+} and the observed scattering suggest the complete reduction to Pd^0 and the formation of metal nanoparticles [32,33].

2.2. Infrared (IR) Spectroscopy

The interaction of the capping agents with Pd precursor was also investigated using Fourier transform infrared (FTIR) spectroscopy. The FTIR spectra of PVA and PVA–Pd precursor (Na$_2$PdCl$_4$) are shown in Figure 3a. The primary bands for the PVA films are: (i) 3431 cm^{-1}, O–H stretching vibrations (the broad nature of the –OH stretching vibration is characteristic of residual water, i.e., KBr is hygroscopic) and (ii) 2918 cm^{-1}, aliphatic C–H stretching vibrations (the peak at 2356 cm^{-1} is due to gas phase CO$_2$). The peak at 1726 cm^{-1} can be assigned to C=O stretching deriving from the residual C=O due to the incomplete hydrolysis of the acetate group [34]. PVA is produced by the polymerization of vinyl acetate to poly(vinyl acetate), and its subsequent hydrolysis to PVA. The peaks between 1380 and 1247 cm^{-1} can be assigned to O–H deformation, whereas the peaks at 1200–1000 cm^{-1} can be attributed to the stretching vibrations of C–O–C linkage, which suggest the presence of cross-linked PVA molecules [35]. The slight modification observed for the peaks at 1380–1247 cm^{-1} in the spectrum of PVA+Na$_2$PdCl$_4$ suggests a weak interaction between the metal precursor and the –OH groups present in PVA.

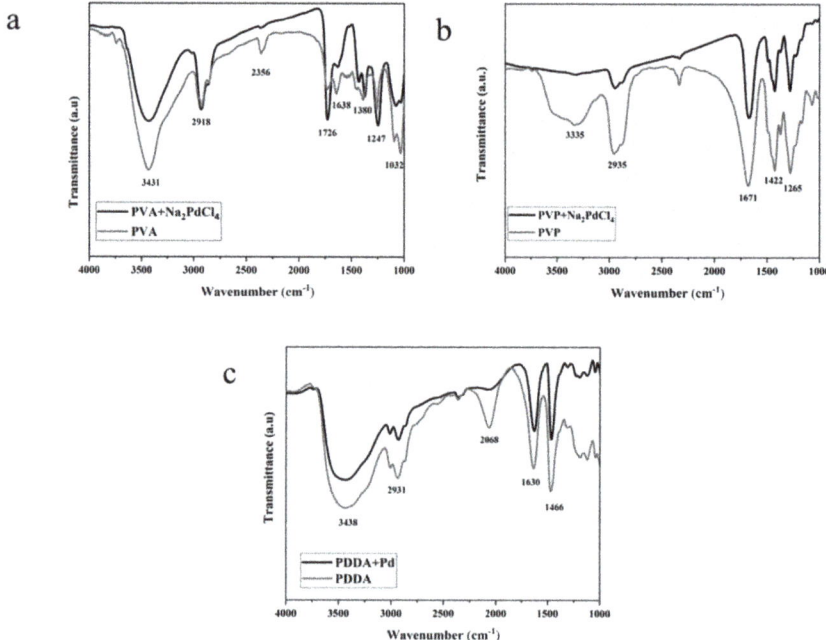

Figure 3. FTIR spectra of (a) PVA and PVA/Pd, (b) PVP and PVP-Pd, and (c) PDDA and PDDA/Pd.

The prominent band related to PVP (Figure 3b) at 3335 cm^{-1} is due to the O–H stretching vibrations of adsorbed water at the surface of particles. The peak at 2951 cm^{-1} could be related to asymmetric CH$_2$ stretching of the ring, while the one at 1671 cm^{-1} is assigned to the stretching vibration of C=O. The absorption peak at 1425 cm^{-1} is due to the C–H$_2$ scissor. The peak at 1281 cm^{-1} can be assigned to the C–N stretching vibrations. Previous studies revealed that PVP can bind to Pd surfaces from the carbonyl group or nitrogen atom of the pyrrolidone units [36,37]. The decrease in intensity of the peak at 1671 cm^{-1} and at 1265 cm^{-1}, observed after the addition of Pd, confirms that both O and N groups present in PVP interact with the metal precursor [38,39].

The characteristic FTIR spectrum of PDDA is illustrated in Figure 3c. The peak at 1635 cm^{-1} is attributed to the bending vibration of the C–N group, and the one at 1466 cm^{-1} is assigned to C=C stretching, in a positively charged environment contributed by the positively charged nitrogen of the polycations [40]. After the addition of Pd, the observed decrease in the intensity of the peaks located at 1635 and 1466 cm^{-1} indicates that the activities of these vibrational modes are modified in the mixture, probably due to the PDDA–Pd interaction. With further inspection of Figure 3c, one can observe a loss in the peak at 2068 cm^{-1} after the addition of the Pd precursor. This result has been already observed in the literature, but the origin of this peak is not clear [40].

2.3. Transmission Electron Microscopy (TEM)

After the immobilization of the Pd nanoparticles on the support, the morphological specification of the prepared catalysts was investigated by HRTEM (Figure 4). It is noticeable that the capping agent has a great influence on the Pd NPs' size and distribution when using activated carbon (Camel (X40S)) as a support. All the catalysts showed an average particle size of 3–4 nm, with the presence of isolated larger particles, in particular for Pd$_{PDDA}$/C (see Figure 4).

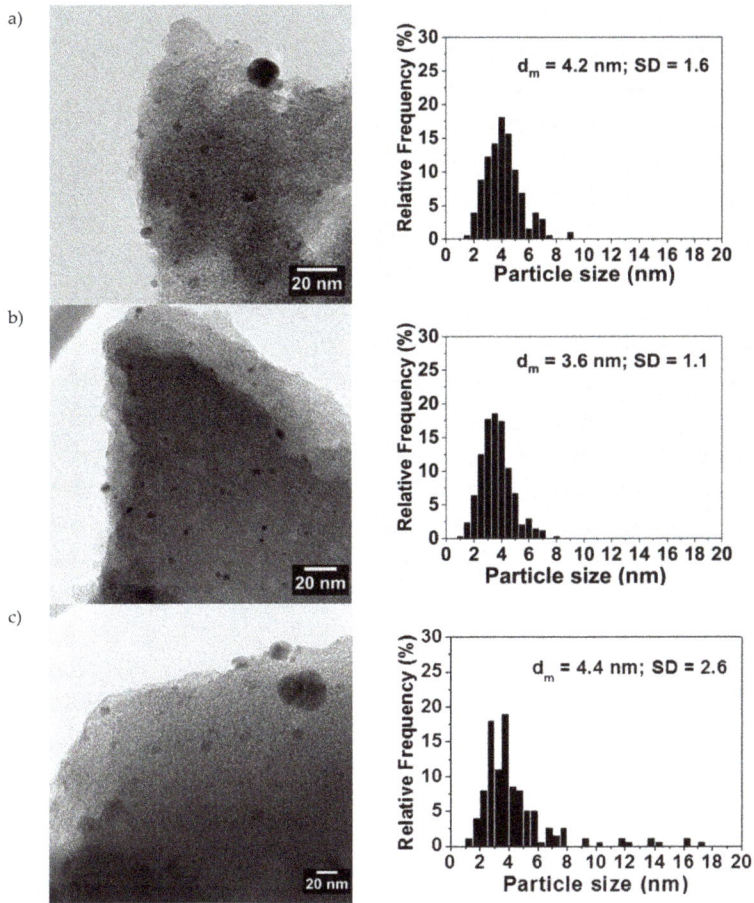

Figure 4. HRTEM micrograph and histogram of particle size distribution of (**a**) Pd$_{PVA}$/C, (**b**) Pd$_{PVP}$/C, and (**c**) Pd$_{PDDA}$/C.

2.4. X-ray Photoelectron Spectroscopy (XPS)

X-ray photoelectron spectroscopy (XPS) of the catalysts was performed to investigate the surface chemistry of the carbon support, the chemical state, and the exposure of supported palladium species. XPS survey data indicated the presence of only Pd, C, N, and O species. No evidence of Na or B residues from NaBH$_4$ was detected (Table 1). Depending on the capping agent used, significant differences were observed in the relative amount of Pd at the surface: Pd$_{PVA}$/C (1.30%) > Pd$_{PDDA}$/C (0.76%) > Pd$_{PVP}$/C (0.50) (Table 1). The data revealed a different oxygen content in the samples. Pd$_{PVA}$/C shows the highest relative amount of O 1s (14.7%) compared to Pd$_{PDDA}$/C (9.70%) and Pd$_{PVP}$/C (9.10%), which show an oxygen content similar to bare carbon (9.17%) (Table 1). The highest oxygen content on the surface of Pd$_{PVA}$/C catalyst is probably due to the presence of oxygen groups contained in PVA, confirming the presence of the capping agent on the surface of the catalyst.

Table 1. XPS data for the relative surface amount of Pd, O, N, C, and O/C.

Catalyst	Content (at %)				O/C
	Pd	O	N	C	
C (X40S)	0	9.17	0	90.82	0.10
Pd$_{PVA}$/AC	1.30	14.70	0	84.23	0.17
Pd$_{PVP}$/C	0.50	9.10	2.95	87.54	0.10
Pd$_{PDDA}$/C	0.76	9.70	1.86	87.84	0.11
Pd$_{PVA}$/AC used	0.32	45.70	0	53.98	1.10
Pd$_{PVA}$/AC used and washed	0.93	18.00	0	81.07	0.22

Pd$_{PVP}$/C evidenced the highest amount of N (2.95%), superior to Pd$_{PDDA}$/C (1.86%), whereas for Pd$_{PVA}$/C nitrogen was not detected on the surface, as expected.

The XPS spectra show two components in the Pd 3d$_{3/2}$ region. The full width at half maximum (FWHM) was set between 1 and 3 eV. The measured spectra were deconvoluted with a few traces of the Gauss (80%)–Lorentz (20%) mixed function.

The XPS spectra show two components in the Pd 3d$_{3/2}$ region. The first peak at a binding energy (BE) of ~335 eV is ascribed to Pd^0, while a second peak centered at ~338 eV was ascribed to Pd$^{\delta+}$ species (Table 2 and Figure 4). Pd is mainly present in its oxidized form. However, the Pd^0/Pd$^{\delta+}$ ratio varies when using different capping agents. A higher content of Pd^0 species was observed in the case of Pd$_{PVA}$/C and Pd$_{PVP}$/C catalysts (30.9% and 23.9%, respectively), compared to Pd$_{PDDA}$/C, which mainly contain oxidized Pd on the surface of the nanoparticles (Table 2, Figure 5c), in agreement with the increasing steric hindrance of the capping agent (PVA < PVP < PDDA).

Table 2. XPS analysis of carbon-supported Pd NPs.

Sample		Pd3d	
		Pd^0	Pd$^{\delta+}$
Pd$_{PVA}$/C	BE [eV]	335.7	337.6
	[%]	30.9	69.1
Pd$_{PVP}$/C	BE [eV]	335.6	337.6
	[%]	23.9	76.1
Pd$_{PDDA}$/C	BE [eV]	335.7	338.0
	[%]	9.4	90.6
Pd$_{PVA}$/C used	BE [eV]	335.4	337.3
	[%]	72.7	27.3
Pd$_{PVA}$/C used and washed	BE [eV]	335.8	337.0
	[%]	65.5	34.5

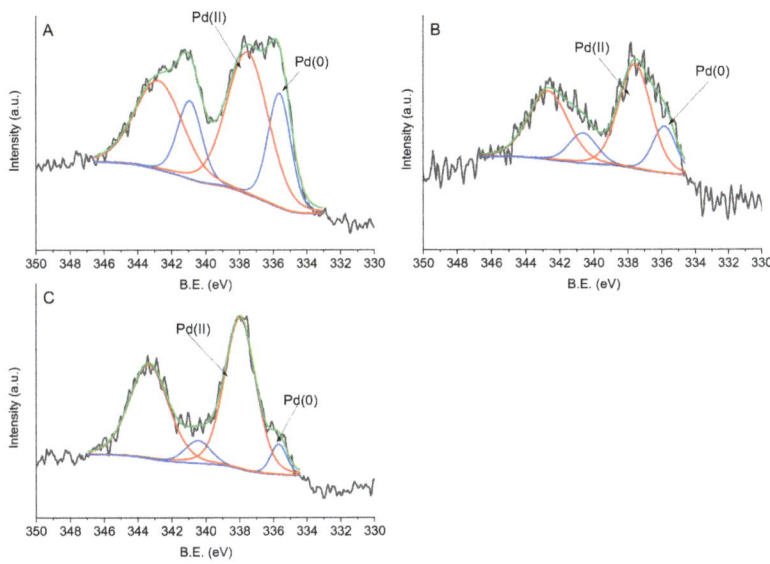

Figure 5. Fitting of XPS envelopes of (**a**) Pd$_{PVA}$/C, (**b**) Pd$_{PVP}$/C, and (**c**) Pd$_{PDDA}$/C for Pd 3d.

In the C 1s region we can identify four peaks. The signal at 284.0–284.6 eV can be assigned to sp^2-hybridised carbon, the one at 285.3–285.8 eV to sp^3-hybridized carbon, the one at 286.6–287.9 eV to the C=O group, and the one at 288.8–292.2 eV to C–OH/C–O–C, as indicated in Table 3 and Figure 6. Four main oxygen groups were identified. Binding Energy (BE) of 529.5–531.4 eV corresponds to a carbon–oxygen double bond and the one at 532.1–532.8 eV to an ether-like single bond, whereas BE at 533.4–534.6 eV can be attributed to carbon–oxygen single bonds in hydroxyl groups [41] (Figure 7). Oxygen species around 536.0 eV refer to the presence of carboxylic groups [42]. Bare X40S carbon contains 30% C=O groups, 41.9% C–O–C, 22.1% C–OH groups, and 6.0% carboxylic groups (Table 3). After the addition of Pd$_{PVP}$ nanoparticles, the oxygen species' relative ratio remains similar, whereas after the addition of Pd$_{PVA}$ and Pd$_{PDDA}$ the percentage of C–O–C increases with a decrease in C–OH species. For Pd$_{PVA}$, we can attribute the unexpected decrease in superficial –OH groups to the formation of C–O–C species between hydroxyl groups of the carbon and of PVA and the crosslinking of PVA chains, supporting the findings of FTIR spectroscopy

Table 3. C species for 1 wt % Pd/C samples.

Sample		C1s			
		C=C	C–C	C=O	C–OH/ C–O–C
C (X40S)	BE [eV]	284.6	285.4	286.7	289.2
	[%]	55.4	22.7	12.9	9.0
Pd$_{PVA}$/C	BE [eV]	284.0	285.4	286.7	289.2
	[%]	42.9	29.9	14.1	13.1
Pd$_{PVP}$/C	BE [eV]	284.6	285.9	287.9	289.9
	[%]	54.7	32.3	9.4	3.6
Pd$_{PDDA}$/C	BE [eV]	284.6	285.3	286.9	289.1
	[%]	28.4	47.5	18.5	5.6
Pd$_{PVA}$/C used	BE [eV]	284.6	285.8	287.8	289.2
	[%]	35.0	36.2	22.5	6.3
Pd$_{PVA}$/C used and washed	BE [eV]	284.6	285.5	286.6	288.8
	[%]	33.1	31.1	23.6	12.2

Table 3. *Cont.*

Sample		O1s			
		C=O	C–O–C	C–OH	C–OOH
C (X40S)	BE [eV]	531.1	532.7	534.0	536.5
	[%]	30.0	41.9	22.1	6.0
Pd$_{PVA}$/C	BE [eV]	531.2	532.9	534.6	536.2
	[%]	33.8	54.8	9.1	2.3
Pd$_{PVP}$/C	BE [eV]	531.0	532.1	533.4	534.8
	[%]	30.9	32.9	24.7	11.1
Pd$_{PDDA}$/C	BE [eV]	531.4	532.8	534.3	535.9
	[%]	28.1	52.5	14.2	5.2
Pd$_{PVA}$/C used	BE [eV]	-	532.6	533.9	535.9
	[%]	-	53.1	42.4	4.5
Pd$_{PVA}$/C used and washed	BE [eV]	529.8	531.0	532.4	533.6
	[%]	42.8	27.3	17.7	12.2

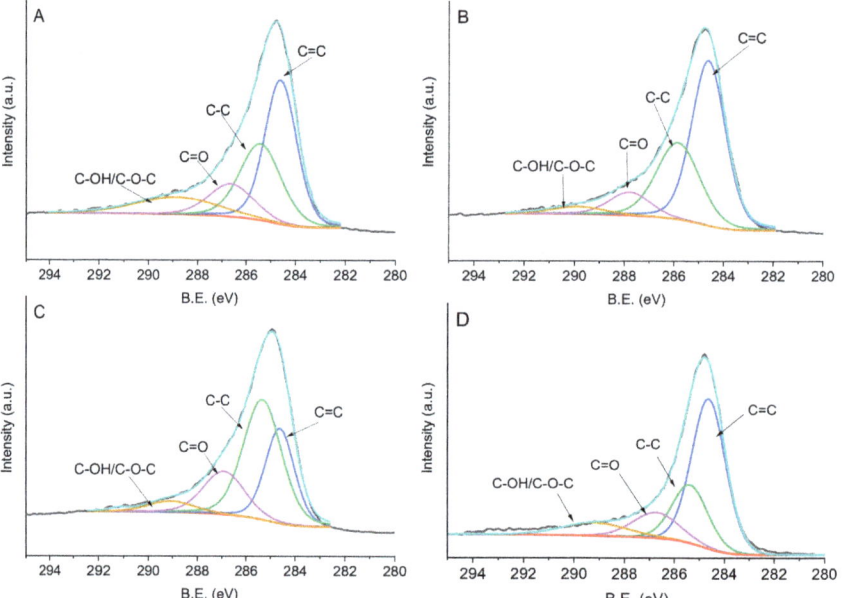

Figure 6. Fitting of XPS envelopes of (**a**) Pd$_{PVA}$/C, (**b**) Pd$_{PVP}$/C, (**c**) Pd$_{PDDA}$/C, and (**d**) X40S carbon for C1s species.

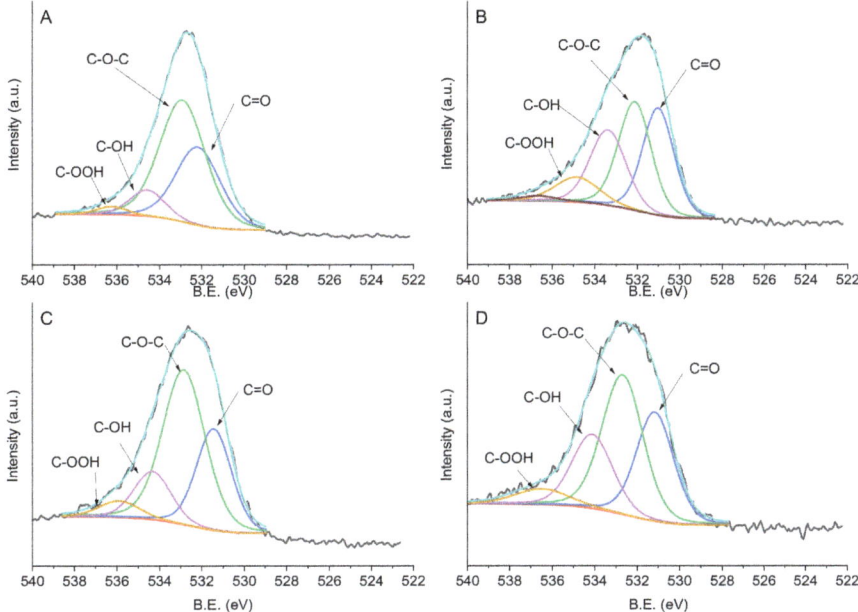

Figure 7. Fitting of XPS envelopes of (**a**) Pd$_{PVA}$/C, (**b**) Pd$_{PVP}$/C, (**c**) Pd$_{PDDA}$/C, and (**d**) X40S carbon for O1s species.

2.5. Catalytic Results

The performance of the prepared catalysts was evaluated for the liquid-phase hydrogenation of furfural (furfural 0.3 M; furfural/metal ratio 500 mol/mol, 5 bar of H_2, in the temperature range 25–75 °C), with 2-propanol as the solvent. As illustrated in Figure 1, the reaction route map is a cascade of reactions that forms various products such as FA, THFA, 2-MF, 2-MTHF, and ethers.

Hydrogenation of the carbonyl group (C=O) in furfural yields FA. The formation of THFA, on the other hand, requires the unselective hydrogenation of the furan double bonds along with the hydrogenation of the carbonyl group. Moreover, hydrogenolysis via ring activation is a primary pathway in the C–OH bond cleavage for the conversion of FA to 2-MF. By using alcohols as the solvent, several other byproducts can be formed, such as acetals [43] and ethers [44–46].

When the reaction was performed at 25 °C and 50 °C, Pd$_{PVA}$/C showed the highest initial activity (calculated after 5 min as 824 h^{-1} and 1775 h^{-1} at 25 °C and 50 °C, respectively), higher than Pd$_{PDDA}$ (624 h^{-1} and 1422 h^{-1} at 25 °C and 50 °C, respectively) and Pd$_{PVP}$/C (304 h^{-1} and 573 h^{-1} at 25 °C and 50 °C, respectively) (Table 4). At 75 °C all the catalysts were very active; it is not possible to precisely calculate the initial activity (conversion >55% after 5 min in all cases). We can exclude any effect of the size of Pd nanoparticles on the activity since all the catalysts had a similar particle size (3–4 nm, Table 4). However, it is possible to correlate the activity to the relative amount of Pd at the surface (Tables 2 and 4): Pd$_{PVA}$/C (1.3%) > Pd$_{PDDA}$/C (0.76%) > Pd$_{PVP}$/C (0.50%). Therefore, the initial activity can be affected by using different capping agents as they can change the available Pd surface sites regardless of any particle size effect. Indeed, as already reported in the literature, the N groups present in PVP blocks the Pd active sites more than in PDDA, whereas the oxygen groups in PVA interact weakly with the Pd surface [38]. By plotting the conversion versus time of reaction (Figure 8), it is possible to observe an induction time in the first 15 min of the reaction, for all the tested catalysts, in particular for the reaction performed at 25 °C. It is worth noting that at 25 °C the activity of Pd$_{PDDA}$/C increased after 1 h, becoming the most active catalyst, whereas Pd$_{PVA}$/C and Pd$_{PVP}$/C showed a similar reaction profile. This behavior can be attributed to the easier removal of PDDA from Pd active sites during the reaction, and the reduction of the superficial oxidized Pd (Table 2). At 75 °C the catalysts show similar reactivity. We ascribed both effects to the partial removal of a capping agent during the reaction, thereby exposing a higher amount of active sites to the reactant [47].

Table 4. Comparison of the activity and selectivity for Pd$_{PVA}$/C, Pd$_{PVP}$/C, and Pd$_{PDDA}$/C in the selective hydrogenation of furfural.

Catalyst	T (°C)	Activity (h^{-1}) [a]	Selectivity (%) [b]						Carbon Balance (%)	Pd (XPS, at%)	Pd size (TEM, nm)
			FA	THFA	2-MF	FA ether	THFA ether	Acetal			
Pd$_{PVA}$/C	25	824	45.2	5.0	1.8	17.3	3.7	19.0	92.0	1.30	4.2
	50	1775	49.6	6.0	4.4	31.0	3.3	2.7	97.0		
	75	2522	46.0	1.7	5.2	32.4	4.1	4.3	93.7		
Pd$_{PVP}$/C	25	304	64.0	6.8	2.9	19.2	5.4	-	98.3	0.50	3.6
	50	573	56.7	7.7	5.4	19.0	3.8	2.9	95.5		
	75	2722	42.0	1.9	9.2	34.2	3.4	5.1	95.8		
Pd$_{PDDA}$/C	25	624	52.8	10.2	4.9	23.3	5.6	-	96.8	0.76	4.4
	50	1422	49.0	5.7	4.6	27.4	5.2	4.6	96.5		
	75	2581	44.1	4.7	4.5	32.2	2.9	3.2	91.2		

Reaction conditions: Furfural 0.3M in isopropanol, metal/substrate 1:500, 5 bar of H2. [a] calculated at 5 min of reaction as mol$_{furfural}$ converted mol$_{metal}$$^{-1}$ h^{-1}. [b] selectivity at 60% of conversion

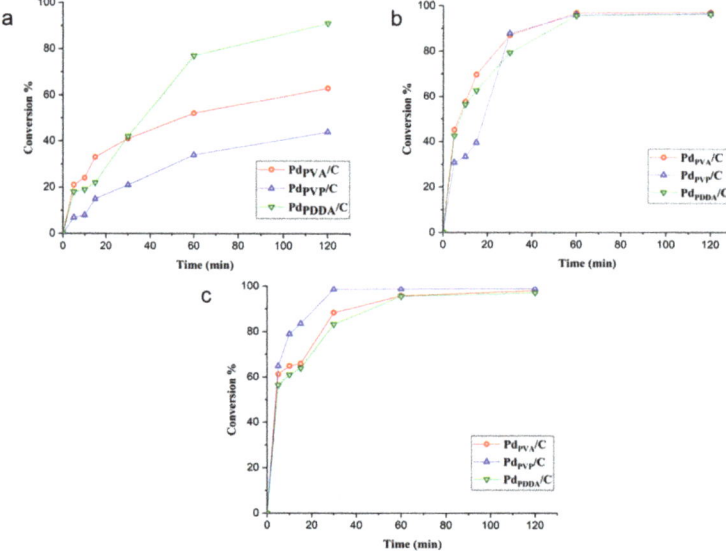

Figure 8. Conversion % vs. time plot for the catalysts at (**a**) 25 °C, (**b**) 50 °C, and (**c**) 75 °C.

Comparing the selectivity at low temperature (25 °C and 50 °C), Pd$_{PVP}$/C exhibits better selectivity towards furfuryl alcohol (64% and 56.7% at 25 °C and 50 °C, respectively) compared to the other catalysts (45.2% and 49.6% at 25 °C and 50 °C, respectively, for Pd$_{PVA}$/C and 52.8% and 49.0% at 25 °C and 50 °C, respectively, for Pd$_{PDDA}$/C) (Table 4). We already reported that the presence of a higher amount of protective agent can enhance the selectivity to furfuryl alcohol, directing the geometry of the adsorption of the furfural on the surface of the catalyst [21,48]. The evolution of the selectivity with the time on stream for the reaction performed at 25 °C is reported in Figure S2. For all catalysts, with a low reaction time, the main product is the acetal. When increasing the conversion rate, the production of acetal decreases, with the formation of furfuryl alcohol as the main product and furfuryl ether. According to the literature, Pd/C can catalyze the formation of ether even in the absence of an acid support [44]. Indeed, the Pd hydride can catalyze the formation of acetal, followed by hydrogenolysis to form the corresponding ether. Figure S2 shows the initial formation of acetals, which are converted to ether at a higher conversion rate, confirming the proposed mechanism. At 75 °C the selectivity is similar due to the removal of the protective agent.

The stability of the most active catalyst (Pd_{PVA}/C) has been studied. Five successive reactions were conducted by simply filtering the catalyst and reusing it either without any treatment or by washing with acetone. As shown in Figure 9a, the conversion starts to decrease immediately after the first run. Washing with acetone was performed to prove if the deactivation is caused by the irreversible adsorption of the products. Pretreating the catalyst after each run, the catalyst maintained almost the same activity and selectivity (Figure 9b), confirming our assumption.

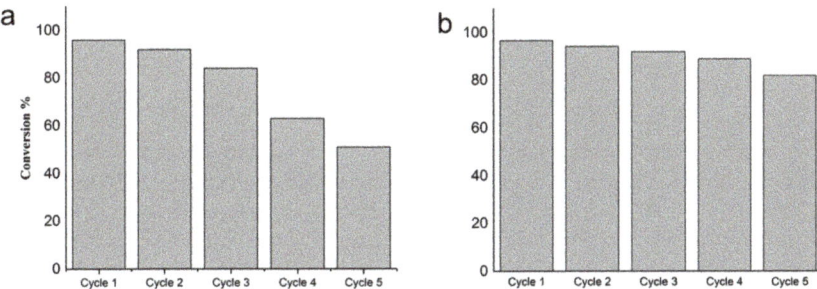

Figure 9. Reusability tests on Pd_{PVA}/C performed at 25 °C (**a**) without any treatment; (**b**) after washing the catalyst with acetone.

We performed XPS of the used catalysts (Table 1, Figures S4–S6). The catalyst used without any treatment showed a high O/C ratio (1.10) (10 times higher than in the fresh one (0.10)), and a low surface Pd content (0.32 and 1.30 at % for the used and fresh Pd_{PVA}/C, respectively; see Table 1), attributed to the adsorption of furfural and the hydrogenation products (Table 1). After washing, the O/C ratio decreased to 0.22, and Pd exposure increased from 0.32 to 0.93 at %, confirming the partial removal of the adsorbed species. The Pd3d spectrum of the used Pd_{PVA}/C catalyst shows that a reduction of $Pd^{\delta+}$ to Pd^0 occurred during the reaction (Pd^0 of 30.9% to 72.7% before and after reaction) (Table 2). This result can also explain why the catalysts become more active after 15 min of reaction. The slight deactivation observed in the recycling tests performed after washing the catalyst after each run can be attributed to the formation of palladium carbide, as already demonstrated by Rogers et al. [21].

3. Materials and Methods

3.1. Materials

Sodium tetrachloropalladate (II) (99.99%, Sigma-Aldrich, St. Louis, MO, USA), sodium borohydride (99.99%, Sigma-Aldrich, St. Louis, MO, USA), poly(vinyl alcohol) (average molar weight 10,000, 87–89% hydrolyzed, Sigma-Aldrich, St. Louis, MO, USA), polyvinylpyrrolidone (average molar weight 10,000, Sigma-Aldrich, St. Louis, MO, USA), poly (diallyldimethylamonium chloride, Sigma-Aldrich, St. Louis, MO, USA) (20 wt % in water), and activated carbon (Camel (X40S)) were used without any pretreatment for the catalyst synthesis All the catalytic tests were carried out using furfural (99 %, Sigma-Aldrich) and 2-propanol (99.8%, Sigma-Aldrich, St. Louis, MO, USA) as the substrate and the solvent, respectively. Furfuryl alcohol (98%, Sigma-Aldrich, St. Louis, MO, USA), tetrahydrofurfuryl alcohol (99%, Sigma-Aldrich, St. Louis, MO, USA), methyl furan (99%, Sigma-Aldrich, St. Louis, MO, USA), and 2-Methyltetrahydrofuran (>99%, Sigma-Aldrich, St. Louis, MO, USA) were used as standards. Isopropyl-furfuryl ether and isopropyl-tetrahydrofurfuryl ether were obtained by reacting isopropanol with furfuryl alcohol and tetrahydrofurfuryl alcohol, respectively, using ZSM5 as a catalyst at 150 °C and as a standard.

3.2. Catalysts Preparation

Supported Pd nanoparticles were synthesized by the sol-immobilization method [27]. Two aqueous solutions, one containing the metal precursor (Na$_2$PdCl$_4$, 1 mL of a 1.26 × 10^{-4} mol L^{-1} solution) and one containing the desired capping agent (1 mL of a 1 wt % solution, capping agents/Pd (w/w) = 1), were added to 100 mL of deionized water. The third solution of NaBH$_4$ (3.7 mL of a 0.1 mol L^{-1} solution; NaBH$_4$/Pd (mol/mol) = 8) was then added to the yellow-brown solution under stirring. After the reduction of Pd species (30 min), the colloidal solution was immobilized on activated carbon (Camel X40S; Surface Area = 1200 m^{2}g^{-1}) under vigorous stirring. The amount of support material required was calculated to give a final metal loading of 1 wt %. The suspension was acidified to pH 2 by sulfuric acid before being stirred for 60 min to accomplish full immobilization of the metal NPs onto the support [49]. The slurry was filtered, washed completely with distilled water, and dried at 80 °C for 2 h.

3.3. Catalytic Tests

The liquid phase hydrogenation of furfural was carried out in a stainless-steel autoclave equipped with a heater and a magnetic stirrer. For a typical experiment, 0.063 g of catalyst (1:500 metal/substrate) was loaded in the reactor and tested at 5 bar of H$_2$ and different temperatures (25 °C, 50 °C, and 75 °C) for 2 h. Before starting the reaction, the reactor was charged with 10 mL of 0.3 M furfural solution using 2-propanol as the solvent; the reactor was then purged with nitrogen and finally charged at 5 bar of H$_2$. The reactor was heated to the desired temperature and the reaction commenced as soon as the stirring speed was 1000 rpm. Samples were collected at regular intervals for gas chromatography analysis (Agilent 6890, equipped with a Zebron ZB5 60 m x 0.32 mm x 1 μm column and a FID detector, Agilent, Santa Clara, CA, USA). Response factors were determined using a known concentration of standard solutions of the pure compounds. To identify unknown products, gas chromatography–mass spectrometry (GC-MS) was used (Thermo Scientific ISQ QD, equipped with an Agilent VF-5ms column, 60 m × 0.32 mm I.D × 1 μm film thickness, Thermo Scientific, Waltham, MA, USA).

3.4. Characterization

The ultraviolet-visible (UV-Vis) absorption spectra of capping agents and the corresponding Pd sols were collected using a UV-Vis spectrophotometer (Shimadzu UV3600, Kyoto, Japan) with quartz cuvettes with 1 cm optical path length, in the wavelength range 150–600 nm.

The FTIR spectra of catalysts were recorded with a JASCO spectrometer (Jasco FT/IR-4100; Jasco Corporation, Tokyo, Japan) in the range of 4000–400 cm^{-1}. For the measurements, 1 mL of a solution of capping agent (1 wt %) was mixed thoroughly with 1 mL of a metal precursor solution (10 mg Pd mL^{-1}); 0.15 g of KBr were then added, and the solution was mixed until all the solid was completely dissolved. The obtained solution was dried in a rotary evaporator and the resulting solid was crushed in a mortar and pressed to obtain a thin film.

The morphological characteristics of the samples were determined using a high-resolution transmission electron microscope (LIBRA 200FE Zeiss at 200 kV equipped with a high-angle annular dark-field detector (HAADF), Zeiss, Oberkochen, Germany). The samples were suspended in isopropanol and sonicated; afterwards, each suspension was dropped onto a carbon-coated copper grid (300 mesh), and the solvent was evaporated. Particle size distribution curves were obtained by counting onto the micrographs at least 300 particles.

X-ray photoelectron spectra (XPS) were acquired in an M-probe apparatus (Surface Science Instruments, Warrington, UK) equipped with an atmospheric reaction chamber. The XPS lines of Pd 3d, O 1s, and C 1s were recorded by applying a Al Kα characteristic X-ray line and (hv = 1486.6 eV) pass energy.

The metal content was checked by atomic absorption spectroscopy (AAS) analysis of the filtrate on a Perkin Elmer 3100 instrument (Perkin Elmer, Waltham, MA, USA). After the catalyst filtration, a 10 mL sample of the filtered solution was collected and analyzed

4. Conclusions

A sol-immobilization methodology was utilized for preparing a series of catalysts with different capping agents (PVA, PVP, and PDDA), using carbon as the support (Camel X40s). UV-Vis and FTIR analyses were applied for studying the interaction between the Pd precursor and the capping agents, and XPS and HRTEM techniques were used to identify the metal surface exposure and particle size/distribution, respectively.

The results showed differences in the relative amount of Pd at the surface of catalysts. Moreover, the use of different capping agents affected the initial activity of the catalysts ($Pd_{PVA}/C > Pd_{PDDA}/C > Pd_{PVP}/C$), depending on the Pd exposure on the catalyst surface. However, in terms of particle size and distribution, the capping agents were not as effective as expected, since all the catalysts revealed a roughly similar range of particle size and distribution when using carbon as the support. We also found that the use of different capping agents affects the product distribution at low temperatures (25 °C and 50 °C).

Supplementary Materials: The following are available online at http://www.mdpi.com/2073-4344/10/1/11/s1, Figure S1: UV–vis spectra of (a) Pd/PVA sols in H_2O, (b) Pd/PVP sols in H_2O, and Pd/PDDA sols in H_2O, Figure S2: Selectivity vs. time of reaction (a) PdPVA/C (b) PdPVP/C, and PdPDDA/C at 25 °C, Figure S3: GC-MS results of the furfuryl isopropyl ether, tetrahydrofurfuryl isopropyl ether and 2-(diisopropoxymethyl)furan, Figure S4: Fitting of XPS envelopes of (a) PdPVA/C used and (b) PdPVA/C, used and washed for Pd 3d, Figure S5: Fitting of XPS envelopes of (a) PdPVA/C used and (b) PdPVA/C, used and washed for C1s, Figure S6: Fitting of XPS envelopes of (a) PdPVA/C used and (b) PdPVA/C, used and washed for O1s.

Author Contributions: S.A. carried out the catalytic evaluation and wrote the article; S.C. (Sofia Capelli) carried out the XPS experiments and helped with the interpretation; C.E. carried out the TEM and helped with the interpretation; S.C. (Stefano Cattaneo) prepared the catalysts; K.M.H.M. carried out the UV and IR experiments; M.S. carried out the GC analysis and helped with the interpretation; A.V., F.T., and P.P.W. were involved in writing and editing the manuscript. All authors have read and agreed to the published version of the manuscript.

Funding: This research received no external funding.

Conflicts of Interest: The authors declare no conflict of interest.

References

1. Gielen, D.; Boshell, F.; Saygin, D.; Bazilian, M.D.; Wagner, N.; Gorini, R. The role of renewable energy in the global energy transformation. *Energy Strategy Rev.* **2019**, *24*, 38–50. [CrossRef]
2. Lanzafame, P.; Centi, G.; Perathoner, S. Catalysis for biomass and CO_2 use through solar energy: Opening new scenarios for a sustainable and low-carbon chemical production. *Chem. Soc. Rev.* **2014**, *43*, 7562–7580. [CrossRef]
3. Chen, S.; Wojcieszak, R.; Dumeignil, F.; Marceau, E.; Royer, S. How Catalysts and Experimental Conditions Determine the Selective Hydroconversion of Furfural and 5-Hydroxymethylfurfural. *Chem. Rev.* **2018**, *118*, 11023–11117. [CrossRef]
4. Ramirez-Barria, C.; Isaacs, M.; Wilson, K.; Guerrero-Ruiz, A.; Rodríguez-Ramos, I. Optimization of ruthenium based catalysts for the aqueous phase hydrogenation of furfural to furfuryl alcohol. *Appl. Catal. A Gen.* **2018**, *563*, 177–184. [CrossRef]
5. Taylor, M.J.; Durndell, L.J.; Isaacs, M.A.; Parlett, C.M.A.; Wilson, K.; Lee, A.F.; Kyriakou, G. Highly selective hydrogenation of furfural over supported Pt nanoparticles under mild conditions. *Appl. Catal. B Environ.* **2016**, *180*, 580–585. [CrossRef]
6. Aldosari, O.F.; Iqbal, S.; Miedziak, P.J.; Brett, G.L.; Jones, D.R.; Liu, X.; Edwards, J.K.; Morgan, D.J.; Knight, D.K.; Hutchings, G.J. Pd-Ru/TiO_2 catalyst—An active and selective catalyst for furfural hydrogenation. *Catal. Sci. Technol.* **2016**, *6*, 234–242. [CrossRef]

7. Mariscal, R.; Maireles-Torres, P.; Ojeda, M.; Sádaba, I.; López Granados, M. Furfural: A renewable and versatile platform molecule for the synthesis of chemicals and fuels. *Energy Environ. Sci.* **2016**, *9*, 1144–1189. [CrossRef]
8. Liu, L.; Lou, H.; Chen, M. Selective hydrogenation of furfural over Pt based and Pd based bimetallic catalysts supported on modified multiwalled carbon nanotubes (MWNT). *Appl. Catal. A Gen.* **2018**, *550*, 1–10. [CrossRef]
9. Gilkey, M.J.; Panagiotopoulou, P.; Mironenko, A.V.; Jenness, G.R.; Vlachos, D.G.; Xu, B. Mechanistic Insights into Metal Lewis Acid-Mediated Catalytic Transfer Hydrogenation of Furfural to 2-Methylfuran. *ACS Catal.* **2015**, *5*, 3988–3994. [CrossRef]
10. O'Driscoll, A.; Leahy, J.J.; Curtin, T. The influence of metal selection on catalyst activity for the liquid phase hydrogenation of furfural to furfuryl alcohol. *Catal. Today* **2017**, *279*, 194–201. [CrossRef]
11. Gupta, K.; Rai, R.K.; Singh, S.K. Metal Catalysts for the Efficient Transformation of Biomass-derived HMF and Furfural to Value Added Chemicals. *ChemCatChem* **2018**, *10*, 2326–2349. [CrossRef]
12. Li, X.; Jia, P.; Wang, T. Furfural: A Promising Platform Compound for Sustainable Production of C4 and C5 Chemicals. *ACS Catal.* **2016**, *6*, 7621–7640. [CrossRef]
13. Giorgianni, G.; Abate, S.; Centi, G.; Perathoner, S.; Van Beuzekom, S.; Soo-Tang, S.H.; Van Der Waal, J.C. Effect of the Solvent in Enhancing the Selectivity to Furan Derivatives in the Catalytic Hydrogenation of Furfural. *ACS Sustain. Chem. Eng.* **2018**, *6*, 16235–16247.
14. Bhogeswararao, S.; Srinivas, D. Catalytic conversion of furfural to industrial chemicals over supported Pt and Pd catalysts. *J. Catal.* **2015**, *327*, 65–77. [CrossRef]
15. Wang, W.; Villa, A.; Kübel, C.; Hahn, H.; Wang, D. Tailoring the 3D Structure of Pd Nanocatalysts Supported on Mesoporous Carbon for Furfural Hydrogenation. *ChemNanoMat* **2018**, *4*, 1125–1132. [CrossRef]
16. Wang, C.; Guo, Z.; Yang, Y.; Chang, J.; Borgna, A. Hydrogenation of furfural as model reaction of bio-oil stabilization under mild conditions using multiwalled carbon nanotube (MWNT)-supported pt catalysts. *Ind. Eng. Chem. Res.* **2014**, *53*, 11284–11291. [CrossRef]
17. Meng, X.; Yang, Y.; Chen, L.; Xu, M.; Zhang, X.; Wei, M. A Control over Hydrogenation Selectivity of Furfural via Tuning Exposed Facet of Ni Catalysts. *ACS Catal.* **2019**, *9*, 4226–4235. [CrossRef]
18. Zhou, X.; Feng, Z.; Guo, W.; Liu, J.; Li, R.; Chen, R.; Huang, J. Hydrogenation and Hydrolysis of Furfural to Furfuryl Alcohol, Cyclopentanone, and Cyclopentanol with a Heterogeneous Copper Catalyst in Water. *Ind. Eng. Chem. Res.* **2019**, *58*, 3988–3993. [CrossRef]
19. Liu, F.; Liu, Q.; Xu, J.; Li, L.; Cui, Y.T.; Lang, R.; Li, L.; Su, Y.; Miao, S.; Sun, H.; et al. Catalytic cascade conversion of furfural to 1,4-pentanediol in a single reactor. *Green Chem.* **2018**, *20*, 1770–1776. [CrossRef]
20. Lam, E.; Luong, J.H.T. Carbon materials as catalyst supports and catalysts in the transformation of biomass to fuels and chemicals. *ACS Catal.* **2014**, *4*, 3393–3410. [CrossRef]
21. Rogers, S.M.; Catlow, C.R.A.; Chan-Thaw, C.E.; Chutia, A.; Jian, N.; Palmer, R.E.; Perdjon, M.; Thetford, A.; Dimitratos, N.; Villa, A.; et al. Tandem Site- and Size-Controlled Pd Nanoparticles for the Directed Hydrogenation of Furfural. *ACS Catal.* **2017**, *7*, 2266–2274. [CrossRef]
22. Huang, R.; Cui, Q.; Yuan, Q.; Wu, H.; Guan, Y.; Wu, P. Total Hydrogenation of Furfural over Pd/Al$_2$O$_3$ and Ru/ZrO$_2$ Mixture under Mild Conditions: Essential Role of Tetrahydrofurfural as an Intermediate and Support Effect. *ACS Sustain. Chem. Eng.* **2018**, *6*, 6957–6964. [CrossRef]
23. Campisi, S.; Schiavoni, M.; Chan-Thaw, C.E.; Villa, A. Untangling the role of the capping agent in nanocatalysis: Recent advances and perspectives. *Catalysts* **2016**, *6*, 185. [CrossRef]
24. Abedini, A.; Bakar, A.A.A.; Larki, F.; Menon, P.S.; Islam, M.S.; Shaari, S. Recent Advances in Shape-Controlled Synthesis of Noble Metal Nanoparticles by Radiolysis Route. *Nanoscale Res. Lett.* **2016**, *11*, 1–13. [CrossRef] [PubMed]
25. Niu, Z.; Li, Y. Removal and utilization of capping agents in nanocatalysis. *Chem. Mater.* **2014**, *26*, 72–83. [CrossRef]
26. Pang, S.H.; Schoenbaum, C.A.; Schwartz, D.K.; Will Medlin, J. Effects of thiol modifiers on the kinetics of furfural hydrogenation over Pd catalysts. *ACS Catal.* **2014**, *4*, 3123–3131. [CrossRef]
27. Campisi, S.; Ferri, D.; Villa, A.; Wang, W.; Wang, D.; Kröcher, O.; Prati, L. Selectivity Control in Palladium-Catalyzed Alcohol Oxidation through Selective Blocking of Active Sites. *J. Phys. Chem. C* **2016**, *120*, 14027–14033. [CrossRef]

28. Rogers, S.M.; Catlow, C.R.A.; Chan-Thaw, C.E.; Gianolio, D.; Gibson, E.K.; Gould, A.L.; Jian, N.; Logsdail, A.J.; Palmer, R.E.; Prati, L.; et al. Tailoring Gold Nanoparticle Characteristics and the Impact on Aqueous-Phase Oxidation of Glycerol. *ACS Catal.* **2015**, *5*, 4377–4384. [CrossRef]
29. Zhao, Y.; Liang, W.; Li, Y.; Lefferts, L. Effect of chlorine on performance of Pd catalysts prepared via colloidal immobilization. *Catal. Today* **2017**, *297*, 308–315. [CrossRef]
30. Chowdhury, S.R.; Roy, P.S.; Bhattacharya, S.K. Room temperature synthesis of polyvinyl alcohol stabilized palladium nanoparticles: Solvent effect on shape and electro-catalytic activity. *Nano-Struct. Nano-Objects* **2018**, *14*, 11–18. [CrossRef]
31. Eisa, W.H.; Shabaka, A.A. Ag seeds mediated growth of Au nanoparticles within PVA matrix: An eco-friendly catalyst for degradation of 4-nitrophenol. *React. Funct. Polym.* **2013**, *73*, 1510–1516. [CrossRef]
32. Liu, J.; Ruffini, N.; Pollet, P.; Llopis-Mestre, V.; Dilek, C.; Eckert, C.A.; Liotta, C.L.; Roberts, C.B. More benign synthesis of palladium nanoparticles in dimethyl sulfoxide and their extraction into an organic phase. *Ind. Eng. Chem. Res.* **2010**, *49*, 8174–8179. [CrossRef]
33. Tang, H.; Pan, M.; Jiang, S.; Wan, Z.; Yuan, R. Self-assembling multi-layer Pd nanoparticles onto Nafion^{TM} membrane to reduce methanol crossover. *Colloids Surfaces A Physicochem. Eng. Asp.* **2005**, *262*, 65–70. [CrossRef]
34. Hassan, C.M.; Peppas, N.A. Structure and applications of poly (vinyl alcohol) hydrogels produced by conventional crosslinking or by freezing/thawing methods. *Adv. Polym. Sci.* **2000**, *153*, 37–65.
35. Luo, L.B.; Yu, S.H.; Qian, H.S.; Zhou, T. Large-scale fabrication of flexible silver/cross-linked poly (vinyl alcohol) coaxial nanocables by a facile solution approach. *J. Am. Chem. Soc.* **2005**, *127*, 2822–2823. [CrossRef] [PubMed]
36. Koczkur, K.M.; Mourdikoudis, S.; Polavarapu, L.; Skrabalak, S.E. Polyvinylpyrrolidone (PVP) in nanoparticle synthesis. *Dalton Trans.* **2015**, *44*, 17883–17905. [CrossRef]
37. Song, Y.J.; Wang, M.; Zhang, X.Y.; Wu, J.Y.; Zhang, T. Investigation on the role of the molecular weight of polyvinyl pyrrolidone in the shape control of high yield silver nanospheres and nanowires. *Nanoscale Res. Lett.* **2014**, *9*, 17. [CrossRef]
38. Chan-Thaw, C.E.; Villa, A.; Veith, G.M.; Prati, L. Identifying the Role of N-Heteroatom Location in the Activity of Metal Catalysts for Alcohol Oxidation. *ChemCatChem* **2015**, *7*, 1338–1346. [CrossRef]
39. Xian, J.; Hua, Q.; Jiang, Z.; Ma, Y.; Huang, W. Size-Dependent Interaction of the Poly(N-vinyl-2-pyrrolidone) Capping Ligand with Pd Nanocrystals. *Langmuir* **2012**, *28*, 6736–6741. [CrossRef]
40. Yang, D.Q.; Rochelte, J.F.; Sacher, E. Spectroscopic evidence for π-π interaction between poly(diallyl dimethylammonium) chloride and multiwalled carbon nanotubes. *J. Phys. Chem. B* **2005**, *109*, 4481–4484. [CrossRef]
41. Figueiredo, J.L.; Pereira, M.F.R.; Freitas, M.M.A.; Órfão, J.J.M. Modification of the surface chemistry of activated carbons. *Carbon* **1999**, *37*, 1379–1389. [CrossRef]
42. Walczyk, M.; Swiatkowski, A.; Pakuła, M.; Biniak, S. Electrochemical studies of the interaction between a modified activated carbon surface and heavy metal ions. *J. Appl. Electrochem.* **2005**, *35*, 123–130. [CrossRef]
43. Mertens, P.G.N.; Cuypers, F.; Vandezande, P.; Ye, X.; Verpoort, F.; Vankelecom, I.F.J.; De Vos, D.E. Ag^0 and Co^0 nanocolloids as recyclable quasihomogeneous metal catalysts for the hydrogenation of α, β-unsaturated aldehydes to allylic alcohol fragrances. *Appl. Catal. A Gen.* **2007**, *325*, 130–139. [CrossRef]
44. Wang, Y.; Cui, Q.; Guan, Y.; Wu, P. Facile synthesis of furfuryl ethyl ether in high yield: Via the reductive etherification of furfural in ethanol over Pd/C under mild conditions. *Green Chem.* **2018**, *20*, 2110–2117. [CrossRef]
45. Padovan, D.; Al-Nayili, A.; Hammond, C. Bifunctional Lewis and Brønsted acidic zeolites permit the continuous production of bio-renewable furanic ethers. *Green Chem.* **2017**, *19*, 2846–2854. [CrossRef]
46. Pizzi, R.; van Putten, R.J.; Brust, H.; Perathoner, S.; Centi, G.; van der Waal, J.C. High-throughput screening of heterogeneous catalysts for the conversion of furfural to bio-based fuel components. *Catalysts* **2015**, *5*, 2244–2257. [CrossRef]
47. Rossi, L.M.; Fiorio, J.L.; Garcia, M.A.S.; Ferraz, C.P. The role and fate of capping ligands in colloidally prepared metal nanoparticle catalysts. *Dalton Trans.* **2018**, *47*, 5889–5915. [CrossRef]

48. Villa, A.; Wang, D.; Veith, G.M.; Vindigni, F.; Prati, L. Sol immobilization technique: A delicate balance between activity, selectivity and stability of gold catalysts. *Catal. Sci. Technol.* **2013**, *3*, 3036. [CrossRef]
49. Villa, A.; Wang, D.; Dimitratos, N.; Su, D.; Trevisan, V.; Prati, L. Pd on carbon nanotubes for liquid phase alcohol oxidation. *Catal. Today* **2010**, *150*, 8–15. [CrossRef]

© 2019 by the authors. Licensee MDPI, Basel, Switzerland. This article is an open access article distributed under the terms and conditions of the Creative Commons Attribution (CC BY) license (http://creativecommons.org/licenses/by/4.0/).

Article

First-Principles-Based Simulation of an Industrial Ethanol Dehydration Reactor

Kristof Van der Borght, Konstantinos Alexopoulos, Kenneth Toch, Joris W. Thybaut, Guy B. Marin and Vladimir V. Galvita *

Laboratory for Chemical Technology, Ghent University, Technologiepark 125, B-9052 Ghent, Belgium; Kristof.Van-der-Borght@eurochem.be (K.V.d.B.); konalex@udel.edu (K.A.); kenneth.toch@arcelormittal.com (K.T.); Joris.Thybaut@UGent.be (J.W.T.); Guy.Marin@UGent.be (G.B.M.)
* Correspondence: Vladimir.Galvita@UGent.be; Tel.: +32-9-331-17-22

Received: 26 August 2019; Accepted: 31 October 2019; Published: 5 November 2019

Abstract: The achievement of new economically viable chemical processes often involves the translation of observed lab-scale phenomena into performance in an industrial reactor. In this work, the in silico design and optimization of an industrial ethanol dehydration reactor were performed, employing a multiscale model ranging from nano-, over micro-, to macroscale. The intrinsic kinetics of the elementary steps was quantified through ab initio obtained rate and equilibrium coefficients. Heat and mass transfer limitations for the industrial design case were assessed via literature correlations. The industrial reactor model developed indicated that it is not beneficial to utilize feeds with high ethanol content, as they result in lower ethanol conversion and ethene yield. Furthermore, a more pronounced temperature drop over the reactor was simulated. It is preferred to use a more H_2O-diluted feed for the operation of an industrial ethanol dehydration reactor.

Keywords: diffusion; ab initio; industrial design; H-ZSM-5; multiscale modeling; adiabatic reactor; zeolite catalysis

1. Introduction

Since their initial discovery in the late 1970s, the conversion processes of oxygenates have been gaining importance rapidly as an alternative route for the production of fuels and chemicals [1–3]. Most industrial focus has been given to the conversion of methanol to hydrocarbons, with products ranging from light olefins to gasoline. Both fixed and fluidized bed reactors are in use in the industry. A fluidized bed reactor with SAPO-34 catalysts offers the advantage of adequately coping with rapid catalyst deactivation and the high exothermicity of the methanol-to-olefins (MTO) reaction. However, the corresponding setbacks are its notable catalyst attrition and low single-pass methanol conversion in addition to its high investment cost. A fixed bed variant, on the other hand, is simple in construction and can easily be operated, certainly in adiabatic operation.

The first records on ethanol dehydration date back to the 18th century, and several plants have been in operation in the course of the 20th century. In contrast to the MTO reaction, ethanol dehydration is an endothermic process typically operated in a multitubular, isothermal reactor at temperatures exceeding 623 K. Such a reactor configuration, which employs indirect heating via a heating fluid, has disadvantages in both its technical and its economic aspects, resulting in a shift towards adiabatic fixed bed reactors [4]. Initially, the catalyst employed was alumina or silica–alumina, while, more recently, also zeolites have been considered for this process [5].

A schematic overview of an ethanol dehydration plant is shown in Figure 1 [6]. When starting from a fermentation broth, i.e., bioethanol, a distillation column (1) is required to partly remove the water from this ethanol–water mixture. The ethanol feedstock is subsequently mixed with unreacted ethanol from the purification section. Next, a heat exchanger (2) allows heat recovery from the reactor effluent, i.e., the latent heat of the effluent is used to vaporize the ethanol feedstock, which is subsequently pressurized (3). In a subsequent heat exchanger (4), the ethanol feedstock is superheated. Finally, a furnace (5) is installed to bring the feed to the temperature of the first ethanol dehydration reactor (6). The effluent from the first reactor is sent to the next ethanol dehydration reactor (8) via an additional furnace (7). The number of reactors in series depends on the reaction conditions and the intended conversion. The effluent of the second reactor undergoes a series of heat exchanges as described above to ensure maximum heat recovery. Downstream of the reactor, the effluent is separated in a distillation column (9) into an ethene top stream and a bottom stream comprising water, side products, and unreacted ethanol. The latter is sent to a second separation column (10) and results in three streams: side products, i.e., by-products (C_{3+} olefins and oxygenates), water, and unconverted ethanol, which can be recycled.

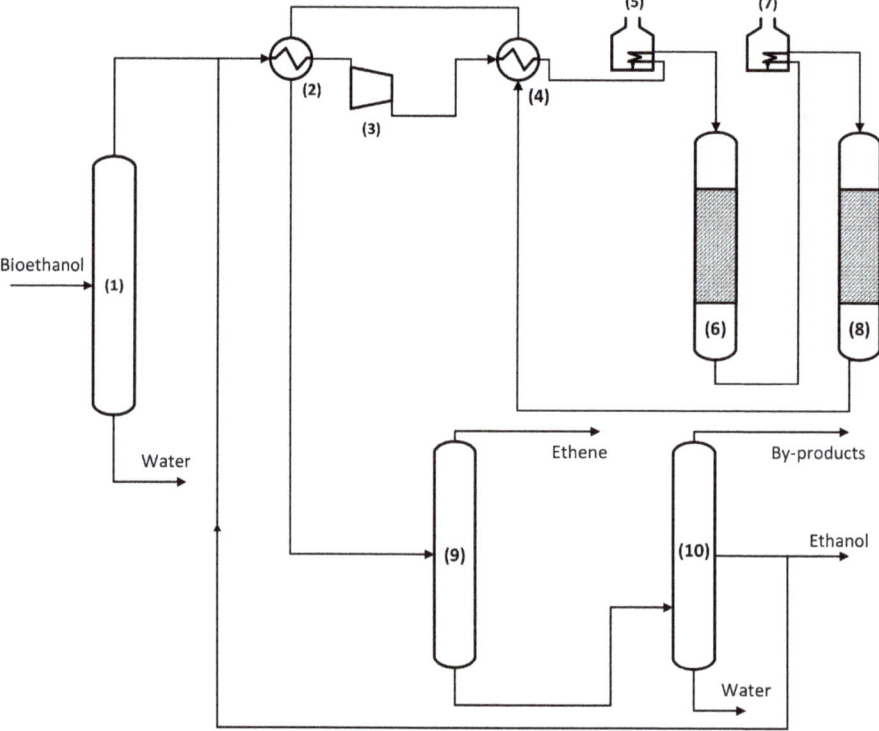

Figure 1. Flow sheet of an ethanol dehydration plant [6] consisting of (1) a pretreatment distillation column, (2) and (4) heat exchangers, (3) a compressor, (5) and (7) heating furnaces, (6) and (8) ethanol dehydration reactors, and (9) and (10) gas/liquid separation columns.

The capability of accurately simulating the behavior of a chemical reaction over a broad range of process conditions opens up perspectives for the design and optimization of industrial chemical reactors. The current reactor models described in literature typically rely on simplified kinetic models [7]. Efforts have already been undertaken to extend such models towards more complex reaction networks based on elementary steps [8–11]. However, employing model parameter values that have been determined by regression to experimental results, potentially jeopardizes the extrapolative capabilities of the model. Indeed, it is not guaranteed that all the kinetically relevant reactions are accounted for in the most adequate manner. In contrast, ab initio developed models incorporate information on the level of the active site and truly represent the intrinsic kinetics of the investigated reactions within the constraints under which the corresponding calculations have been performed.

In multiscale industrial reactor modeling, the chemical reaction rates are described via a kinetic model embedded within a suitable reactor model, which accounts for all relevant physical transport phenomena. Ab initio reactor modeling has already been successfully applied for thermal processes [12], but because of the complexity of heterogeneous catalyzed reactions, only few examples of simulations of catalytic processes solely based on ab initio obtained rate and equilibrium coefficients are reported, e.g., NH_3 synthesis [13] and benzene hydrogenation [14]. For zeolite catalysis, a successful simulation of an industrial reactor would provide a proof of principle that reliable ab initio modeling of catalytic reactions is possible from molecular to industrial scale.

A reactor model provides guidelines for the design, optimization, and operation of industrial reactors. Alwahabi and Froment [15] developed a conceptual reactor design for the MTO reaction using a SAPO-34 catalyst and compared three different configurations: a multi-tubular quasi-isothermal reactor, a multi-bed adiabatic reactor with intermediate heat exchangers, and a bubbling fluidized bed reactor with internal heat exchanger. The advantages of relying on a fundamental kinetic model was already demonstrated by Park and Froment [16], who explored the use of a multi-bed adiabatic reactor for maximizing the propylene yield over H-ZSM-5 yield during the MTO reaction. CFD-ased models for a fixed bed [17] and a fluidized bed [18] using lumped kinetics have also been proposed. However, so far, no industrial reactor simulation model has been developed for the dehydration of ethanol on zeolites.

In the present work, a multi-bed adiabatic reactor model was developed for the dehydration of ethanol on H-ZSM-5 with Si/Al ratio 140 and acid site concentration of 0.003 mol kg^{-1}. The model also accounts for intermediate heat exchange between the fixed beds. The kinetics implemented in the reactor model are solely based on quantum chemically obtained rate and equilibrium coefficients. A comparison of the ab initio modeling-based reactor simulation results with data found in patent literature provides the ultimate test of the model validity and methodology presented in this work. The benefits of accurate reaction and reactor model are illustrated by exploration of the water content effect.

2. Assessment of Internal and External Mass and Heat Transfer Limitations

A key factor in the development of an adequate reactor model is to assess the extent to which mass and heat transfer inside the catalyst particle and between the fluid bulk in the reactor and the catalyst surface impact on the overall performance, i.e., a determination of the occurrence of internal and external heat and mass transport limitations [19].

The most extensively investigated catalyst for ethanol dehydration is H-ZSM-5, which is composed of pentasil units. It consists of elliptical straight channels (0.53 nm × 0.56 nm) and near-circular sinusoidal channels (0.51 nm × 0.55 nm) that perpendicularly intersect [20]. This pore network is located in small crystallites with a diameter (d_c) ranging between 10^{-7} and 10^{-5} m. These crystallites are typically embedded in a binder when applied industrially, to increase the mechanical strength and allow the formation of larger pellets ($d_p = 10^{-3} - 10^{-2}$ m), so to limit the pressure drop over the catalyst bed. Therefore, two different length scales for internal mass transport limitations exist. An assessment of the relative importance of these limitations can be made using the Weisz–Prater criterion [21]:

$$\frac{(n+1)}{2} \frac{d^2 \rho_p R_i^{obs}}{6 D_{e,i} C_i^s} < 0.08 \qquad (1)$$

in which n is the apparent order of reaction, d is the diameter of either the catalyst crystallite (d_c) or the catalyst pellet (d_p), ρ_p is the density, $D_{e,i}$ is the effective diffusion coefficient of component i (m^2 s^{-1}), and C_i^s is the concentration of component i at the catalyst surface.

The Carberry number [22] (Ca) allows verifying the absence of external mass transfer limitations. It expresses the fractional concentration difference between the concentration of component i in the bulk phase, C_i^{bl}, and the concentration of component i on the external surface, C_i^s:

$$Ca = \frac{C_i^{bl} - C_i^s}{C_i^{bl}} = \frac{R_i^{obs}}{k_{fi} a_s C_i^{bl}} < \frac{0.05}{n} \quad (2)$$

where R_i^{obs} is the observed reaction rate per unit of catalyst mass, k_{fi} is the external mass transfer coefficient of component i which can be calculated via correlations, a_s is the specific external surface area of the catalyst, i.e., $6/d_p$ for spherical particles, C_i^{bl} an C_i^s refer to the bulk and surface concentration of component i, respectively, and n is the reaction order.

Mears [22] proposes criteria to assess external (Equation (3)) and internal heat transfer limitations (Equation (4)), similar to external mass transfer limitations, stating that the observed reaction rate should not deviate more than 5% from the rate under isothermal conditions:

$$\Delta T_{film} = \frac{R_{w,i}^{obs} \rho_p d_p |-\Delta H_r|}{6 \alpha T_{bl}} \frac{E_a}{R T_{bl}} < 0.05 \quad (3)$$

$$\Delta T_{pellet} = \frac{R_{w,i}^{obs} \rho_p d_p^2 |-\Delta_r H|}{60 \lambda_p T_{bl}} \frac{E_a}{R T_{bl}} < 0.05 \quad (4)$$

where $|-\Delta H_r|$ corresponds to the reaction enthalpy, α is the heat transfer coefficient inside the film, i.e., the boundary layer between fluid and catalyst surface, λ_p is the catalyst pellet heat conductivity, T_{bl} is the bulk temperature, and E_a is the apparent activation energy of the reaction.

Table 1 shows the results of the transport limitations assessment for ethanol dehydration in an industrial reactor using H-ZSM-5. It can be seen that the catalyst particle is practically isothermal, which is consistent with the results of Froment et al. [23]. In addition, external transport limitations can also be neglected.

Table 1. External and internal heat and mass transport limitations in an industrial ethanol dehydration reactor.

		Heat Transport Limitations		
External	Equation (3)	$|\Delta T_{film}|$	0.012	<2.35
Internal	Equation (4)	$|\Delta T_{pellet}|$	0.736	<2.35
		Mass transport limitations		
External	Equation (2)	Ca	0.00764	<0.05
Internal	Equation (1)		See Figure 2	

The results obtained by applying the Weisz–Prater criterion are shown in Figure 2 for a wide range of pellet and crystallite diameters and effective diffusion coefficients. The area below the black line, which indicates the limit of 0.08, is the region where internal diffusion limitations will occur. Above that line, no internal diffusion limitations will occur. It can be seen that, under the conditions and catalyst studied in this work, internal mass transfer limitations are only expected at the pellet scale.

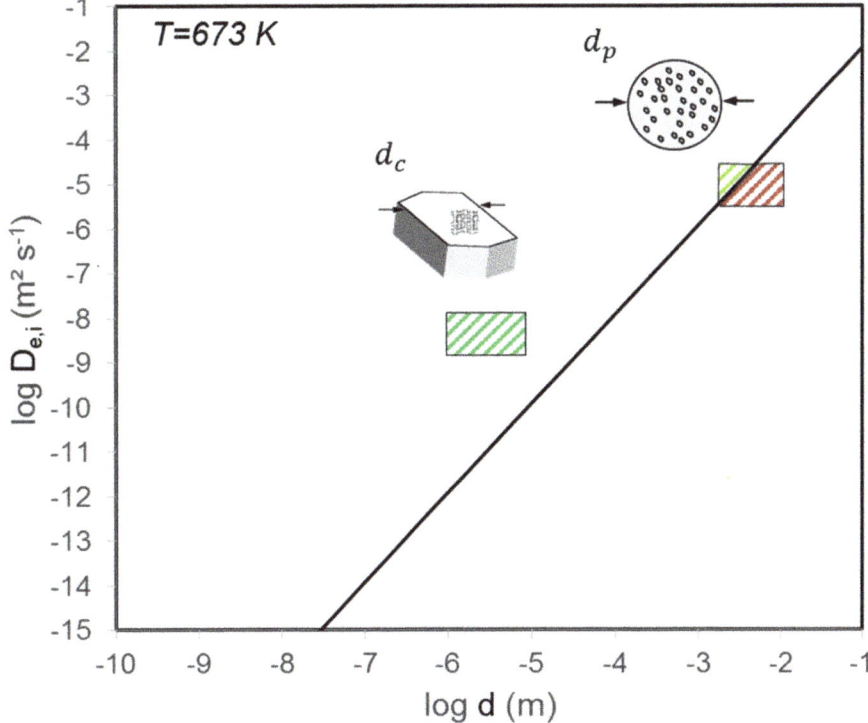

Figure 2. Internal mass transfer limitations assessed by the Weisz–Prater criterion (Equation (1)) in an industrial ethanol dehydration reactor as a function of particle diameter d, which can correspond either to the crystallite diameter, i.e., d_c, or to the pellet diameter, i.e., d_p, and the effective diffusion coefficient $D_{e,i}$. The black line indicates the criterion limit of 0.08. Boxes indicate the typical ranges of diffusion coefficient and diameter for either crystallite or pellet. (Green: no internal mass transport limitations; red: internal mass transport limitations).

3. Industrial Reactor Model for Ethanol Dehydration

3.1. Reactor Model

A graphical representation of the reactor model and the phenomena that are taken into consideration are given in Figure 3. The reactor model consists of a tubular fixed bed reactor with specified length and diameter, i.e., L_r and d_r. The molar inlet flow rate of ethanol and water, as well as the inlet temperature and pressure, are specified. The reactor is operated in adiabatic mode. The pressure drop over the fixed bed along the axial reactor coordinate is also taken into account. Further, the reactor model explicitly includes intraparticle mass transfer limitations which lead to a typical concentration profile, as shown below the catalyst pellet.

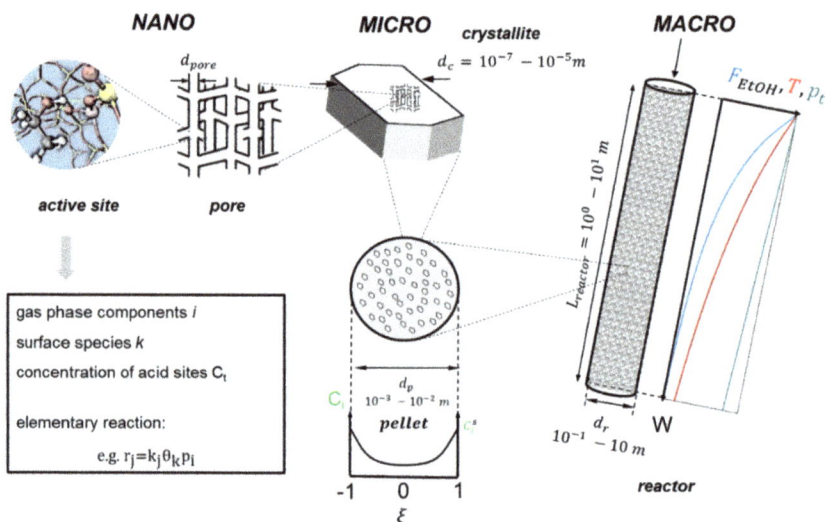

Figure 3. Graphical representation of the fixed bed reactor for ethanol dehydration.

3.1.1. Macroscale: The Reactor

The reactor is described by continuity equations for the conservation of mass, energy, and momentum. The reactor is considered to be in steady state and, hence, no accumulation term has to be added. A one-dimensional heterogeneous reactor model with ideal plug flow was considered. The continuity equation for component i in the gas phase, e.g., ethanol, is given by:

$$\frac{dF_i}{dW} = \overline{R}_i \qquad (5)$$

in which F_i is the molar flow rate of gas phase component i (mol s^{-1}), W is the catalyst mass (kg), $\overline{R}_i$ is the net rate of formation of gas phase component i (mol s^{-1} kg^{-1}).

As the reactor is operated adiabatically, no heat exchange with the wall is occurring, and thus, the energy equation for the gas phase is given by:

$$\frac{dT}{dW} = \frac{1}{Gc_p} \sum_{i=1}^{ncomp} \Delta H_{f,i} \overline{R}_i \qquad (6)$$

T is the temperature (K), $\Delta H_{f,i}$ is the standard formation enthalpy of component i (J mol^{-1}), G is the total mass flow rate (kg s^{-1}), c_p is the heat capacity of the gas (J kg^{-1} K^{-1}). The latter is determined via the method of Chung et al. [24]. The standard formation enthalpy can be determined via a group additivity method such as Benson's or directly taken from the literature [24].

Momentum can be lost throughout the reactor because of friction of the gas with the packed bed and is accounted for by:

$$\frac{dp_t}{dW} = -f \frac{G^2}{\rho_b \rho_{fl} A_r^3 d_p} \qquad (7)$$

where p_t is the total pressure in the reactor (Pa), ρ_{fl} is the density of the fluid (kg m^{-3}), ρ_b is the bed density (kg m^{-3}), A_r is the cross-sectional surface area of the reactor tube (m^2), and d_p is the diameter of the catalyst pellet (m).

The friction factor f is determined by a correlation proposed by Hicks [25]:

$$f = 6.8 \frac{(1-\varepsilon_B)^{1.2}}{\varepsilon_B^3} Re_p^{-0.2} \qquad (8)$$

where ε_B is the bed porosity, and Re_p is the pellet Reynolds number, which is given by:

$$Re_p = \frac{\rho_b u_s d_p}{\mu(1-\varepsilon_B)} \qquad (9)$$

where u_s is the superficial velocity (m s^{-1}), and μ is the dynamic viscosity of the gas phase mixture (Pa s), which is determined according to the method of Chung et al. [24]. The bed porosity ε_B can be found via the correlation of Haughey and Beveridge [26]:

$$\varepsilon_B = 0.38 + 0.073 \left[1 + \frac{\left(\frac{d_t}{d_p} - 2\right)^2}{\left(\frac{d_t}{d_p}\right)^2} \right] \qquad (10)$$

where d_t is the diameter of the reactor (m).

The initial conditions for this set of differential equation (Equations (6)–(8)) are given by:

$$\left. \begin{array}{l} F_i = F_i^0 \\ T = T^0 \\ P_t = P_t^0 \end{array} \right\} \text{at } W = 0 \qquad (11)$$

3.1.2. Microscale: The Catalyst Pellet

A one-dimensional mass balance for each gas phase component i over an infinitesimal volume of the catalyst pellet is considered:

$$\frac{\partial C_i}{\partial t} = R_i \rho_s - \frac{4}{d_p^2} \left(\frac{s}{\xi} D_{e,i} \frac{\partial C_i}{\partial \xi} + \frac{\partial D_{e,i}}{\partial \xi} \frac{\partial C_i}{\partial \xi} + D_{e,i} \frac{\partial^2 C_i}{\partial \xi^2} \right) \qquad (12)$$

Here, ρ_s is the solid density of the catalyst (kg m^{-3}), C_i is the concentration of the gas phase component i inside the catalyst pellet (mol m^{-3}), ξ is the position coordinate within the pellet, s is the pellet shape factor, i.e., 0, 1 or 2 for, respectively, a slab, a cylinder, and a sphere, R_i is the net rate of formation of component i (mol s^{-1} kg^{-1}), and $D_{e,i}$ is the effective diffusion coefficient for gas phase component i (m^2 s^{-1}).

For this set of differential equations, the following initial conditions were considered:

$$\begin{array}{ll} C_i = C_i^s & \xi = 1 \\ \frac{dC_i}{d\xi} = 0 & \xi = 0 \end{array} \qquad (13)$$

In contrast to a homogeneous medium, the porous pellets consist of interconnected non-uniform pores, inside which the gaseous components move. This internal void fraction of the porous material and the tortuous nature of the pores are taken into account by using the effective diffusivity for component i, i.e., $D_{e,i}$:

$$D_{e,i} = \frac{\varepsilon_p}{\tau_p} D_i \qquad (14)$$

where ε_p is the porosity, i.e., the fraction of the volume occupied by the pores, and τ_p is the tortuosity.

The diffusion coefficient D_i is given as the sum of two resistances by the so-called Bosanquet equation [27], which is composed of the diffusion coefficient, corresponding to intermolecular collisions, i.e., $D_{i,m}$, and the Knudsen diffusion coefficient, i.e., $D_{i,K}$, corresponding to the collisions of the molecules with the pore wall:

$$\frac{1}{D_i} = \frac{1}{D_{i,m}} + \frac{1}{D_{i,K}} \tag{15}$$

The molecular diffusion coefficient $D_{i,m}$ is preferably calculated using the rigorous Stefan–Maxwell model [28,29], but this can be computationally demanding. The bulk diffusivity of gas phase component i in a gas mixture, $D_{m,i}$, is therefore calculated from the individual binary diffusion coefficients, using the Wilke equation [30]:

$$D_{i,m} = \left(\sum_{\substack{j=1 \\ j \neq i}} \frac{y_j}{D_{ij}} \right)^{-1} \tag{16}$$

where y_i is the molar fraction of component i in the gas phase.

The Wilke equation assumes diffusion in a stagnant mixture and is valid when using dilute systems. Solsvik and Jakobsen [31,32] compared the rigorous Stefan–Maxwell model to the simpler Wilke model and concluded that it is appropriate to use the Stefan–Maxwell model in the simulation of a fixed packed-bed methanol synthesis reactor. Good results have been obtained by applying the Wilke–Bosanquet combination for the determination of diffusivity in multicomponent gas mixtures at low pressures in combination with complex reactions such as the MTO reaction [33] and hydrodesulphurization [9].

The molecular binary diffusion coefficient of component i in component j, $D_{i,j}$, is calculated using the Füller–Schettler–Giddings relation [34], which is recommended by Reid et al. [35]:

$$D_{i,j} = 1 \times 10^{-7} \frac{T^{1.75}}{P_t \left(\frac{1}{M_i} + \frac{1}{M_j}\right)^{-1/2} \left((\Sigma_v)_i^{1/3} + (\Sigma_v)_j^{1/3}\right)^2} \tag{17}$$

where M_i is the molecular mass of component i (mol kg^{-1}), and $(\Sigma_v)_i$ is the atomic diffusion volume for component i, which was found to be 51.77 cm^3 for ethanol, 41.04 cm^3 for ethene, 92.81 cm^3 for di-ethyl ether, and 13.10 cm^3 for water.

The Knudsen diffusion coefficient of component i, $D_{i,K}$, is given by:

$$D_{i,K} = \frac{2}{3} \frac{d_{pore}}{2} \sqrt{\frac{8RT}{\pi M_i}} \tag{18}$$

Due to the second-order nature of the balances to be solved over the catalyst pellet, see Equation (12), a meaningful solution is not guaranteed. Therefore, the set of differential equations originating from Equation (12) was solved by integration from an initial to a steady state, rather than by directly solving the steady-state mass balance. A finite difference method was used for solving these second-order differential equations, i.e., the pellet diameter was discretized over a user-defined number of mesh points, n_{mesh}. Every partial differential equation was rewritten as a set of n_{mesh} ordinary differential equations.

The net production rate of component i in case of diffusion limitations, i.e., $\overline{R}_i$, can be determined via:

$$\overline{R}_i = \int_0^V R_i dV \tag{19}$$

In practice, its value was obtained by averaging the pointwise net rate of formation of component i at position ξ of the catalyst pellet. A number of equidistant grid points was defined, and a trapezoidal discretization produce was followed for integration:

$$\overline{R_i} = \int_0^V R_i dV = \frac{s+1}{2n_{grid}} \sum_{j=1}^{n_{grid}} \left[R_i(r_{p,j}) r_{p,j}^s + R_i(r_{p,j+1}) r_{p,j+1}^s \right] \quad (20)$$

where n_{grid} is the number of grid points, $R_i(r_{p,j})$ is the net production rate of component i at a location $r_{p,i}$ inside the pellet, and V is the pellet volume.

The catalyst effectiveness factor is calculated as the ratio of the reaction rate in the presence of pore diffusion resistance to the reaction rate in the absence of diffusion limitations, i.e., at gas bulk concentrations:

$$\eta = \frac{\overline{R_i}}{R_i^s} = \frac{\int_0^V R_i dV}{R_i^s} \quad (21)$$

The catalyst effectiveness factor as a function of the number of mesh points is nearly constant after 25 mesh points. In this work, 35 mesh points were used for the simulations.

3.1.3. Nanoscale: The Active Site

A fully ab initio reaction network [36,37] consisting of 15 elementary steps was used for describing the intrinsic kinetics of ethanol dehydration and is shown in Figure 4. Three different reaction pathways were identified and are given below along with the corresponding reaction enthalpies:

$$C_2H_5OH \rightarrow C_2H_4 + H_2O \quad \Delta H_r = 46 \text{ kJ mol}^{-1}_{EtOH} \quad (22)$$

$$2\, C_2H_5OH \rightarrow (C_2H_5)_2O + H_2O \quad \Delta H_r = -12 \text{ kJ mol}^{-1}_{EtOH} \quad (23)$$

$$(C_2H_5)_2O \rightarrow C_2H_4 + C_2H_5OH \quad \Delta H_r = 70 \text{ kJ mol}^{-1}_{EtOH} \quad (24)$$

The monomolecular pathway (Equation (22)) describes the direct dehydration of ethanol to ethene, which is endothermic. The alternative route towards ethene comprises the bimolecular dehydration of ethanol to di-ethyl ether (Equation (23)) and its subsequent decomposition into ethanol and ethene (Equation (24)). The former is slightly exothermic, while the latter is endothermic. The mechanism for the production of C_{3+} hydrocarbons from ethanol is still a matter of debate [38–40]. Therefore, it was opted to include the dimerization of ethene to 1-butene, which serves as a representation of higher hydrocarbons formation:

$$2\, C_2H_4 \rightarrow C_4H_8 \quad \Delta H_r = -53 \text{ kJ mol}^{-1}_{C2H4} \quad (25)$$

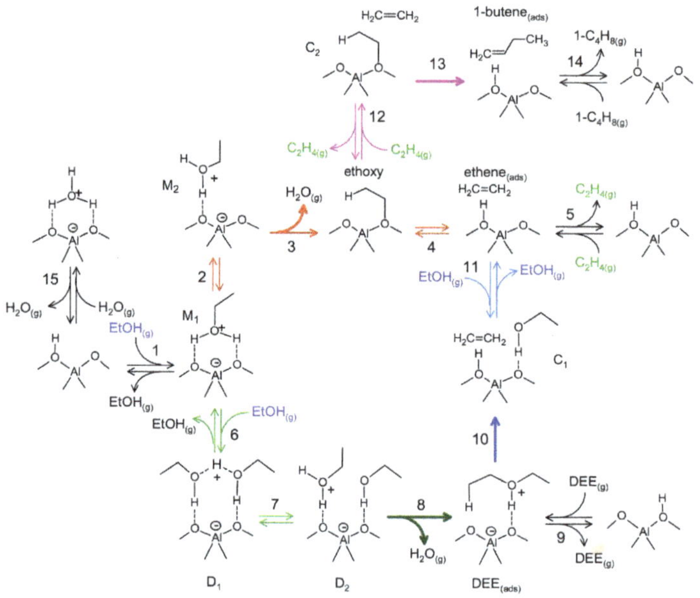

Figure 4. Reaction network used for the simulation of the industrial reactor (red: monomolecular dehydration, green: bimolecular dehydration, blue: di-ethyl ether decomposition, magenta: ethene dimerization). Modified from Alexopoulos et al. [36].

The following net rates of formation were applied for the surface species k and gas phase components i, complemented with a site balance:

$$R_k = C_t \sum_j v_{jk} r_j = 0 \qquad (26)$$

$$R_i = C_t \sum_j v_{ji} r_j \qquad (27)$$

$$\theta_{H^+} + \sum_k \theta_k = 1 \qquad (28)$$

where r_j is the turnover frequency of elementary step j, v_{ji} and v_{jk} are the stoichiometric coefficient of gas phase component i or surface species k in the elementary step j. The forward reaction rate of a typical elementary step j can be written as:

$$r_j = k_j \theta_k^n p_i^m \qquad (29)$$

where θ_k is the fractional occupancy of surface species k, and p_i is the partial pressure of gas phase component i.

Equilibrium coefficients for each elementary reaction were obtained using the following formula:

$$K_j = \exp\left(-\frac{\Delta H^0 - T\Delta S^0}{RT}\right) = \exp\left(-\frac{\Delta G^{0,\#}}{RT}\right) \qquad (30)$$

where R is the universal gas constant, ΔH^0 is the standard enthalpy of the reaction, ΔS^0 is the standard entropy of the reaction, and ΔG^0 is the standard Gibbs free energy of the reaction. The rate coefficients for each elementary reaction were calculated on the basis of the transition state theory:

$$k_j = \frac{k_B T}{h} \exp\left(\frac{\Delta S^{0,\ddagger}}{R}\right) \exp\left(\frac{\Delta H^{0,\ddagger}}{RT}\right) = \frac{k_B T}{h} \exp\left(-\frac{\Delta G^{0,\ddagger\#}}{RT}\right) \tag{31}$$

where k_B is the Boltzmann constant, h is the Planck constant, $\Delta H^{0,\ddagger}$ is the standard enthalpy of activation, $\Delta S^{0,\ddagger}$ is the standard entropy of activation, and $\Delta G^{0,\ddagger}$ is the standard Gibbs free energy of activation. Arrhenius pre-exponential factors (A_f) and activation energies ($E_{a(f)}$), as well as values for $\Delta S°$ and $\Delta H°$, were determined on the basis of the computational work discussed by Alexopoulous et al. [36] and reported in Table 2.

Table 2. Standard reaction enthalpy (ΔH_r^0 in kJ mol^{-1}), standard reaction entropy (ΔS_r^0 in J mol^{-1} K^{-1}), activation energy ($E_{a(f)}$ in kJ mol^{-1}), and pre-exponential factor (A_f in s^{-1} or 10^{-2} kPa^{-1} s^{-1}) of the forward reaction for the elementary steps, numbered as indicated in Figure 4. The activation steps are indicated in bold.

	Elementary Steps	ΔH_r^0	ΔS_r^0	$E_{a(f)}$	A_f
1	EtOH$_{(g)}$ + * ↔ M$_1$	−122	−167	-	-
2	M$_1$ ↔ M$_2$	14	7	-	-
3	**M$_2$ ↔ Ethoxy + H$_2$O$_{(g)}$**	77	146	118	4.0 10^{13}
4	**Ethoxy ↔ Ethene$_{(ads)}$**	44	60	106	9.4 10^{12}
5	Ethene$_{(ads)}$ ↔ C$_2$H$_{4(g)}$ + *	48	99	-	-
6	M$_1$ + EtOH$_{(g)}$ ↔ D$_1$	−99	−162	-	-
7	D$_1$ ↔ D$_2$	44	24	-	-
8	**D$_2$ ↔ DEE$_{(ads)}$ + H$_2$O$_{(g)}$**	16	125	92	3.5 10^{12}
9	DEE$_{(ads)}$ ↔ DEE$_{(g)}$	139	165	-	-
10	**DEE$_{(ads)}$ ↔ C$_1$**	114	51	145	4.6 10^{13}
11	C$_1$ ↔ Ethene* + EtOH$_{(g)}$	59	175	-	-
12	Ethoxy + Ethene ↔ C$_2$	−33	−113	-	-
13	**C$_2$ ↔ 1-butene$_{(ads)}$**	−82	−25	81	1.7 10^{12}
14	1-butene$_{(ads)}$ ↔ 1-butene + *	90	159	-	-
15	W ↔ H$_2$O$_{(g)}$ + *	83	151	-	-

4. Multi-Scale Reactor Model Validation

A survey of publicly available information yielded the following patent US 2013/0090510 [6] as the most relevant one for assessing the adequacy of the model developed in this work. The operating conditions and catalyst properties for this design case are presented in Table 3. The process configuration comprised two adiabatic reactors in series with intermediate heating, having a combined catalyst mass amounting to 6 ton. The inlet temperature and pressure for the first adiabatic reactor amounted to 673 K and 590 kPa, while 679 K and 530 kPa were used for the second one. The considered feedstock was an aqueous ethanol mixture containing 26 wt.% ethanol which considerably exceeded the ethanol content of the fermentation broth, i.e., 10 wt.%. The yearly ethanol processing capacity was estimated at 360 kton.

Table 3. Experimental operating conditions: catalyst mass (W_t), inlet temperature (T_0), and pressure ($p_{t,0}$) for each adiabatic reactor and the annual ethene production capacity (G_{C2H4}) and inlet water content for the first reactor ($x_{EtOH,0}$).

Operating condition	Reactor 1	Reactor 2
W (ton)	3	3
T_0 (K)	673	679
$p_{t,0}$ (kPa)	590	530
$F_{C2H5OH,0}$ (kton y^{-1})	360	
$x_{EtOH,0}$	0.26	
Catalyst property		
d_p (m)	4 10^{-3}	
ε_p (-)	0.6	
τ (-)	5	
ρ_p (kg m^{-3})	700	
C_t (mol kg^{-1})	0.003	

The performance results for the configuration comprising two adiabatic reactors are given in Table 4. Herein, the ethanol conversion (X_{EtOH}) and yield of gas phase component i (Y_i) are defined as:

$$X_{EtOH} = \frac{F^0_{EtOH} - F_{EtOH}}{F^0_{EtOH}} \tag{32}$$

$$Y_i = \frac{F_i}{F^0_{EtOH}} \tag{33}$$

in which F^0_{EtOH} is the molar inlet flow rate of ethanol, and F_i is the molar flow rate of gas phase component i.

Table 4. Performance results, i.e., conversion (X_{EtOH}), ethene, oxygenates, and C_{3+} olefin yield (respectively, Y_{C2H4}, Y_{oxy}, Y_{ole}), temperature (T), and pressure (p_t), as described in Coupard et al. [6].

	X_{EtOH} (-)	Y_{C2H4} (-)	Y_{oxy} (-)	Y_{ole} (-)	T (K)	p_t (kPa)
Outlet reactor 1	0.71	0.69	0.02	0.00	591	560
Outlet reactor 2	0.99	0.97	0.00	0.01	653	500

According to the results described in patent US 2013/0090510 [6], an ethanol conversion amounting to 0.71 was observed, with a corresponding ethene yield of 0.69 in the first reactor. The by-product at the reactor outlet was said to consist of oxygenates, which are represented in the kinetic model by di-ethyl ether. A temperature drop over the first catalyst bed of more than 80 K was observed. At the end of the second reactor, almost complete ethanol conversion was achieved, with a high yield of ethene (0.97). Table 4 indicates that in this case, the by-products were higher olefins, represented in the kinetic model employed in this work by 1-butene. A less pronounced temperature drop of 26 K was observed over the second catalyst bed.

Figure 5 shows the calculated conversion and yield profiles along the axial reactor position. Ethene was the most abundant product throughout the reactor. At the end of the first catalyst bed, around 2% oxygenates product was observed, which is accurately described by the kinetic model with di-ethyl ether as a representative product. At the end of the second bed, no more di-ethyl ether was present as a consequence of its decomposition into ethene and ethanol, while the formation of higher hydrocarbon by-products, here represented by 1-butene, was observed. The reactor model hence provides a detailed picture of the product evolution throughout the reactor and allows assessing the effects of temperature and pressure.

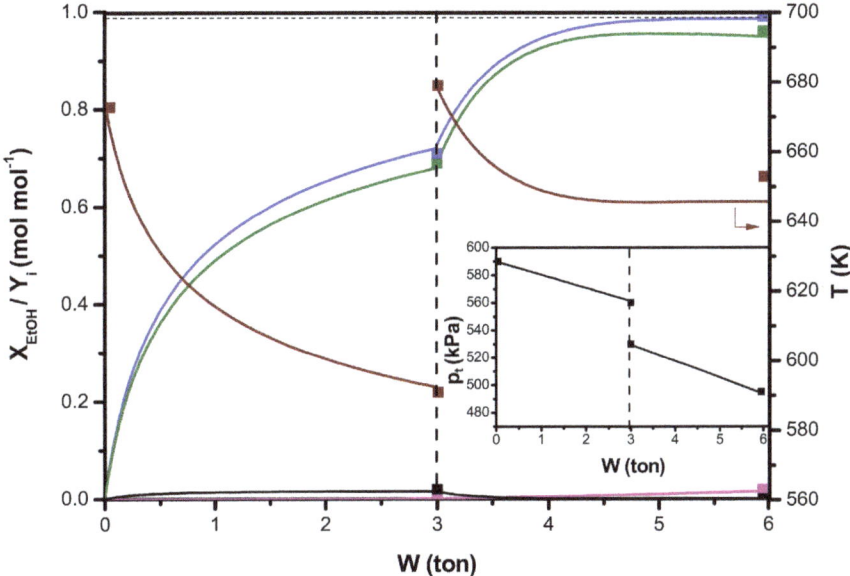

Figure 5. Ethanol conversion (X_{EtOH}, blue), ethene, di-ethyl ether, and butene yield (green: Y_{C2H4}; black : Y_{DEE}; magenta; Y_{C4H8}), and temperature profiles (T) as a function of catalyst mass. The inset shows the pressure drop (p_t) as a function of catalyst mass. Calculations were made by integration of Equations (5)–(7) and (12) and simultaneously solving Equations (26) and (28) with the corresponding net production rates as defined in Equation (27), with parameters taken from Table 2 and the experimental conditions given in Table 3. Square symbols indicate the experimental points from [4] given in Table 4.

A monotonous temperature decrease with increasing catalyst mass was observed in Figure 5, indicating that the monomolecular pathway (Equation (18)) was the most dominant along the entire reactor axis. Downstream of the first bed, the temperature of the outlet flow was increased via intermediate heating prior to sending the effluent to the subsequent reactor. Although the temperature showed good agreement at the end of the first reactor, a discrepancy between the simulated and the reported temperature was observed at the end of the second reactor. A total temperature drop of 116 K was simulated, while a temperature drop of only 107 K was observed. This can be compared to the total maximum adiabatic temperature drop calculated by:

$$\Delta T_{ad,max} = \frac{F^0_{EtOH}(-\Delta H^0_r)}{G\, c_p} \qquad (34)$$

This maximum adiabatic temperature drop was found to be 119 K and was closer to the simulated temperature drop than to the experimentally determined one. The pressure drop was also described adequately, as shown in the inset in Figure 5.

The catalyst effectiveness factor along the first reactor bed is shown in Figure 6 and was found to increase from 0.21 to 0.42. A concentration profile along the dimensionless catalyst pellet diameter is shown in the inset, indicating that severe diffusion limitations existed at the catalyst pellet scale.

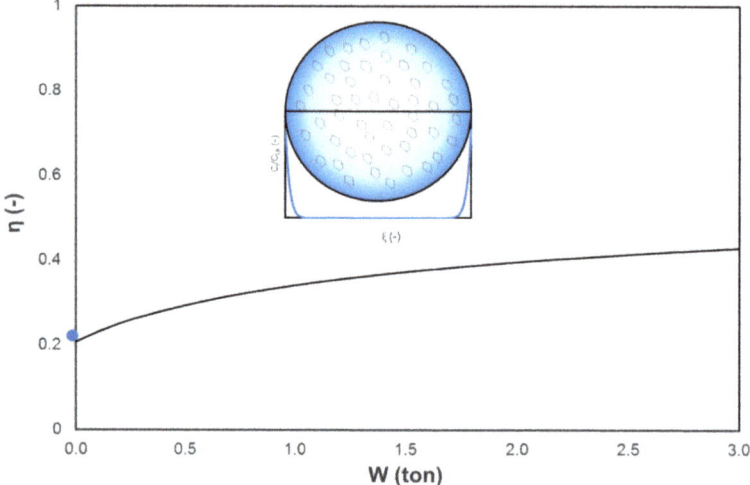

Figure 6. Catalyst effectiveness factor, calculated by Equation (21), as a function of catalyst mass. The inset shows the relative concentration profile along the dimensionless catalyst pellet diameter. Calculations were made by integration of Equations (5)–(7) and (12) and simultaneously solving Equations (26) and (28) with the corresponding net production rates as defined in Equation (27), with parameters taken from Table 2 and the experimental conditions given in Table 3.

5. Optimization of an Industrial Ethanol Dehydration Reactor

As good agreement was achieved between the model and the reported values, the model was considered to provide reliable predictions and, hence, was used to investigate and optimize an industrial ethanol dehydration reactor. Key process parameters for industrial operation are the amount of water added to the feed and the operating temperature. The thermodynamic equilibrium composition as a function of temperature was investigated by minimization of the Gibbs free energy. Three cases were considered:

(I). Dehydration of pure ethanol, i.e., no additional water added in the feed, which considers ethanol, ethene, di-ethyl ether, and water in the product mixture.
(II). Dehydration of aqueous ethanol, i.e., 90 mol% water contained in the feed, which corresponds to the lower limit of ethanol content obtained via biomass fermentation. This case also considers ethanol, ethene, di-ethyl ether, and water in the product mixture.
(III). Dehydration of aqueous ethanol, i.e., 90 mol% water in the feed, with dimerization of ethene included as a model reaction for the formation of higher hydrocarbons. In addition to the compounds mentioned above, 1-butene was also added to the calculation.

The dehydration of pure ethanol was found to be complete at 600 K, as shown in Figure 7. At low temperatures, the major product was di-ethyl ether, which gradually decreased with the temperature in favor of ethene. It was possible to achieve 100% selectivity towards ethene, i.e., no thermodynamic constraints were encountered in the industrial implementation of this process. The addition of water at the reactor inlet resulted in lower ethanol conversion and higher selectivity towards ethene at lower temperatures. Nevertheless, the ethanol conversion was still complete at temperatures exceeding 650 K.

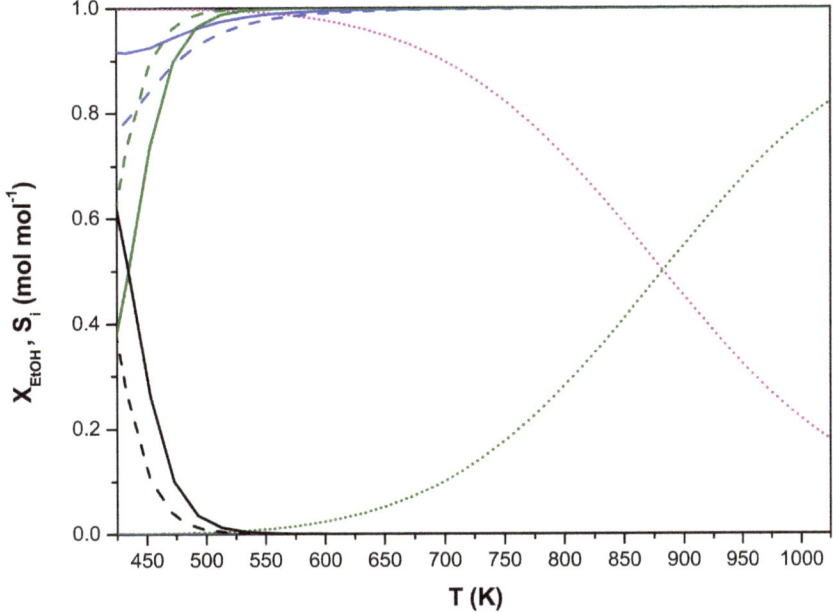

Figure 7. Thermodynamic equilibrium of ethanol conversion (blue) and selectivity towards ethene (green), di-ethyl ether (black), and 1-butene (magenta) as a function of temperature. Full line: ethanol dehydration (Equations (22)–(24)) with no additional water (case I); dashed line: ethanol dehydration (Equations (22)–(24)) with 90 mol% water (case II); dotted line: ethanol dehydration with ethene dimerization (Equations (22)–(25)) including 1-butene as a product (case III).

Taking into account ethene dimerization as a representative for the formation of higher hydrocarbons, it was shown that ethanol conversion remained complete over the entire temperature range shown in Figure 7. However, 1-butene was now the major product when solely considering thermodynamics. Only at temperatures exceeding 650 K, ethene became the principal product. This illustrates the existing competition between ethanol dehydration and the subsequent formation of higher hydrocarbons.

The effect of the ethanol content in the feed on the maximum adiabatic temperature drop is illustrated in Figure 8. The higher the ethanol content, the higher the maximum adiabatic temperature drop over the reactor. This is related to changes in the mixture heat capacity due to changes in feed composition. At 673 K, a pure ethanol feed would result in a total temperature drop amounting to 400 K, while the aqueous conditions investigated in this work only resulted in a temperature drop of 119 K. As heat is consumed along the reactor with increasing ethanol conversion due to the endothermicity of monomolecular ethanol dehydration, a higher water content has a higher heat buffering effect that can be utilized throughout the reaction. At low ethanol content, the reactor inlet temperature effect on the maximum adiabatic temperature drop can be neglected. However, at high ethanol content, a substantial difference can be observed: a temperature difference of 70 K was found between 573 K and 773 K for a feed with no additional water. Hydrous ethanol is a particularly attractive feedstock because the production of anhydrous ethanol is very energy- and cost-intensive. In the context of zeolites, water can influence the kinetics of alcohol dehydration, potentially by competing with alcohol reactants for Brønsted acid sites, by shifting the dehydration–hydration equilibrium, and by inducing potentially different solvation strengths in all states along the reaction coordinates in zeolite [41]. Previously, it was reported that water in the ethanol feed enhances the steady-state catalytic activity of

H-ZSM-5 and the selectivity for ethylene formation by possibly moderating the acidity of the catalytic sites, resulting in less extensive deactivation due to coking [42,43].

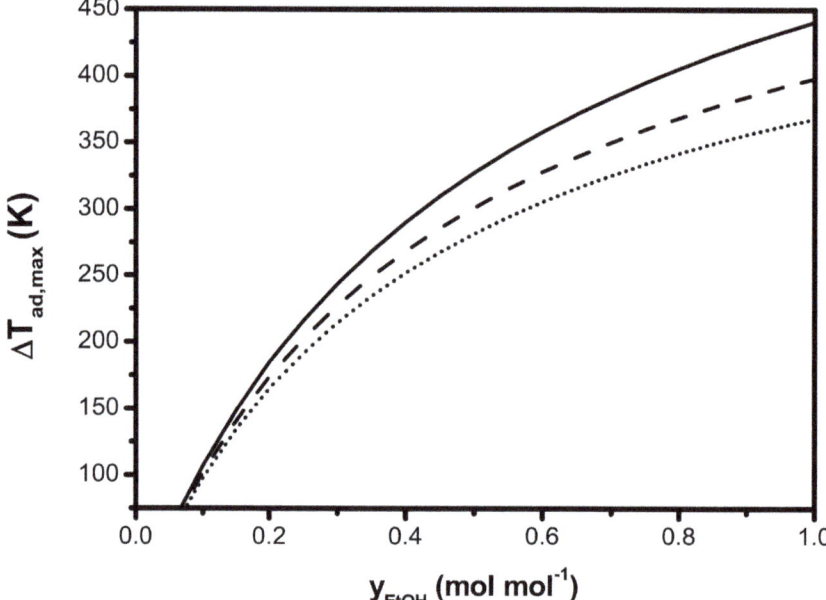

Figure 8. Maximum adiabatic temperature drop (Equation (34)) as a function of molar ethanol fraction in the feed for three different reactor inlet temperatures (full line: 573 K, dashed line: 673 K, dotted line: 773 K) and the process conditions indicated in Table 3.

The effect of varying water contents on conversion, ethene yield, and temperature is shown in Figure 9. The highest conversion and ethene yield were obtained in the range of high water content. This high water content also resulted in the lowest adiabatic temperature drop, as shown in Figure 8, and, hence, in overall higher reaction rates compared to less diluted ethanol feeds. However, this can only be assessed when also the size and cost of the other pieces of equipment (compressors) are taken into account.

Decreasing the water content resulted in a decreased ethanol conversion and, remarkably, also a decreased ethene yield (Figure 9). At high water content, the ratio between ethene yield and ethanol conversion was close to one, but decreasing the water content decreased this ratio. Instead, di-ethyl ether was produced in higher quantities, which lowered the ethene production. Higher ethanol partial pressures favored the formation of di-ethyl ether and decreased ethene selectivity. Dimerization of ethene to 1-butene was not observed to a significant extent in any of the case studies. The temperature drop observed in the reactor was not as pronounced as shown in Figure 9, which is related to ethanol conversion to di-ethyl ether, i.e., a mildly exothermic reaction. A less pronounced temperature drop due to high water content automatically resulted in a higher conversion, as can be seen in Figure 9.

These simulation results are in line with patent literature showing the necessity of introducing a heat carrying fluid in the reactor when working with a pure ethanol feed. The use of water vapor has been proposed. This water may come from an external source or be produced internally in the process and recycled from the effluent [4]. Without effluent separation into ethane and water, the latter is not advisable as ethane recycling towards the reactor inlet affects the thermodynamic equilibrium of the dehydration reaction. Ethene also participates in the subsequent conversion to higher hydrocarbons, which will increase the yield of secondary products.

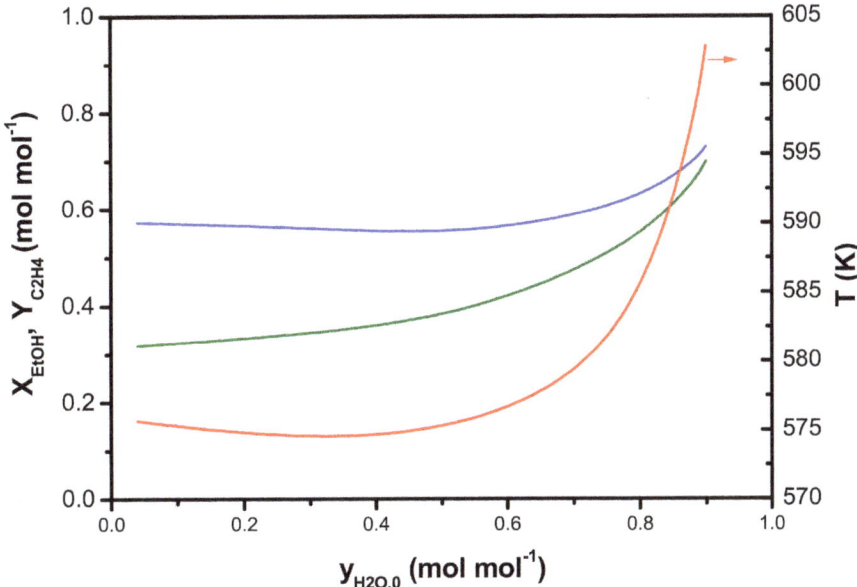

Figure 9. Ethanol conversion (X_{EtOH}, blue), ethene yield (Y_{C2H4}, green), and temperature (T, red) as a function of water inlet content. Calculations were made by integration of Equations (5)–(7) and (12) and simultaneously solving Equations (26) and (28) with the corresponding net production rates as defined in Equation (27), with parameters taken from Table 2 and the experimental conditions given in Table 3.

6. Conclusions

A fully ab initio reaction network for ethanol dehydration on H-ZSM-5 was used to simulate an industrial multi-bed adiabatic reactor. Internal transport limitations inside the catalyst pellet needed to be explicitly accounted for, while no external transport limitations nor internal heat transport limitations were diagnosed at the conditions investigated in this work. Good agreement was found with the literature results, i.e., temperature, pressure, and outlet flow rates. The industrial reactor model developed indicated that it is not beneficial to utilize feeds with high ethanol content, as they result in a lower ethanol conversion and ethene yield. Furthermore, a more pronounced temperature drop over the reactor was simulated. It is preferred to use a vapor feed more diluted with water for the operation of the industrial ethanol dehydration reactor. Of course, in order to properly assess an economically optimal configuration and reasonable feedstock properties, the investment and operating costs of the entire facility should be taken into account.

Author Contributions: Conceptualization, K.V.d.B., G.B.M. and V.V.G.; methodology, K.A., J.W.T. and K.T.; software, J.W.T. and K.T.; formal analysis, K.V.d.B.; data curation, K.V.d.B.; writing—original draft preparation, K.V.d.B.; writing—review and editing, K.V.d.B., J.W.T., K.A., K.T., V.V.G. and G.B.M; supervision, G.B.M and V.V.G.

Funding: This research was funded by e 'Long-Term Structural Methusalem Funding with Grant No. BOF09/01M00409.

Conflicts of Interest: The authors declare no conflict of interest.

List of Symbols

Roman symbols

*	free acid sitessur
A_r	face area of the cross section of the reactor [m^2]
c_p	specific heat capacity [J kg^{-1} K^{-1}]
C_p	specific molar heat capacity [J mol^{-1} K^{-1}]
C_i	concentration of component i in the pellet [mol kg$_{cat}^{-1}$]
C_t	acid site concentration [mol$_{H+}$ kg^{-1}]
d	diameter [m]
$D_{e,i}$	effective diffusion coefficient of component i [m^2 s^{-1}]
f	friction factor [-]
F_i	molar flow rate of gas phase component i [mol s^{-1}]
G	mass flow rate [kg s^{-1}]
ΔG	Gibbs free energy of reaction [J mol^{-1}]
h	Planck constant = 6.63. 10^{-34} m^2 kg s^{-1}
ΔH	enthalpy of reaction [J mol^{-1}]
k_B	Boltzmann's constant = 1.38. 10^{-23} m^2 kg s^{-2} K^{-1}
k_j	rate coefficient of elementary step j [variable]
M	molecular mass [kg mol^{-1}]
n	apparent order of reaction
n_{comp}	number of components
N_i	molar flux of i with respect to a fixed plane [mol m^{-2} s^{-1}]
p_i	partial pressure of component i [Pa]
p_t	total reactor pressure [Pa]
u_s	superficial velocity [m s^{-1}]
r	radius of the catalyst pellet [m]
R	universal gas constant = 8.31 J mol^{-1} K^{-1}
R_i	net production rate of component i [mol mol$_{H+}^{-1}$ s^{-1}]
S_i	selectivity of component i [mol mol^{-1}]
ΔS	entropy of reaction [J mol^{-1} K^{-1}]
T	temperature [K]
v	stoichiometric coefficient
V	volume [m^3]
W	catalyst mass [kg]
X_i	conversion of component i [mol mol^{-1}]
y_i	molar fraction of component i in the gas phase [mol mol^{-1}]
Y_i	yield of component i [mol mol^{-1}]

Greek symbols

ε	porosity [-]
η	catalyst effectiveness [-]
μ	dynamic viscosity [Pa s]
ρ	density [kg m^{-3}]
ξ	dimensionless distance [-]
θ_k	fractional coverage of surface species k [-]

Subscripts

b	catalyst bed
bl	bulk
f	formation
fl	fluid
i	gas phase species
j	elementary step
k	surface species
p	catalyst pellet
pore	pore
m	mixture
r	reaction
r	reactor
s	surface
v	volumetric

Subscripts

‾	average
‡	activation
∘	inlet
∘	standard
C_2H_4	ethene
EtOH	ethanol

References

1. Derouane, E.G.; Nagy, J.B.; Dejaifve, P.; Vanhooff, J.H.C.; Spekman, B.P.; Vedrine, J.C.; Naccache, C. Elucidation of Mechanism of Conversion of Methanol and Ethanol to Hydrocarbons on a New Type of Synthetic Zeolite. *J. Catal.* **1978**, *53*, 40–55. [CrossRef]
2. Sheldon, R.A. Green and sustainable manufacture of chemicals from biomass: State of the art. *Green Chem.* **2014**, *16*, 950–963. [CrossRef]
3. Donnis, B.; Egeberg, R.G.; Blom, P.; Knudsen, K.G. Hydroprocessing of Bio-Oils and Oxygenates to Hydrocarbons. Understanding the Reaction Routes. *Top. Catal.* **2009**, *52*, 229–240. [CrossRef]
4. Barrocas, H.V.V.; da Silva, J.B.d.M.; de Assis, R.C. Process for Preparing Ethene. U.S. Patent 4232179 A, 4 November 1980.
5. Phung, T.K.; Hernández, L.P.; Lagazzo, A.; Busca, G. Dehydration of Ethanol over Zeolites, Silica Alumina and Alumina: Lewis Acidity, Brønsted Acidity and Confinement EffectsAppl. *Catal. A: General* **2015**, *493*, 77–89. [CrossRef]
6. Coupard, V.; Touchais, N.; Fleurier, S.; Penas, H.G.; de Smedt, P.; Vermeiren, W.; Adam, C.; Minoux, D. Process for Dehydration of Dilute Ethanol into Ethylene with Low Energy Consumption without Recycling of Water. U.S. Patent 20130090510 A1, 11 April 2013.
7. Jiang, B.; Feng, X.; Yan, L.; Jiang, Y.; Liao, Z.; Wang, J.; Yang, Y. Methanol to Propylene Process in a Moving Bed Reactor with Byproducts Recycling: Kinetic Study and Reactor Simulation. *Ind. Eng. Chem. Res.* **2014**, *53*, 4623–4632. [CrossRef]
8. Lee, W.L.; Froment, G.F. Ethylbenzene Dehydrogenation into Styrene: Kinetic Modeling and Reactor Simulation. *Ind. Eng. Chem. Res.* **2008**, *47*, 9183–9194. [CrossRef]
9. Froment, G.F.; Depauw, G.A.; Vanrysselberghe, V. Kinetic Modeling and Reactor Simulation in Hydrodesulfurization of Oil Fractions. *Ind. Eng. Chem. Res.* **1994**, *33*, 2975–2988. [CrossRef]
10. Dewachtere, N.V.; Santaella, F.; Froment, G.F. Application of a single-event kinetic model in the simulation of an industrial riser reactor for the catalytic cracking of vacuum gas oil. *Chem. Eng. Sci.* **1999**, *54*, 3653–3660. [CrossRef]
11. Batchu, R.; Galvita, V.V.; Alexopoulos, K.; Glazneva, T.S.; Poelman, H.; Reyniers, M.-F.; Marin, G.B. Ethanol dehydration pathways in H-ZSM-5: Insights from temporal analysis of products. *Catal. Today* **2019**. [CrossRef]

12. Sabbe, M.K.; van Geem, K.M.; Reyniers, M.F.; Marin, G.B. First Principle-Based Simulation of Ethane Steam Cracking. *AiChe J.* **2011**, *57*, 482–496. [CrossRef]
13. Jacobsen, C.J.H.; Dahl, S.; Boisen, A.; Clausen, B.S.; Topsøe, H.; Logadottir, A.; Nørskov, J.K. Optimal Catalyst Curves: Connecting Density Functional Theory Calculations with Industrial Reactor Design and Catalyst Selection. *J. Catal.* **2002**, *205*, 382–387. [CrossRef]
14. Sabbe, M.K.; Canduela-Rodriguez, G.; Reyniers, M.-F.; Marin, G.B. DFT-based modeling of benzene hydrogenation on Pt at industrially relevant coverage. *J. Catal.* **2015**, *330*, 406–422. [CrossRef]
15. Alwahabi, S.M.; Froment, G.F. Conceptual Reactor Design for the Methanol-to-Olefins Process on SAPO-34. *Ind. Eng. Chem. Res.* **2004**, *43*, 5112–5122. [CrossRef]
16. Park, T.-Y.; Froment, G.F. Analysis of Fundamental Reaction Rates in the Methanol-to-Olefins Process on ZSM-5 as a Basis for Reactor Design and Operation. *Ind. Eng. Chem. Res.* **2004**, *43*, 682–689. [CrossRef]
17. Zhuang, Y.-Q.; Gao, X.; Zhu, Y.-P.; Luo, Z.-H. CFD modeling of methanol to olefins process in a fixed-bed reactor. *Powder Technol.* **2012**, *221*, 419–430. [CrossRef]
18. Zhao, Y.; Li, H.; Ye, M.; Liu, Z. 3D Numerical Simulation of a Large Scale MTO Fluidized Bed Reactor. *Ind. Eng. Chem. Res.* **2013**, *52*, 11354–11364. [CrossRef]
19. Berger, R.J.; Stitt, E.H.; Marin, G.B.; Kapteijn, F.; Moulijn, J.A. Eurokin. Chemical Reaction Kinetics in Practice. *CATTECH* **2001**, *5*, 36–60. [CrossRef]
20. Kokotailo, G.T.; Lawton, S.L.; Olson, D.H.; Olson, D.H.; Meier, W.M. Structure of Synthetic Zeolite ZSM-5. *Nature* **1978**, *272*, 437–438. [CrossRef]
21. Weisz, P.B.; Prater, C.D. Interpretation of Measurements in Experimental Catalysis. In *Advances in Catalysis*; Frankenburg, V.I.K.W.G., Rideal, E.K., Eds.; Academic Press: New York, NY, USA, 1954; pp. 143–196.
22. Mears, D.E. Diagnostic Criteria for Heat Transport Limitations in Fixed Bed Reactors. *J. Catal.* **1971**, *20*, 127–132. [CrossRef]
23. Froment, G.F.; de Wilde, J.; Bischoff, K.B. *Chemical Reactor Analysis and Design*, 3rd ed.; Wiley: Hoboken, NJ, USA, 2011.
24. Reid, R.C.; Prausnitz, J.M.; Sherwood, T.K. *The Properties of Gases and Liquids*, 3rd ed.; McGraw-Hill: New York, NY, USA, 1977.
25. Hicks, R.E. Pressure Drop in Packed Beds of Spheres. *Ind. Eng. Chem. Fundam.* **1970**, *9*, 500–502. [CrossRef]
26. Haughey, D.P.; Beveridge, G.S.G. Structural properties of packed beds—A review. *Can. J. Chem. Eng.* **1969**, *47*, 130–140. [CrossRef]
27. Bosanquet, C.H. *British TA Report BR-507*; Springer: London, UK, 1944.
28. Maxwell, J.C. On the Dynamical Theory of Gases. *Philos. Trans. R. Soc. Lond.* **1867**, *157*, 49–88.
29. Stefan, J. Über das Gleichgewicht un die Bewegung, insbesondere die Diffusion von Gasgemengen. *Sitzungsberichte Akad Wiss Wien* **1871**, *63*, 63–124.
30. Wilke, C. Diffusional properties of multicomponent gases. *Chem. Eng. Prog.* **1950**, *46*, 95–104.
31. Solsvik, J.; Jakobsen, H.A. Modeling of multicomponent mass diffusion in porous spherical pellets: Application to steam methane reforming and methanol synthesis. *Chem. Eng. Sci.* **2011**, *66*, 1986–2000. [CrossRef]
32. Solsvik, J.; Jakobsen, H.A. Multicomponent mass diffusion in porous pellets: Effects of flux models on the pellet level and impacts on the reactor level. Application to methanol synthesis. *Can. J. Chem. Eng.* **2013**, *91*, 66–76. [CrossRef]
33. Thybaut, J.W.; Marin, G.B. Single-Event MicroKinetics: Catalyst design for complex reaction networks. *J. Catal.* **2013**, *308*, 352–362. [CrossRef]
34. Fuller, E.N.; Schettle, P.; Giddings, J.C. A New Method for Prediction of Binary Gas-Phase Diffusion Coeffecients. *Ind. Eng. Chem.* **1966**, *58*, 19.
35. Reid, R.C.; Prausnitz, J.M.; Poling, B.E. *The Properties of Gases and Liquids*, 4th ed.; McGraw-Hill: New York, NY, USA, 1987.
36. Reyniers, M.-F.; Marin, G.B. Experimental and Theoretical Methods in Kinetic Studies of Heterogeneously Catalyzed Reactions. *Annu. Rev. Chem. Biomol. Eng.* **2014**, *5*, 563–594. [CrossRef]
37. Alexopoulos, K.; John, M.; van der Borght, K.; Galvita, V.; Reyniers, M.-F.; Marin, G.B. DFT-based microkinetic modeling of ethanol dehydration in H-ZSM-5. *J. Catal.* **2016**, *339*, 173–185. [CrossRef]
38. Johansson, R.; Hruby, S.L.; Rass-Hansen, J.; Christensen, C.H. The Hydrocarbon Pool in Ethanol-to-Gasoline over HZSM-5 Catalysts. *Catal. Lett.* **2009**, *127*, 1–6. [CrossRef]

39. Madeira, F.F.; Gnep, N.S.; Magnoux, P.; Vezin, H.; Maury, S.; Cadran, N. Mechanistic insights on the ethanol transformation into hydrocarbons over HZSM-5 zeolite. *Chem. Eng. J.* **2010**, *161*, 403–408. [CrossRef]
40. Aguayo, A.T.; Gayubo, A.G.; Atutxa, A.; Olazar, M.; Bilbao, J. Catalyst deactivation by coke in the transformation of aqueous ethanol into hydrocarbons. Kinetic modeling and acidity deterioration of the catalyst. *Ind. Eng. Chem. Res.* **2002**, *41*, 4216–4224. [CrossRef]
41. Macht, J.; Janik, M.J.; Neurock, M.; Iglesia, E. Mechanistic Consequences of Composition in Acid Catalysis by Polyoxometalate Keggin Clusters. *J. Am. Chem. Soc.* **2008**, *130*, 10369–10379. [CrossRef] [PubMed]
42. Phillips, C.B.; Datta, R. Production of ethylene from hydrous ethanol on H-ZSM-5 under mild conditions. *Ind. Eng. Chem. Res.* **1997**, *36*, 4466–4475. [CrossRef]
43. Zhi, Y.C.; Shi, H.; Mu, L.Y.; Liu, Y.; Mei, D.H.; Camaioni, D.M.; Lercher, J.A. Dehydration Pathways of 1-Propanol on HZSM-5 in the Presence and Absence of Water. *J. Am. Chem. Soc.* **2015**, *137*, 15781–15794. [CrossRef]

© 2019 by the authors. Licensee MDPI, Basel, Switzerland. This article is an open access article distributed under the terms and conditions of the Creative Commons Attribution (CC BY) license (http://creativecommons.org/licenses/by/4.0/).

Article

Tandem Hydrogenation/Hydrogenolysis of Furfural to 2-Methylfuran over a Fe/Mg/O Catalyst: Structure–Activity Relationship

Carlo Lucarelli [1], Danilo Bonincontro [2], Yu Zhang [2,3], Lorenzo Grazia [2], Marc Renom-Carrasco [3], Chloé Thieuleux [3], Elsje Alessandra Quadrelli [3], Nikolaos Dimitratos [2], Fabrizio Cavani [2] and Stefania Albonetti [2,*]

1. Dipartimento di Scienza e Alta tecnologia, Università dell'Insubria, Via Valleggio 9, 22100 Como, Italy; carlo.lucarelli@uninsubria.it
2. Dipartimento di Chimica Industriale "Toso Montanari", University of Bologna Viale del Risorgimento 4, 40136 Bologna, Italy; danilo.bonincontro2@unibo.it (D.B.); niobium224@gmail.com (Y.Z.); Lorenzo.Grazia@polynt.com (L.G.); nikolaos.dimitratos@unibo.it (N.D.); fabrizio.cavani@unibo.it (F.C.)
3. Université de Lyon, C2P2—UMR 5265 (CNRS—Université de Lyon 1—CPE Lyon), Équipe Chimie; Organométallique de Surface CPE Lyon, 43 Boulevard du 11 Novembre 1918, CEDEX, 69616 Villeurbanne, France; mrenomc@gmail.com (M.R.-C.); thieuleux@cpe.fr (C.T.); quadrelli@cpe.fr (E.A.Q.)
* Correspondence: stefania.albonetti@unibo.it; Tel.: +39-051-2093681

Received: 8 October 2019; Accepted: 23 October 2019; Published: 27 October 2019

Abstract: The hydrodeoxygenation of furfural (FU) was investigated over Fe-containing MgO catalysts, on a continuous gas flow reactor, using methanol as a hydrogen donor. Catalysts were prepared either by coprecipitation or impregnation methods, with different Fe/Mg atomic ratios. The main product was 2-methylfuran (MFU), an important highly added value chemical, up to 92% selectivity. The catalyst design helped our understanding of the impact of acid/base properties and the nature of iron species in terms of catalytic performance. In particular, the addition of iron on the surface of the basic oxide led to (i) the increase of Lewis acid sites, (ii) the increase of the dehydrogenation capacity of the presented catalytic system, and (iii) to the significant enhancement of the FU conversion to MFU. FTIR studies, using methanol as the chosen probe molecule, indicated that, at the low temperature regime, the process follows the typical hydrogen transfer reduction, but at the high temperature regime, methanol dehydrogenation and methanol disproportionation were both presented, whereas iron oxide promoted methanol transfer. FTIR studies were performed using furfural and furfuryl alcohol as probe molecules. These studies indicated that furfuryl alcohol activation is the rate-determining step for methyl furan formation. Our experimental results clearly demonstrate that the nature of iron oxide is critical in the efficient hydrodeoxygenation of furfural to methyl furan and provides insights toward the rational design of catalysts toward C–O bonds' hydrodeoxygenation in the production of fuel components.

Keywords: furfural; 2 methyl-furan; hydrodeoxygenation; catalyst design; iron; magnesium oxide; catalytic hydrogen transfer reduction; methanol

1. Introduction

The use of biomass, particularly lignocellulosic materials for fuels and chemical production, aims at reducing the exploitation of non-renewable resources. However, the industry is facing the challenge of developing new chemical processes for converting these renewable feedstocks containing highly functionalized carbohydrates into platform molecules with reduced oxygen content [1]. Many of those systems involve molecular hydrogen as the reductant; at the same time, H-transfer processes, where an organic molecule (e.g., an alcohol) behaves as the hydrogen donor, are a promising alternative [2]. Avoiding the use of H$_2$ for substrate reduction could induce safer and more selective chemical processes. Indeed, the lower hydrogenating capability of most hydrogen donors promotes a higher degree of control, especially when partially hydrogenated molecules are needed [3]; this is the case in hydrodeoxygenation (HDO) processes, where C–O bonds are selectively cleaved, leaving the nearby C=C and C–C bonds unchanged.

Within bio-derived platform molecules, furan derivatives are important intermediates because of their rich chemistry. For this reason, many efforts were made in the conversion of furfural (FU), which can be large-scale produced from hemi-cellulose into furan-based compounds in the form of furfuryl alcohol (FAL) and 2-methylfuran (MFU) [4]. FAL can be formed by selective hydrogenation of the FU carbonyl group (see Scheme 1). MFU is often produced through further hydrogenolysis of FAL [5–7] and has drawn the attention of researchers as a gasoline alternative due to its very attractive combustion performance in engines [8].

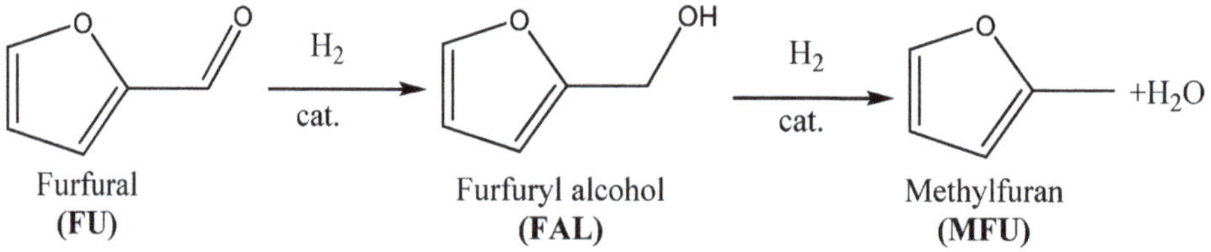

Scheme 1. Sequential reduction of furfural to 2-methylfuran.

Catalytic transfer HDO of furfural was investigated over heterogeneous catalysts using different hydrogen donors in the recent years [9–12]. In our group, the liquid-phase reduction of FU into FAL was carried out using methanol as a hydrogen donor and MgO as a heterogeneous basic catalyst. Furfural was completely reduced into its corresponding alcohol through a Meerwein–Ponndorf–Verley (MPV)-like mechanism [13]. More recently, Hermans and co-workers obtained a 62% yield of 2-methylfuran (and 2-methyltetrahydrofuran) over Pd/Fe$_2$O$_3$, using 2-propanol as the H-donor in continuous liquid phase (180 °C, 25 bar) [4]. Vlachos et al. showed that Ru/RuO$_x$/C catalysts were also active for the liquid-phase transfer hydrogenation of FU to MFU. Thorough mechanistic studies to better understand the reaction mechanism and the role of active sites, supported by DFT calculations, they proposed three different catalytic sites: Lewis acidity in the RuO$_2$ promotes the transfer hydrogenation of FU to FAL, RuO$_2$ oxygen vacancies catalyze the C–O scission to form MFU, and metallic Ru helps form hydride species that regenerate the RuO$_2$ vacancies [14–16].

In the gas-phase furfural hydrogenation reaction, the reaction of furfural with hydrogen over an SiO$_2$-supported transition metal catalyst was investigated by D.E. Resasco [17]. They specified that η^1 (O) aldehyde is the mostly likely species adsorbed on the Cu surface, which will favor furfuryl alcohol formation. Furthermore, they indicated that surface η^2 (C=O) adsorption geometry facilitated the formation of methyl furan [18,19]. Other groups have shown that Fe promotes methyl furan formation in Ni–Fe systems [20] and Fe–Cu systems [6,21]. The authors concluded that the addition of Fe suppresses the decarbonization and promotes the C=O hydrogenation at low temperatures and the C–O hydrogenolysis at high temperatures. Moreover, it reveals that the partial reduction of Fe^{3+} to Fe^{2+} played the role of promoter. However, it is still unclear what the exact role of Fe in the hydrogen-transfer processes is.

Recently, we observed high MFU yield (79%) in the gas-phase reduction of furfural, using methanol as the hydrogen donor over an iron–magnesium mixed oxide catalyst (Fe/Mg/O) [22]. In that work, we showed that the introduction of Fe^{3+} cations into the magnesia structure leads to the formation of higher quantities of MFU, derived from FAL hydrogenolysis, without presenting a detailed structure–activity correlation study. Since iron is known to have both redox and acid–base properties [23], we focused on studying the role of Fe in the Fe/MgO catalytic system, in order to determine structure–activity relationships. Therefore, in the current study, catalysts with different iron content were prepared, in order to investigate this relationship in MFU formation. Fe/Mg/O catalysts were synthesized either by coprecipitation or incipient wetness impregnation methods, to understand how the preparation routes affect the activity and product distribution of FU transformation. The synthesized materials were thoroughly characterized and compared based on their redox and acid–base properties and their crystalline phase. Finally, we carried out in situ DRIFT studies to elucidate (i) the activity and selectivity determining factors, (ii) the influence of the basicity, and (iii) the nature of FeO_x species for the hydrogenation/hydrogenolysis of furfural with the purpose of exploring the role of FeO_x/MgO in terms of furfural activation and production of MFU.

2. Results and Discussion

2.1. Physical–Chemical Properties of the Iron-Containing MgO Catalyst

Several Fe/Mg/O catalysts containing different Fe/Mg atomic ratios were prepared both by coprecipitation and by incipient wetness impregnation methods, as described in the experimental section. Characterizations of these materials are summarized in Table 1.

Table 1. Physicochemical properties (specific surface area, crystalline phase, Lewis acidity, Brønsted acidity and basicity) of MgO, Fe_2O_3 and Fe/Mg/O coprecipitated and synthesized by incipient wetness impregnation samples.

Catalyst	Fe:Mg	Surface Area ($m^2 \cdot g^{-1}$)	Crystalline Phase (XRD)	Lewis Acidity ($mmol \cdot g^{-1}$) [a]	Brønsted Acidity ($mmol \cdot g^{-1}$) [a]	Basicity ($mmol \cdot g^{-1}$) [b]
MgO		172	Periclase MgO	0	0	7.51
Fe_2O_3		51	Hematite	_ [d]	_ [d]	1.38
Fe/Mg/O_1_10	1:10	102	Periclase MgO	0	0	2.62
Fe/Mg/O_1_2	1:2	140	Periclase MgO and Magnetite/Magnesioferrite	0.43 [0.20 + 0.07 + 0.16] [c]	0	2.34
Fe/Mg/O_1_1	1:1	74	Periclase MgO and Magnetite/Magnesioferrite	0.62 [0.25 + 0.24 + 0.13] [c]	0	1.09
FeOx/MgO_1_100	1:100	150	Periclase MgO	0	0	2.62
FeOx/MgO_1_20	1:20	129	Periclase MgO and Hematite	0	0	2.34
FeOx/MgO_1_10	1:10	94	Periclase MgO and Hematite	0.15 [0.05 + 0.06 + 0.04] [c]	0	2.67
FeOx/MgO_1_2	1:2	33	Periclase MgO and Hematite	0.29 [0.04 + 0.17 + 0.08] [c]	0	0.96

[a] Quantification of surface Lewis and Brønsted acid sites was obtained from Pyridine-FTIR analysis. [b] Basicity measurements were performed by TPD analysis using CO_2 as probe molecule. [c] Distribution of the Lewis acid sites among weak, medium, and strong sites, respectively, based on the temperature at which pyridine desorption is observed, i.e., 20–200 °C (weak), 200–400 °C (medium), and >400 °C (strong). [d] Not possible to analyze by this technique.

The addition of Fe generally caused the decrease of surface area both in the case of coprecipitated and impregnated samples. In the case of impregnation, the decrease of surface area is more pronounced because, in this case, the total amount of iron oxides is located on the surface of MgO; in the case of the coprecipitated samples, the amount of superficial Fe may be less significant (Table 1). This assumption may be supported by the XRD analysis, which revealed that, in the case of the impregnated samples, the iron-containing phases are detected, even at a low concentration (i.e., from 20:1 Mg:Fe molar ratio on, Figure 1).

In the case of coprecipitated samples (Figure 1a), the main crystalline phase detected is ascribed to the MgO-periclase phase. There is no indication on the Fe intercalation in the MgO lattice, but a new crystalline phase arises by increasing the Fe content, and such a phase could be ascribed to Magnetite or Magnesioferrite; among the two, the latter is more likely formed since periclase diffraction peak intensities decrease, because the incorporation of trivalent Fe^{3+} cation in the periclase lattice is known to generate cationic defects and to produce a low crystalline degree [24–26]. Conversely, the samples prepared by impregnation (Figure 1b) show periclase as the main phase, with no evidences of mixed Mg/Fe phases, since only the Hematite crystalline phase was detected alongside the one of periclase. This confirms that addition of Fe by impregnation on the MgO does not influence periclase crystallinity.

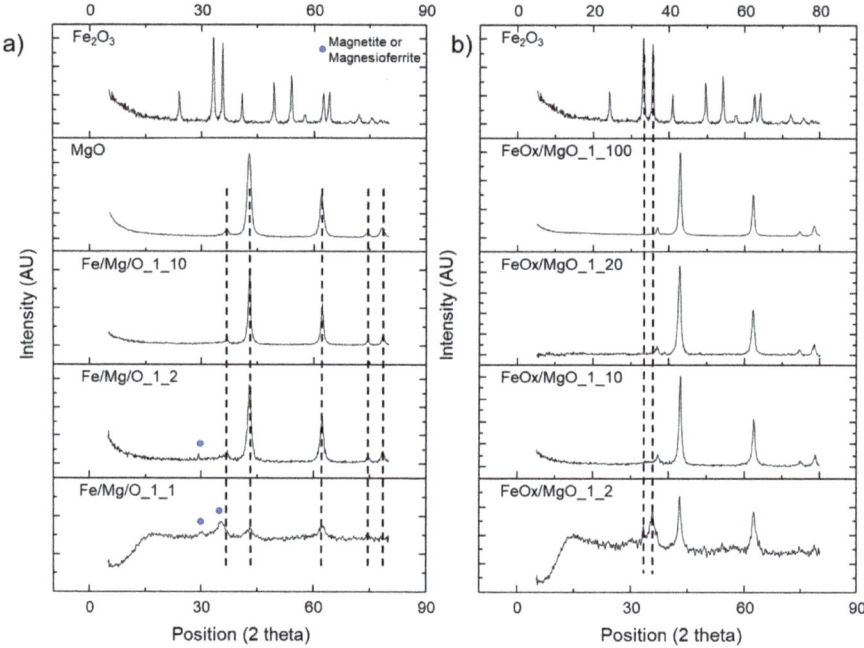

Figure 1. XRD patterns of (**a**) coprecipitated samples and (**b**) impregnated samples.

Fe addition is expected to introduce both acid and redox properties in the mixed oxide [23]. FTIR studies were carried out using pyridine as the chosen probe molecule. The pyridine adsorption desorption FTIR (Py-FTIR) spectra were recorded and reported in Figures S1 and S2 for the coprecipitated samples, and in Figures S3 and S4 for the impregnated ones; the relative acidity measurements are summarized in Table 1. In all cases, the absence of band at 1540 cm^{-1}, which is the characteristic band of Brønsted acid sites, and the presence of a band at 1444 cm^{-1}, corresponding to the adsorption of Pyridine at the Lewis acid sites, showed that only Lewis acid sites were present in Fe/Mg/O that showed acidity. The specific amounts of weak, medium, and strong basic sites are reported in Table 1. The weak sites are defined as the ones from which pyridine is removed by evacuation at 200 °C; the medium weak sites correspond to pyridine evacuation between 200 and 400 °C; and, finally, the strong sites correspond to pyridine evacuation above 400 °C [27]. Semi-quantitative evaluation of the surface acid sites was obtained by peak integration. The integration of the bands allowed for a few comparisons to be made to determine the effect of the Fe content and the effect of the synthesis method. The increase of the Fe amount led to an increase in total acidity and change in the distribution among different-strength acid sites. In the case of the impregnation method, the pyridine was adsorbed on the samples with Fe/Mg 1/10 molar ratio; in the case of the coprecipitation method, the adsorption of the probe molecule occurred only on the samples with an Fe/Mg of 1/1 and 1/2 molar ratio. Moreover, samples with the same Fe/Mg molar ratio, but prepared with different experimental methods, showed different characteristics; in particular, the coprecipitated ones displayed higher total acidity than the ones obtained by impregnation.

TPD analysis, using CO_2 as the probe molecule, was performed to determine basic properties of the samples. The results are presented in Table 1 and Figure 2. Based on the desorption temperature of CO_2, the adsorption is usually classified into three categories: weak adsorption (25–125 °C), medium adsorption (125–225 °C), and strong adsorption (>225 °C), which are assigned to surface hydroxyl groups, oxygen in Mg^{2+}-O^{2-} pairs, and low coordination oxygen anions, respectively [28].

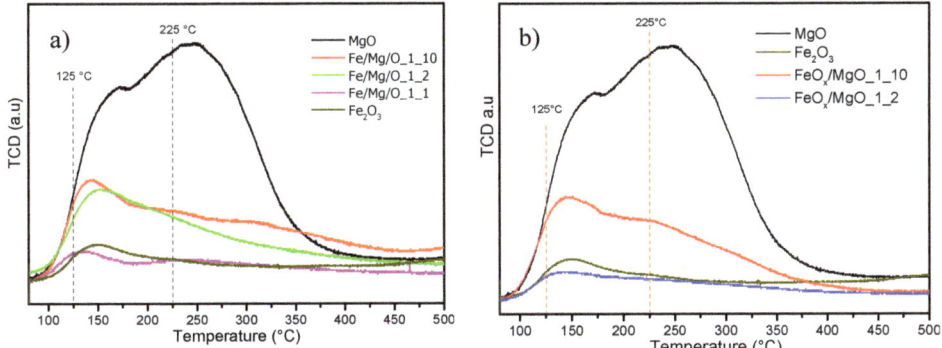

Figure 2. CO_2-TPD curves of: (a) coprecipitated Fe/Mg/O samples, MgO, and Fe_2O_3; and (b) impregnated FeO_x/MgO samples, MgO, and Fe_2O_3.

Catalysts with a higher Fe/Mg molar ratio showed lower basicity values because of the higher electronegativity of Fe^{3+} species, with respect to Mg^{2+} species. As a consequence, the charge density decreases and makes the O^{2-} less electrophilic than in pure MgO [29].

The higher basicity of the Fe/Mg 1/2 molar ratio coprecipitated sample with respect to the analogous impregnated one may be attributed to the different crystalline phases formed at such a high Fe concentration.

2.2. Reactivity Tests of Iron/Magnesium Oxide Catalysts in the Hydrodeoxygenation of Furfural

In our previous work, FU was converted to MFU, with methanol as the hydrogen source, through a tandem hydrogenation/hydrogenolysis sequence, using MgO and Mg/Fe/O_1_2 catalysts (Scheme 2) [22].

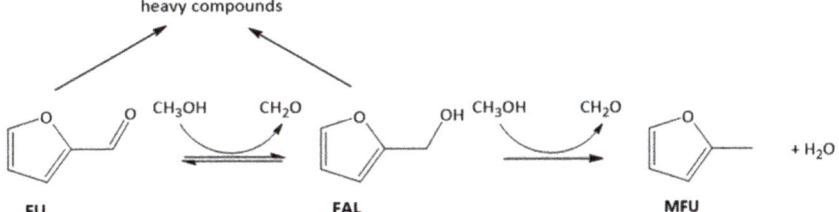

Scheme 2. Reaction pathways for furfural hydrodeoxygenation.

The catalytic systems in the reported operative conditions (optimized in a previous work [22]) were both active in FU conversion; however, their different chemical–physical properties led to different product selectivity. MgO was selective to FAL, while the mixed oxide produced preferentially MFU. We report hereafter a detailed study aiming at explaining the role of Fe in this change of selectivity.

The catalytic performances of the coprecipitated Fe/Mg/O catalysts described in Table 1 for furfuryl alcohol (FAL) and methyl furan (MFU) production from furfural (FU) are reported in Figure 3.

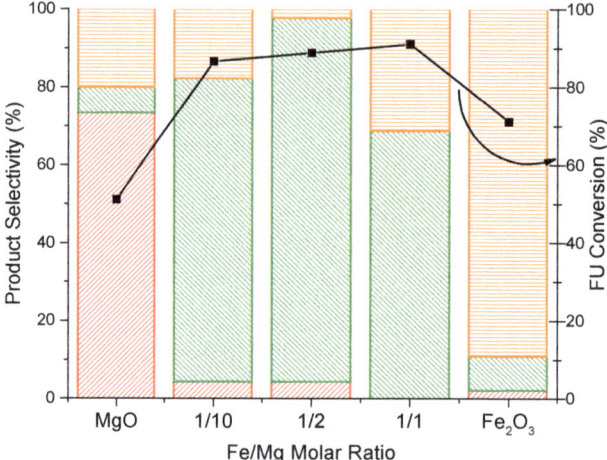

Figure 3. Profiles of coprecipitated Fe/Mg/O mixed oxide catalysts with different Fe content levels. Feed composition: 5% FU, 50% CH$_3$OH, 45% N$_2$, 1 atm, overall gas contact time 1.1 s, reaction time 1 h, 380 °C. Legend: FAL ▨ MFU ▧ Carbon loss ▭ FU conversion ■.

In the reaction with MgO, the H-transfer hydrogenation occurred with a high selectivity and moderate conversion toward FAL. However, under these experimental conditions, MgO exhibited limited activity in the further hydrogenolysis to MFU. The formation of some side products, consistent with the C-loss (20%), was also observed. Compared with pure MgO, the presence of Fe improved furfural conversion and methyl furan selectivity. Notably, when a low amount of iron (Fe/Mg = 1/10) was introduced, the selectivity of MFU was significantly increased from 5% to 80%, while furfural conversion increased from 52% to 88%. The maximum MFU selectivity (92%) was achieved when using Fe/Mg/O with Fe/Mg = 1/2 molar ratio. A further increase of the Fe content (Fe/Mg molar ratio 1/1) significantly decreased the MFU formation (70% selectivity) and led to a poor carbon balance, probably due to higher heavy compounds' deposition on the catalyst surface [22].

Therefore, Fe introduction favored the formation of the targeted MFU. FAL hydrogenolysis to form MFU was strongly influenced by the amount of Fe introduced in the catalyst and by the changes in its chemical–physical properties.

When the basicity density decreased from MgO (7.51 mmol/g) to Fe/Mg/O_1_2 (2.34 mmol/g), a clear enhancement in MFU yield was observed, indicating that MFU formation was not related to the presence of basic sites. Although the good selectivity showed by Fe/Mg/O_1_2 could be also ascribed to the presence of acidic sites, an excess in acidity will be detrimental for the selectivity toward MFU. Indeed, as suggested by the high conversion but poor selectivity obtained with pure iron oxide, a high level of acidity could promote side reactions (from FU and/or FAL).

Further catalytic tests in the same experimental reaction conditions were carried out using the impregnated samples (Figure 4). For this series of catalysts, the formation of MFU was enhanced by the addition of Fe; high selectivities (74%–88%) were observed with Fe/Mg molar ratios in the range of 1/20 to 1/2.

The catalytic results obtained with the two series of Fe/Mg/O samples show that the best performances were obtained with the Fe/Mg ratio of 1/2 and of 1/10, for the coprecipitated and impregnated samples, respectively. This suggests that the key parameter for the high selectivity to MFU is the interaction between Fe and the MgO surface. Indeed, in the coprecipitated samples, the Fe is likely present as Magnesioferrite, while, in the impregnated one, Fe_2O_3 does not change its phase in coexistence with the magnesium-based periclase phase.

In order to better understand the role of acid–base properties of these samples, the effect of acidity and basicity was studied in samples prepared by replacing (i) Fe with Al or (ii) MgO with SiO_2, respectively.

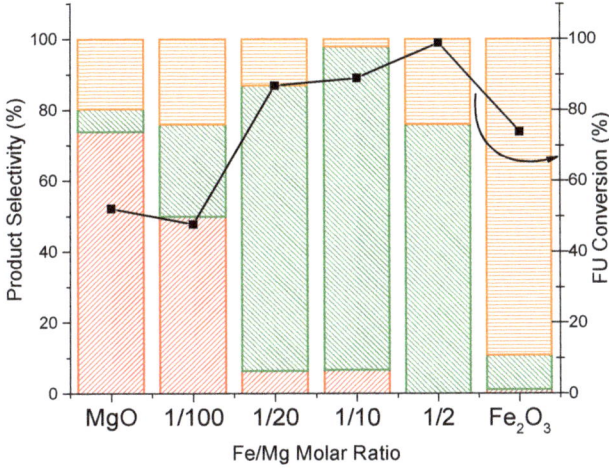

Figure 4. Reaction profiles of impregnated FeO_x/MgO mixed oxide catalysts with different Fe content levels. Feed composition: 5% FU, 50% CH_3OH, 45% N_2, 1 atm, overall gas contact time 1.1 s, reaction time 1 h, 380 °C. Legend: FAL ▨ MFU ▨ Carbon loss ▨ FU conversion ■.

2.3. Effect of the Properties of the Synthesized Catalysts on the Product Distribution of FU Reaction

In the literature, the hydrodeoxygenation capacity of a catalyst is generally associated with the presence of Lewis acid sites in the catalytic system; typical examples are niobium oxide [30] and zeolites [31], which were used in liquid phase. Indeed, it was reported that the acid functionalities catalyze the dehydration of alcohol to form intermediates, which will be substituted by surface hydrides. On metal oxides, it was reported that the electron-rich oxygen anions show basic properties and electron-donating character, while the electron-deficient metal cations show acidic character. Basic and hydrogen-abstracting properties of MgO can be modulated with the introduction of host cations, typically trivalent metal cations. While Fe/Mg/O catalysts exhibited both Lewis acid properties and a moderate to strong redox capacity, the Al/Mg/O system has no redox capacity [23]. In order to verify the contribution of the Lewis acid properties on the reaction network, Al^{3+} was chosen as a dopant metal to modify MgO. Indeed, Al^{3+} was reported to be a typical Lewis acid [32]. Therefore, in order to investigate the reaction pathways and product distribution influenced by Lewis acid properties, the catalytic behavior of the best catalysts for the two series (coprecipitated Fe/Mg/O_1_2 and impregnated FeO_x/MgO_1_10) were compared by substituting Fe with Al, leading to coprecipitated Al/Mg/O_1_2 and impregnated Al_2O_3/MgO_1_10. Such synthesized catalysts were studied using the same experimental conditions as for Fe/Mg/O. The main properties of Al-containing MgO samples are summarized in Table 2. It is worth noting that the big difference in surface area, between the impregnated and the coprecipitated samples, is due to the specific method of synthesis and composition. Both samples present only the periclase phase for MgO.

The acidity and basicity for these samples were analyzed, and the results are presented in Table 2. Al-containing samples showed a higher degree of acidity compared to the corresponding Fe/Mg/O samples, independently of the choice of the preparation method (see Table 1).

Table 2. Physicochemical properties (specific surface area, crystalline phase, Lewis acidity, Brønsted acidity and basicity) of different Al-containing MgO samples.

Catalyst	Surface Area (m^2·g^{-1})	Crystalline Phase (XRD)	Lewis Acidity (mmol·g^{-1}) [a]	Brønsted Acidity (mmol·g^{-1}) [a]	Basicity (mmol·g^{-1}) [b]
Al/Mg/O_1_2	132	Periclase MgO	0.80 [059 + 0.03 + 0.17] [c]	0	4.48
AlO$_x$/MgO_1_10	28	Periclase MgO	0.59 [0.40 + 0.13 + 0.06] [c]	0	2.54

[a] Quantification of Lewis and Brønsted acid sites was obtained from Pyridine-FTIR analysis. [b] Basicity measurements were performed by TPD analysis, using CO$_2$ as a probe molecule. [c] Distribution of the Lewis acid sites among weak, medium, and strong sites, respectively, based on the temperature at which pyridine desorption is observed; i.e., 20–200 °C (weak), 200–400 °C (medium), and >400 °C (strong).

Catalytic results are summarized in Table 3, where Al-based materials are compared to the analogous MgO and Fe-containing catalysts. Al/Mg/O_1_2 reached a conversion of 63% with a 37% of carbon loss and an MFU selectivity of 22%. The comparison between Al/Mg/O_1_2 and Fe/Mg/O_1_2 indicated that the former catalyst converted less FU and showed a greater carbon loss, probably due to the increased acidity. Indeed, Py-FTIR analysis (Tables 1 and 2) showed that the total amount of acid sites in Al/Mg/O_1_2 was twice as much as the one of Fe/Mg/O_1_2. This seems to indicate that the presence of Lewis acid sites is not the only feature leading to MFU formation. Moreover, the high acidity clearly increased by-product formation, as demonstrated by the higher carbon loss observed (37%).

Table 3. Summary of the catalytic performance over different Al, Fe doping MgO catalysts.

Catalyst	M/Mg Molar Ratio	FU Conv (%)	FAL Sel (%)	MFU Sel. (%)	C-Loss (%)
MgO	-	52	75	5	20
Al/Mg/O_1_2	1:2	63	41	22	37
Fe/Mg/O_1_2	1:2	93	5	92	3
AlO$_x$/MgO_1_10	1:10	40	76	5	19
FeO$_x$/MgO_1_10	1:10	89	5	88	7

Feed composition: 5% FU, 50% CH$_3$OH, 45% N$_2$, 1 atm, overall gas contact time 1.1 s, reaction time 1 h, 380 °C.

The same catalytic trend was observed for the impregnated catalysts. AlO$_x$/MgO_1_10 showed a catalytic performance similar to that of MgO. Comparison with FeO$_x$/MgO_1_10 was not relevant, since a very low amount of MFU was formed.

In order to illustrate the importance of basicity and its contribution in the reaction system, we prepared a catalyst containing iron oxide as the main component, but with a support that did not present any basicity, i.e., SiO$_2$, with a Fe/Si atomic ratio equal to 1/10.

The FeO$_x$/SiO$_2$ _1_10 catalyst showed low furfural conversion (19%), which led principally to decomposition products and low selectivity to MFU (25%) (Table 4). In comparison with Fe$_2$O$_3$, a similar product distribution was observed, but with a lower furfural conversion, due to the dilution of FeO$_x$ with SiO$_2$.

All these experiments led us to conclude that (i) FAL formation is related to surface basicity, (ii) Lewis acidity favors MFU formation—although it does not seem to be the key factor—(iii) the presence of Fe is crucial for MFU production, and (iv) an excess of acidity can enhance the formation of degradation products. For these reasons, it is evident that the precise choice of Fe/Mg molar ratio is crucial to obtain a high MFU selectivity. It is also important to notice that MFU is the main product in FeO$_x$-based catalysts, even at very low FU conversion; in other words, the almost exclusive formation of MFU is not due to the higher reactant conversion—which would imply the transformation of the intermediately formed FAL—but is strictly related to catalysts' surface properties.

Table 4. Summary of the catalytic performances of silica-supported iron oxide and comparison with bulk Fe_2O_3 and MgO.

Entry	Catalyst	Fe/Si Molar Ratio	FU Conv (%)	FAL Sel (%)	MFU Sel. (%)	C-Loss (%)
1	MgO	-	52	75	5	20
2	Fe_2O_3	-	73	2	10	88
3	SiO_2	-	0	0	0	0
4	$FeO_x/SiO_2\_1\_10$	1:10	19	1	25	74

Feed composition: 5% FU, 50% CH_3OH, 45% N_2, 1 atm, overall gas contact time 1.1 s, and reaction time 1 h, 380 °C.

2.4. Mechanistic Insights

In the previous section, it was shown that variations in Fe content for Fe/Mg/O catalysts led to very different product distributions. The key factor influencing the methyl furan selectivity was demonstrated to be the presence of FeO_x species. This section aims to provide some mechanistic insights to explain the observed selectivity trend through in situ DRIFT studies. When dealing with hydrogen transfer processes, two main cycles should be considered: (i) the activation of the hydrogen donor—methanol, in this study—and (ii) the activation of the substrate, furfural or furfuryl alcohol. In order to understand the reaction pathways, the three main components of reaction (methanol, furfural and furfuryl alcohol) were investigated separately on three different model catalytic surfaces, with different properties, as we explained in the previous sections, and based on our current catalytic observations. MgO, Fe_2O_3, and the mixed oxide obtained by the impregnation method, FeO_x/MgO_1_10. The IR spectra of the different catalysts were acquired at different temperatures in the range from 25 to 400 °C.

2.4.1. Methanol Adsorption

The spectra of methanol adsorbed on bulk MgO, bulk Fe_2O_3, and FeO_x/MgO 1/10, after outgassing at different temperatures, are shown in Figure 5a–c, respectively.

When methanol was adsorbed on MgO, at room temperature, two sets of peaks could be observed in the C–H region: one at 2942 and 2835 cm^{-1}, corresponding to physisorbed methanol, and the other at 2917 and 2800 cm^{-1}, assigned to mono-coordinated methoxy groups [33]. The $\nu_s(C=O)$ bands corresponding to those species could be found at 1058 and 1108 cm^{-1}, respectively [34]. When the temperature was increased from room temperature to 150 °C, new peaks appeared at 2809 and 1092 cm^{-1}, which were attributed to bridged methoxy species. Further increase of the temperature caused a rapid formation of a species, which could be attributed to a formate isomer, since it displayed a $\nu_s(CH)$ peak at 2846 cm^{-1}, a characteristic $\nu_{as}(COO)$ at 1600–1610 cm^{-1}, and the $\nu_s(COO)$ peaks in the 1379–1339 cm^{-1} region [35,36].

At 300 °C, only formate and bridged methoxy species could be observed, and at 380 °C (the reaction temperature), only formate was present on the MgO surface.

Exposure of α-Fe_2O_3 to methanol stream at room temperature gave rise to different IR absorption bands: similarly to what was observed in the case of MgO surface, the peaks at 2942, 2832 cm^{-1} were due to physical adsorbed methanol, and the ones at 2902, 2802, and 1071 cm^{-1} corresponded to the presence of methoxy species. At 300 °C, the methoxy bands completely disappeared and a substantial amount of formate was formed. This formate was activated in two different modes, with $\nu_s(CH)$ peaks at 2853 and 2804 cm^{-1}, $\nu_{as}(COO)$ at 1644 and 1611 cm^{-1}, and the $\nu_s(COO)$ around 1300 cm^{-1}. Outgassing at 380 °C caused the disappearance of the formate species (no CH vibrations observed), giving way to the presence of new species, presumably carbonates, suggested by the absence of C–H stretching bands at around 2800 cm^{-1} while still observing ν_s and $\nu_{as}(COO)$ bands.

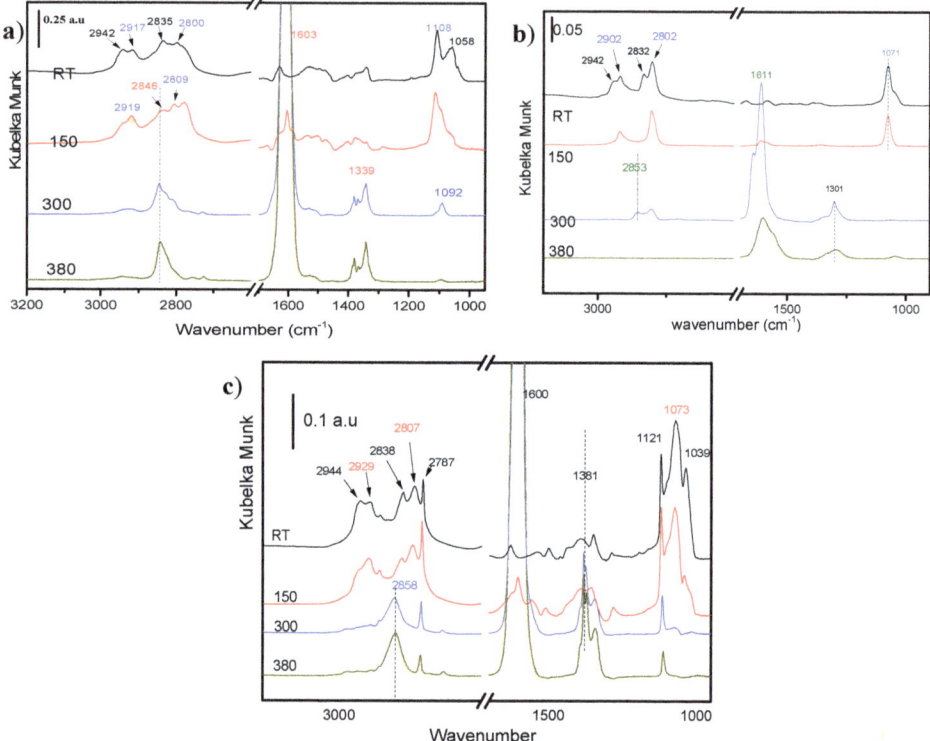

Figure 5. Methanol desorption spectra on (**a**) MgO, (**b**) Fe$_2$O$_3$, and (**c**) impregnated FeO$_x$/MgO_1_10 samples.

Similar to what was observed with the other two surfaces, at room temperature, only physisorbed methanol and methoxy species were present on the surface of the FeO$_x$/MgO_1_10 sample (C–H vibrations between 2944 and 2807 cm^{-1}, and C–O stretching at 1073 and 1039 cm^{-1}). When increasing the temperature, physisorbed methanol rapidly dissociated to give methoxy species, which, at 300 °C was already completely converted into formate (peaks 1600 and between 1385 and 1330 cm^{-1}).

From these series of experiments, it is possible to conclude that methanol adsorbs in a similar way on the three oxides, with the only differences being the presence of two modes of adsorption of the methoxy species in case of pure MgO.

2.4.2. Furfural and Furfuryl Alcohol Adsorption

Understanding the adsorption geometry of furfural and furfuryl alcohol on the catalytic surface is useful for the comprehension of the reaction network. From the current literature [17,20,37–45], it is possible to propose different coordination modes of the furfural with the surface: either through its aldehyde group or through the furan ring, as shown in Scheme 3.

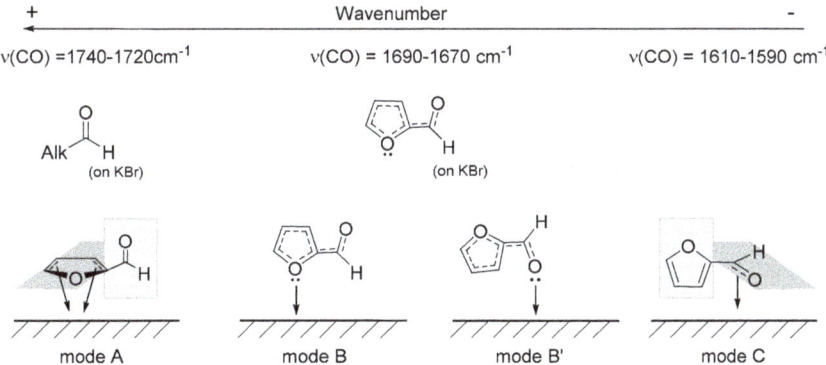

Scheme 3. Proposed rationalization of the literature observed ν(CO) in free furfural and physisorbed furfural, depending on the possible coordination modes.

When furfural was absorbed on MgO, at room temperature (Figure S5), three peaks arising from the C=O stretching could be observed: at 1720, 1672, and 1600 cm^{-1}, corresponding to the modes A, B, and C (Scheme 3, respectively). The peak at 1569 cm^{-1} arises from the C=C of the furan ring. As the temperature increased, the peaks at 1720 and 1672 cm^{-1} decreased, and at 380 °C only the η^2(C=O) mode of activation was present (peak at 1600 cm^{-1}). This suggests that, at this temperature, we have an optimum activation of the C=O bond, which might favor the formation of the furfuryl alcohol [18,46,47]. Furthermore, this experiment demonstrates that adsorbed furfural on MgO is stable even at high temperatures, as was already confirmed by Albonetti et al. [22].

The FTIR adsorption of furfural over Fe_2O_3 could not be performed due to the opacity of the obtained pellets; thus, the experiment was directly carried out with FeO_x/MgO_1_10 (see Figure S6). The obtained spectra were very similar to those of MgO, with the only exception that no peak at 1720 cm^{-1} (corresponding to the furan ring adsorption on the surface) was observed. This may suggest that the presence of FeO_x on the MgO causes a stronger interaction furfural-surface at lower temperatures.

At room temperature, the spectrum of adsorbed furfuryl alcohol over MgO was very similar with respect to the spectrum taken over KBr, meaning that very weak interactions were taking place. When the temperature increased, the furfuryl alcohol bands slowly decreased giving place to the appearance of a band at 1587 cm^{-1}, which, at 400 °C, overlapped with a band at 1602 cm^{-1}. This latter band is the same that was observed for the adsorption of furfural on MgO (see Figure S7). This result proves that, although high temperatures are required, furfuryl alcohol can be dehydrogenated on the MgO surface. Furthermore, it also shows the stability of these two compounds on MgO. A very similar spectrum was obtained when adsorbing furfuryl alcohol at room temperature on FeO_x/MgO_1_10. (Figure S8). However, the dehydrogenation of the alcohol was initiated at lower temperatures (starting at 300 °C) than in the case of MgO. At 380 °C, the band corresponding to furfuryl alcohol almost disappeared and was converted into furfural, as it can be followed by the band at 1505 cm^{-1}.

This section may be divided by subheadings. It should provide a concise and precise description of the experimental results, their interpretation, and the experimental conclusions that can be drawn.

3. Materials and Methods

3.1. Catalyst Preparation

Coprecipitated samples—The MgO and Fe/Mg/O mixed oxide catalysts were prepared, via the precipitation method, from an aqueous solution containing the corresponding metal nitrates $Mg(NO_3)_2 \cdot 6H_2O$ and $Fe(NO_3)_3 \cdot 9H_2O$ (Sigma-Aldrich, St. Louis, MO, USA) in different atomic ratios (procedure described in our previous work [24]). All samples were then dried at 120 °C for 2 h and then calcined in static air at 450 °C for 5 h. Samples were labeled as Fe/Mg/O_n_m, where n_m refers to the Fe-to-Mg molar ratio. The Al/Mg/O sample was prepared via the same experimental procedure, using Al(NO$_3$)$_3$·9H$_2$O (Sigma-Aldrich, St. Louis, MO, USA) as the precursor and named similarly. Bulk Fe$_2$O$_3$ and MgO were synthesized by precipitation of the corresponding nitrates.

Impregnated samples—FeO$_x$ supported over MgO samples were prepared by incipient wet impregnation, using Fe(NO$_3$)$_3$·9H$_2$O. The amount of nitrate required to obtain samples with a nominal Fe/Mg molar range of 1:100 to 1:10. All samples were then dried at 120 °C for 2 h and then calcined in static air at 450 °C for 5 h. Samples were labeled as FeO$_x$/MgO_n_m, where n_m refers to the Fe to Mg molar ratio. AlO$_x$/MgO and FeO$_x$/SiO$_2$ samples were prepared via the same experimental procedure, using Al(NO$_3$)$_3$·9H$_2$O or silica (W. R. Grace & Co.-Conn., Columbia, MD, USA) as support, and named accordingly.

3.2. Characterization of Catalysts

The BET specific surface area of each catalyst was determined by N$_2$ absorption–desorption at liquid N$_2$ temperature, using a Sorpty 1750 Fison instrument (Milan, Italy). Then, 0.3 g of the sample was typically used for the measurement; the sample was outgassed at 150 °C before N$_2$ absorption.

X-ray diffraction analyses (XRD) of the catalysts were recorded with Ni-filtered Cu Kα radiation (λ = 1.54178 Å) on a Philips X'Pert vertical diffractometer equipped with a pulse height analyzer and a secondary curved graphite–crystal monochromator.

Chemisorption experiments were carried out on a BELSORB-max from BEL JAPAN. Typically, 100 mg of catalyst was degassed at 450 °C for 3 h under a 50 mL·min^{-1} flow of pure helium. After cooling to 80 °C, CO$_2$ was adsorbed by flowing the catalysts under 50% CO$_2$-He gas mixture for 30 min (50 mL·min^{-1}), followed by He treatment at 80 °C for 15 min, to remove physisorbed molecules. The catalysts were then heated under He flow (50 mL·min^{-1}), up to 500 °C, at a heating rate of 10 °C·min^{-1}.

FTIR measurements were carried out in Perkin Elmer Spectrum spectrophotometer (Waltham, MA, USA), between 4000 and 400 cm^{-1}. Self-supported wafers of the samples containing around 35 mg (13 mm diameter) were evacuated at 10^{-5} mbar and 450 °C for 1 h. After cooling to room temperature, the spectrum was recorded and used as background for all subsequent spectra. The sample wafer was then exposed to pyridine vapors at room temperature for 30 min, until equilibrium was reached, and a second spectrum was recorded. Then, the wafer was subjected to evacuation for 10 min, and the spectrum was recorded and labeled as RT. Subsequent evacuations were performed at 100, 200, 300, and 400 °C for 10 min, followed by spectral acquisitions at room temperature [37,48–50].

3.3. Catalytic Tests

Catalytic tests were carried out in a continuous-flow fixed-bed micro-reactor (Pyrex, length 38 cm, internal diameter 1/3 inch), already used for the FU reduction [22,51]. The catalyst (30–60 mesh particles) was placed in the reactor in order to have the contact time equal to 1.1 s, and then it was heated to 380 °C under N$_2$ flow (26 mL·min^{-1}). The catalytic reaction was initiated by the vaporization of methanol and furfural (Sigma-Aldrich, St. Louis, MO, USA) in a 10/1 molar ratio, using the N$_2$ flow as the carrier gas (26 mL·min^{-1}). Furfural was purified via distillation prior to being fed into the flowing gas stream. The total volumetric flow rate through the catalytic bed was held constant at 60 mL·min^{-1}, and the concentration of furfural, methanol, and nitrogen were respectively 5%, 50%, and 45%. An analysis of reactants and products was carried out as follows: the outlet stream was scrubbed for 1 h, using a cold-trap glass tube (acetonitrile solution, which was maintained at −26 °C by a F32 Julabo Thermostat, Seelbach, Baden-Württemberg, Germany). The condensed products were analyzed by HPLC, using an Agilent Technologies 1260 Infinity instrument (Santa Clara, CA, USA) equipped with a DAD UV–Vis detector and an Agilent PORO shell 120 C-18 column. An external calibration method was used for the identification and quantification of reactants and products, using reference commercial samples.

4. Conclusions

The MFU production from FU was demonstrated to be strongly dependent on the nature and strength of the acidic sites coexisting on a basic support. The introduction of Fe as Lewis acid on MgO support led to the enhancement of the conversion of FU. The different methodologies employed to synthesize the catalysts and the different Fe/Mg ratios allowed to understand how the acidity is involved, not only in the FU activation, but also in its selective conversion to MFU. In fact, even if the simple impregnation of Fe_2O_3 on MgO led to satisfactory selectivity and activity, best results were obtained by coprecipitation, since this methodology led to the formation of Fe-containing mixed phases. Indeed, the presence of such highly dispersed phases was fundamental to modulate the distribution of acidic sites, which, in turn, allowed us to reach the targeted product (MFU) in high selectivity. Spectroscopic studies helped to understand the different activation modes of both methanol and furfural, indicating a possible reaction pathway.

Supplementary Materials: The following are available online at http://www.mdpi.com/2073-4344/9/11/895/s1. Figure S1: Pyridine-FTIR spectra of Fe/Mg/O_1_1 obtained after evacuation at different temperatures: (a) room temperature; (b) 100 °C; (c) 200 °C; (d) 300 °C; and (e) 400 °C. Figure S2: Pyridine-FTIR spectra of Fe/Mg/O_1_2 obtained after evacuation at different temperatures: (a) room temperature; (b) 100 °C; (c) 200 °C; (d) 300 °C; and (e) 400 °C. Figure S3: Pyridine-FTIR spectra of FeO$_x$/MgO_1_10 obtained after evacuation at different temperatures: (a) room temperature; (b) 100 °C; (c) 200 °C; (d) 300 °C; and (e) 400 °C. Figure S4: Pyridine-FTIR spectra of FeO$_x$/MgO_1_2 obtained after evacuation at different temperatures: (a) room temperature; (b) 100 °C; (c) 200 °C; (d) 300 °C; and (e) 400 °C. Figure S5: Furfural adsorption and desorption over MgO sample from RT to 400 °C. Figure S6: Furfural adsorption/desorption over FeO$_x$/MgO_1_10 sample from RT to 400 °C. Figure S7: FTIR spectra of Furfuryl alcohol adsorption/desorption over MgO at different temperature. Figure S8: Furfuryl alcohol adsorption/desorption over FeO$_x$/MgO_1_10 within different temperature.

Author Contributions: L.G., F.C., E.A.Q., and S.A. designed the different experiments and supported the interpretation of catalytic and catalyst characterization; L.G., D.B., and Y.Z. synthesized the catalysts and carried out catalytic evaluation and characterization of materials (XRD, TPD, and BET); Y.Z. and M.R.-C. carried out FTIR studies; C.L. supported FTIR interpretation; C.L., C.T., E.Q., N.D., and S.A. were involved in the writing and editing the manuscript.

Funding: This work was funded by SINCHEM Joint Doctorate Programme–Erasmus Mundus Action (framework agreement No. 2013-0037; specific grant agreement No. 2015-1600/001-001-EMJD).

Conflicts of Interest: The authors declare no conflict of interest.

References

1. Dusselier, M.; Sels, B.F. Selective Catalysis for Cellulose Conversion to Lactic Acid and Other α-Hydroxy Acids. In *Selective Catalysis for Renewable Feedstocks and Chemicals*; Nicholas, K.M., Ed.; Springer International Publishing: Cham, Switzerland, 2014; pp. 85–125.
2. Lolli, A.; Zhang, Y.; Basile, F.; Cavani, F.; Albonetti, S. Beyond H_2: Exploiting H-Transfer Reaction as a Tool for the Catalytic Reduction of Biomass. In *Chemicals and Fuels from Bio-Based Building Blocks*; Wiley-VCH Verlag GmbH & Co. KGaA: Weinheim, Germany, 2016; pp. 349–378.
3. Gilkey, M.J.; Xu, B. Heterogeneous Catalytic Transfer Hydrogenation as an Effective Pathway in Biomass Upgrading. *Acs Catal.* **2016**, *6*, 1420–1436. [CrossRef]
4. Scholz, D.; Aellig, C.; Hermans, I. Catalytic Transfer Hydrogenation/Hydrogenolysis for Reductive Upgrading of Furfural and 5-(Hydroxymethyl)furfural. *ChemSusChem* **2014**, *7*, 268–275. [CrossRef] [PubMed]
5. Sulmonetti, T.P.; Pang, S.H.; Claure, M.T.; Lee, S.; Cullen, D.A.; Agrawal, P.K.; Jones, C.W. Vapor phase hydrogenation of furfural over nickel mixed metal oxide catalysts derived from layered double hydroxides. *Appl. Catal. A Gen.* **2016**, *517*, 187–195. [CrossRef]
6. Manikandan, M.; Venugopal, A.K.; Nagpure, A.S.; Chilukuri, S.; Raja, T. Promotional effect of Fe on the performance of supported Cu catalyst for ambient pressure hydrogenation of furfural. *Rsc Adv.* **2016**, *6*, 3888–3898. [CrossRef]
7. Lee, J.; Burt, S.P.; Carrero, C.A.; Alba-Rubio, A.C.; Ro, I.; O'Neill, B.J.; Kim, H.J.; Jackson, D.H.K.; Kuech, T.F.; Hermans, I.; et al. Stabilizing cobalt catalysts for aqueous-phase reactions by strong metal-support interaction. *J. Catal.* **2015**, *330*, 19–27. [CrossRef]

8. Wang, C.; Xu, H.; Daniel, R.; Ghafourian, A.; Herreros, J.M.; Shuai, S.; Ma, X. Combustion characteristics and emissions of 2-methylfuran compared to 2,5-dimethylfuran, gasoline and ethanol in a DISI engine. *Fuel* **2013**, *103*, 200–211. [CrossRef]
9. Xu, Y.; Qiu, S.; Long, J.; Wang, C.; Chang, J.; Tan, J.; Liu, Q.; Ma, L.; Wang, T.; Zhang, Q. In situ hydrogenation of furfural with additives over a RANEY Ni catalyst. *Rsc Adv.* **2015**, *5*, 91190–91195. [CrossRef]
10. Villaverde, M.M.; Garetto, T.F.; Marchi, A.J. Liquid-phase transfer hydrogenation of furfural to furfuryl alcohol on Cu–Mg–Al catalysts. *Catal. Commun.* **2015**, *58*, 6–10. [CrossRef]
11. Gong, L.-H.; Cai, Y.-Y.; Li, X.-H.; Zhang, Y.-N.; Su, J.; Chen, J.-S. Room-temperature transfer hydrogenation and fast separation of unsaturated compounds over heterogeneous catalysts in an aqueous solution of formic acid. *Green Chem.* **2014**, *16*, 3746–3751. [CrossRef]
12. Panagiotopoulou, P.; Martin, N.; Vlachos, D.G. Effect of hydrogen donor on liquid phase catalytic transfer hydrogenation of furfural over a Ru/RuO$_2$/C catalyst. *J. Mol. Catal. A: Chem.* **2014**, *392*, 223–228. [CrossRef]
13. Pasini, T.; Lolli, A.; Albonetti, S.; Cavani, F.; Mella, M. Methanol as a clean and efficient H-transfer reactant for carbonyl reduction: Scope, limitations, and reaction mechanism. *J. Catal.* **2014**, *317*, 206–219. [CrossRef]
14. Mironenko, A.V.; Vlachos, D.G. Conjugation-Driven "Reverse Mars–van Krevelen"-Type Radical Mechanism for Low-Temperature C–O Bond Activation. *J. Am. Chem. Soc.* **2016**, *138*, 8104–8113. [CrossRef] [PubMed]
15. Panagiotopoulou, P.; Vlachos, D.G. Liquid phase catalytic transfer hydrogenation of furfural over a Ru/C catalyst. *Appl. Catal. A: Gen.* **2014**, *480*, 17–24. [CrossRef]
16. Gilkey, M.J.; Panagiotopoulou, P.; Mironenko, A.V.; Jenness, G.R.; Vlachos, D.G.; Xu, B. Mechanistic Insights into Metal Lewis Acid-Mediated Catalytic Transfer Hydrogenation of Furfural to 2-Methylfuran. *Acs Catal.* **2015**, *5*, 3988–3994. [CrossRef]
17. Sitthisa, S.; Sooknoi, T.; Ma, Y.; Balbuena, P.B.; Resasco, D.E. Kinetics and mechanism of hydrogenation of furfural on Cu/SiO$_2$ catalysts. *J. Catal.* **2011**, *277*, 1–13. [CrossRef]
18. Sitthisa, S.; Resasco, D.E. Hydrodeoxygenation of Furfural Over Supported Metal Catalysts: A Comparative Study of Cu, Pd and Ni. *Catal. Lett.* **2011**, *141*, 784–791. [CrossRef]
19. Sitthisa, S.; Pham, T.; Prasomsri, T.; Sooknoi, T.; Mallinson, R.G.; Resasco, D.E. Conversion of furfural and 2-methylpentanal on Pd/SiO$_2$ and Pd–Cu/SiO$_2$ catalysts. *J. Catal.* **2011**, *280*, 17–27. [CrossRef]
20. Sitthisa, S.; An, W.; Resasco, D.E. Selective conversion of furfural to methylfuran over silica-supported NiFe bimetallic catalysts. *J. Catal.* **2011**, *284*, 90–101. [CrossRef]
21. Sheng, H.; Lobo, R.F. Iron-Promotion of Silica-Supported Copper Catalysts for Furfural Hydrodeoxygenation. *ChemCatChem* **2016**, *8*, 3402–3408. [CrossRef]
22. Grazia, L.; Lolli, A.; Folco, F.; Zhang, Y.; Albonetti, S.; Cavani, F. Gas-phase cascade upgrading of furfural to 2-methylfuran using methanol as a H-transfer reactant and MgO based catalysts. *Catal. Sci. Technol.* **2016**, *6*, 4418–4427. [CrossRef]
23. Crocella, V.; Cerrato, G.; Magnacca, G.; Morterra, C.; Cavani, F.; Maselli, L.; Passeri, S. Gas-phase phenol methylation over Mg/Me/O (Me = Al, Cr, Fe) catalysts: mechanistic implications due to different acid-base and dehydrogenating properties. *Dalton Trans.* **2010**, *39*, 8527–8537. [CrossRef] [PubMed]
24. Valente, J.S.; Figueras, F.; Gravelle, M.; Kumbhar, P.; Lopez, J.; Besse, J.P. Basic Properties of the Mixed Oxides Obtained by Thermal Decomposition of Hydrotalcites Containing Different Metallic Compositions. *J. Catal.* **2000**, *189*, 370–381. [CrossRef]
25. Sato, T.; Wakabayashi, T.; Shimada, M. Adsorption of various anions by magnesium aluminum oxide of (Mg$_{0.7}$Al$_{0.3}$O$_{1.15}$). *Ind. Eng. Chem. Prod. Res. Dev.* **1986**, *25*, 89–92. [CrossRef]
26. Tichit, D.; Lhouty, M.H.; Guida, A.; Chiche, B.H.; Figueras, F.; Auroux, A.; Bartalini, D.; Garrone, E. Textural Properties and Catalytic Activity of Hydrotalcites. *J. Catal.* **1995**, *151*, 50–59. [CrossRef]
27. Zhang, Y.; Wang, J.; Ren, J.; Liu, X.; Li, X.; Xia, Y.; Lu, G.; Wang, Y. Mesoporous niobium phosphate: an excellent solid acid for the dehydration of fructose to 5-hydroxymethylfurfural in water. *Catal. Sci. Technol.* **2012**, *2*, 2485–2491. [CrossRef]
28. Wang, F.; Ta, N.; Shen, W. MgO nanosheets, nanodisks, and nanofibers for the Meerwein–Ponndorf–Verley reaction. *Appl. Catal. A Gen.* **2014**, *475*, 76–81. [CrossRef]
29. Ballarini, N.; Cavani, F.; Maselli, L.; Montaletti, A.; Passeri, S.; Scagliarini, D.; Flego, C.; Perego, C. The transformations involving methanol in the acid- and base-catalyzed gas-phase methylation of phenol. *J. Catal.* **2007**, *251*, 423–436. [CrossRef]

30. Shao, Y.; Xia, Q.; Liu, X.; Lu, G.; Wang, Y. Pd/Nb$_2$O$_5$/SiO$_2$ Catalyst for the Direct Hydrodeoxygenation of Biomass-Related Compounds to Liquid Alkanes under Mild Conditions. *ChemSusChem* **2015**, *8*, 1761–1767. [CrossRef]
31. Hong, D.-Y.; Miller, S.J.; Agrawal, P.K.; Jones, C.W. Hydrodeoxygenation and coupling of aqueous phenolics over bifunctional zeolite-supported metal catalysts. *Chem. Commun.* **2010**, *46*, 1038–1040. [CrossRef]
32. Di Cosimo, J.I.; Díez, V.K.; Xu, M.; Iglesia, E.; Apesteguía, C.R. Structure and Surface and Catalytic Properties of Mg-Al Basic Oxides. *J. Catal.* **1998**, *178*, 499–510. [CrossRef]
33. Routray, K.; Zhou, W.; Kiely, C.J.; Wachs, I.E. Catalysis Science of Methanol Oxidation over Iron Vanadate Catalysts: Nature of the Catalytic Active Sites. *Acs Catal.* **2011**, *1*, 54–66. [CrossRef]
34. Badri, A.; Binet, C.; Lavalley, J.-C. Use of methanol as an IR molecular probe to study the surface of polycrystalline ceria. *J. Chem. Soc. Faraday Trans.* **1997**, *93*, 1159–1168. [CrossRef]
35. Tabanelli, T.; Passeri, S.; Guidetti, S.; Cavani, F.; Lucarelli, C.; Cargnoni, F.; Mella, M. A cascade mechanism for a simple reaction: The gas-phase methylation of phenol with methanol. *J. Catal.* **2019**, *370*, 447–460. [CrossRef]
36. Tabanelli, T.; Cocchi, S.; Gumina, B.; Izzo, L.; Mella, M.; Passeri, S.; Cavani, F.; Lucarelli, C.; Schütz, J.; Bonrath, W.; et al. Mg/Ga mixed-oxide catalysts for phenol methylation: Outstanding performance in 2,4,6-trimethylphenol synthesis with co-feeding of water. *Appl. Catal. A Gen.* **2018**, *552*, 86–97. [CrossRef]
37. Lucarelli, C.; Galli, S.; Maspero, A.; Cimino, A.; Bandinelli, C.; Lolli, A.; Velasquez Ochoa, J.; Vaccari, A.; Cavani, F.; Albonetti, S. Adsorbent–Adsorbate Interactions in the Oxidation of HMF Catalyzed by Ni-Based MOFs: A DRIFT and FT-IR Insight. *J. Phys. Chem. C* **2016**, *120*, 15310–15321. [CrossRef]
38. Chandramohan, P.; Srinivasan, M.P.; Velmurugan, S.; Narasimhan, S.V. Cation distribution and particle size effect on Raman spectrum of CoFe$_2$O$_4$. *J. Solid State Chem.* **2011**, *184*, 89–96. [CrossRef]
39. Nowicka, E.; Hofmann, J.P.; Parker, S.F.; Sankar, M.; Lari, G.M.; Kondrat, S.A.; Knight, D.W.; Bethell, D.; Weckhuysen, B.M.; Hutchings, G.J. In situ spectroscopic investigation of oxidative dehydrogenation and disproportionation of benzyl alcohol. *Phys. Chem. Chem. Phys.* **2013**, *15*, 12147–12155. [CrossRef]
40. Villa, A.; Ferri, D.; Campisi, S.; Chan-Thaw, C.E.; Lu, Y.; Kröcher, O.; Prati, L. Operando Attenuated Total Reflectance FTIR Spectroscopy: Studies on the Different Selectivity Observed in Benzyl Alcohol Oxidation. *ChemCatChem* **2015**, *7*, 2534–2541. [CrossRef]
41. Shekhar, R.; Barteau, M.A.; Plank, R.V.; Vohs, J.M. Adsorption and Reaction of Aldehydes on Pd Surfaces. *J. Phys. Chem. B* **1997**, *101*, 7939–7951. [CrossRef]
42. Davis, J.L.; Barteau, M.A. Spectroscopic identification of alkoxide, aldehyde, and acyl intermediates in alcohol decomposition on Pd(111). *Surf. Sci.* **1990**, *235*, 235–248. [CrossRef]
43. Socrates, G. *Infrared and Raman Characteristic Group Frequencies: Tables and Charts*; John Wiley & Sons: Hoboken, NJ, USA, 2004.
44. Dimas-Rivera, G.L.; De la Rosa, J.R.; Lucio-Ortiz, C.J.; De los Reyes Heredia, J.A.; González, V.G.; Hernández, T. Desorption of Furfural from Bimetallic Pt-Fe Oxides/Alumina Catalysts. *Materials* **2014**, *7*, 527–541. [CrossRef] [PubMed]
45. Available online: https://sdbs.db.aist.go.jp/sdbs/cgi-bin/direct_frame_top.cgi (accessed on 27 October 2019).
46. Rogojerov, M.; Keresztury, G.; Jordanov, B. Vibrational spectra of partially oriented molecules having two conformers in nematic and isotropic solutions: furfural and 2-chlorobenzaldehyde. *Spectrochim. Acta Part A Mol. Biomol. Spectrosc.* **2005**, *61*, 1661–1670. [CrossRef] [PubMed]
47. Shen, J.; Wang, M.; Wu, Y.-n.; Li, F. Preparation of mesoporous carbon nanofibers from the electrospun poly(furfuryl alcohol)/poly(vinyl acetate)/silica composites. *Rsc Adv.* **2014**, *4*, 21089–21092. [CrossRef]
48. Cesari, C.; Mazzoni, R.; Matteucci, E.; Baschieri, A.; Sambri, L.; Mella, M.; Tagliabue, A.; Basile, F.L.; Lucarelli, C. Hydrogen Transfer Activation via Stabilization of Coordinatively Vacant Sites: Tuning Long-Range π-System Electronic Interaction between Ru(0) and NHC Pendants. *Organometallics* **2019**, *38*, 1041–1051. [CrossRef]
49. Albonetti, S.; Boanini, E.; Jiménez-Morales, I.; Lucarelli, C.; Mella, M.; Molinari, C.; Vaccari, A. Novel thiotolerant catalysts for the on-board partial dehydrogenation of jet fuels. *Rsc Adv.* **2016**, *6*, 48962–48972. [CrossRef]

50. Lucarelli, C.; Giugni, A.; Moroso, G.; Vaccari, A. FT-IR Investigation of Methoxy Substituted Benzenes Adsorbed on Solid Acid Catalysts. *J. Phys. Chem. C* **2012**, *116*, 21308–21317. [CrossRef]
51. Grazia, L.; Bonincontro, D.; Lolli, A.; Tabanelli, T.; Lucarelli, C.; Albonetti, S.; Cavani, F. Exploiting H-transfer as a tool for the catalytic reduction of bio-based building blocks: the gas-phase production of 2-methylfurfural using a FeVO$_4$ catalyst. *Green Chem.* **2017**, *19*, 4412–4422. [CrossRef]

© 2019 by the authors. Licensee MDPI, Basel, Switzerland. This article is an open access article distributed under the terms and conditions of the Creative Commons Attribution (CC BY) license (http://creativecommons.org/licenses/by/4.0/).

Article

One-Pot Catalytic Conversion of Cellobiose to Sorbitol over Nickel Phosphides Supported on MCM-41 and Al-MCM-41

Wipark Anutrasakda [1], Kanyanok Eiamsantipaisarn [2], Duangkamon Jiraroj [2], Apakorn Phasuk [2], Thawatchai Tuntulani [2], Haichao Liu [3] and Duangamol Nuntasri Tungasmita [4,*]

[1] Green Chemistry for Fine Chemical Productions STAR, Department of Chemistry, Faculty of Science, Chulalongkorn University, Payathai Road, Patumwan, Bangkok 10330, Thailand; wipark.a@chula.ac.th

[2] Department of Chemistry, Faculty of Science, Chulalongkorn University, Pathumwan, Bangkok 10330, Thailand; kanyanok.e@gmail.com (K.E.); jiraroj_d@hotmail.com (D.J.); apakorn.phasuk@gmail.com (A.P.); thawatchai.t@chula.ac.th (T.T.)

[3] College of Chemistry and Molecular Engineering, Peking University, Chengfu Road, Haidian, Beijing 100871, China; hcliu@pku.edu.cn

[4] Center of Excellence in Catalysis for Bioenergy and Renewable Chemicals (CBRC), Department of Chemistry, Faculty of Science, Chulalongkorn University, Pathumwan, Bangkok 10330, Thailand

* Correspondence: duangamol.n@chula.ac.th; Tel.: +66-2-218-7619

Received: 4 December 2018; Accepted: 15 January 2019; Published: 16 January 2019

Abstract: MCM-41- and Al-MCM-41-supported nickel phosphide nanomaterials were synthesized at two different initial molar ratios of Ni/P: 10:2 and 10:3 and were tested as heterogeneous catalysts for the one-pot conversion of cellobiose to sorbitol. The catalysts were characterized by X-ray diffractometer (XRD), N2 adsorption-desorption, scanning electron microscope (SEM), transmission electron microscope (TEM), ^{27}Al-magnetic angle spinning-nuclear magnetic resonance spectrometer (^{27}Al MAS-NMR), temperature programmed desorption of ammonia (NH_3-TPD), temperature-programmed reduction (H_2-TPR), and inductively coupled plasma optical emission spectrophotometer (ICP-OES). The characterization indicated that nickel phosphide nanoparticles were successfully incorporated into both supports without destroying their hexagonal framework structures, that the catalysts contained some or all of the following Ni-containing phases: Ni^0, Ni_3P, and $Ni_{12}P_5$, and that the types and relative amounts of Ni-containing phases present in each catalyst were largely determined by the initial molar ratio of Ni/P as well as the type of support used. For cellobiose conversion at 150 °C for 3 h under 4 MPa of H_2, all catalysts showed similarly high conversion of cellobiose (89.5–95.0%). Nevertheless, sorbitol yield was highly correlated to the relative amount of phases with higher content of phosphorus present in the catalysts, giving the following order of catalytic performance of the Ni-containing phases: $Ni_{12}P_5$ > Ni_3P > Ni. Increasing the reaction temperature from 150 °C to 180 °C also led to an improvement in sorbitol yield (from 43.5% to 87.8%).

Keywords: nickel phosphide; cellobiose; sorbitol; MCM-41; hydrolytic hydrogenation

1. Introduction

The use of biomass as a renewable feedstock to produce fuels and chemicals has gained much attention due to the depletion of fossil fuels and global warming [1,2]. The most abundant biomass is lignocellulose, which is composed of cellulose, hemicellulose, and lignin. The proportion of the three components may differ between plants, but cellulose generally makes up the largest proportion [3].

Cellulose is a non-edible saccharide polymer containing glucose units linked together via β-1,4-glycosidic bonds. The intra- and intermolecular hydrogen bondings are formed within and between the cellulose chains, respectively. Due to its highly rigid structure, high crystallinity index, and insolubility in common solvents including water, the conversion of cellulose to valuable chemicals has been a challenge [4]. A number of attempts have been made regarding the use of catalysts to convert cellulose or cellobiose, a model compound of cellulose, to renewable feedstocks such as glucose [5], methyl glucosides [6], 5-HMF [7], gluconic acid [8], ethylene glycol [9] and sorbitol [10].

Sorbitol is one of the most industrially important renewable feedstocks; it is the most used sugar alcohol and has wide applications in food, pharmaceutical and cosmetic industries [11]. For instance, sorbitol is used as low-calorie sweetener, a component in toothpaste, and a feedstock for the production of vitamin C. Typically, sorbitol is commercially produced by the one-step catalytic hydrogenation of glucose [11]. Additionally, since glucose may be hydrolyzed from cellulose or cellobiose, many researches have investigated the two-step catalytic conversion of cellulose or cellobiose to sorbitol under hydrolytic hydrogenation [12–16]. A catalytic hydrolytic hydrogenation system for the conversion of cellulose or cellobiose to sorbitol normally requires two components: (i) soluble acid, such as phosphoric acid, or solid acid support, such as zeolite and acid-functionalized silica, for the catalytic hydrolysis of cellulose/cellobiose to glucose and (ii) noble transition metals, such as platinum, ruthenium, and palladium for the catalytic hydrogenation of glucose to sorbitol.

Catalytic systems based on nickel, a non-noble metal, have also been found to be efficient and cost effective for hydrogenation processes to produce sugar-related compounds [17–19]. It is therefore of interest to develop nickel-based catalysts that contain both acidic and metallic functions in order to accomplish the one-pot hydrolytic hydrogenation for the conversion of cellulose/cellobiose to sorbitol. One way to achieve such bifunctional properties is to form a compound between nickel and phosphorus, namely nickel phosphide. The presence of phosphorus would play an important role in creating Brönsted acid sites from P–OH groups [20]. Nickel phosphide catalysts are otherwise well known for their good catalytic activity in processes such as hydrodesulfurization (HDS) and hydrodenitrogenation (HDN) [21,22]. Furthermore, due to the fact that different stoichiometric ratios of nickel to phosphorus can lead to the formation of different phases of nickel phosphides [23,24], it is also of interest to investigate the bifunctional catalytic properties of different phases of nickel phosphide in the one-pot conversion of cellulose/cellobiose to sorbitol. The use of nickel phosphide nanoparticles as heterogeneous catalyst without support, however, may result in poor catalytic activity due primarily to particle aggregation. Support materials that can help improve the nanoparticle dispersion and therefore circumvent this problem include the widely used mesoporous MCM-41 and Al-MCM-41 materials, owing to their facile synthesis, high specific surface area, uniform pore size distribution, and ordered mesoporous channels [25,26].

In the present study, we report the synthesis and full characterization of supported nickel phosphide catalysts prepared with two different ratios of nickel to phosphorus where MCM-41 and Al-MCM-41 were used as supports in order to enhance the dispersion of nickel phosphide nanoparticles. The synthesized catalysts were then tested for their bifunctional catalytic performance in the one-pot conversion of cellobiose to sorbitol. The catalytic performance of different phases of nickel phosphides was comparatively investigated. The effect of different supports on the formation of nickel phosphide phases and on the catalytic performance was also studied.

2. Results and Discussion

2.1. Characterization of the Catalysts

The low-angle XRD patterns ($2\theta = 1.5°–6.0°$) are shown in Figure 1a,b. At low angles, the XRD patterns of MCM-41 and Al-MCM-41 samples exhibited a sharp peak at $2\theta \approx 2.3°$, and two small peaks at $2\theta \approx 3.9°$ and $4.5°$, which could be indexed as the (100), (110), and (200) planes, respectively [25]. This finding indicates that both samples exhibited hexagonal structure with a high degree of structural ordering. After nickel phosphide doping, the XRD patterns of the M-xNiyP and Al-M-xNiyP samples exhibited similar characteristics to those of the undoped counterparts, indicating that the incorporation of nickel phosphide nanoparticles into the mesoporous structures of MCM-41 and Al-MCM-41 did not destroy the underlying framework structure of the supports. Nevertheless, the XRD peaks of the M-xNiyP and Al-M-xNiyP samples were less intense and slightly shifted to higher angles, suggesting a decrease in their lattice parameters with respect to those of the undoped counterparts.

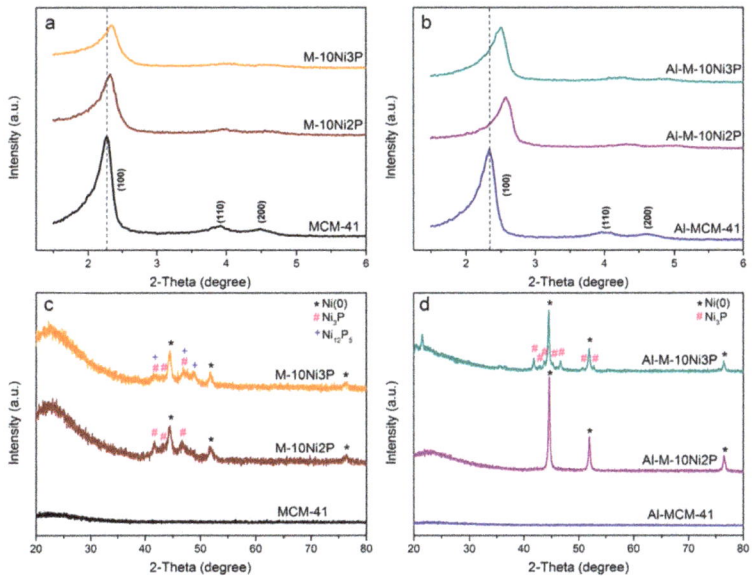

Figure 1. X-ray diffraction patterns of the synthesized materials at low angles (**a**–**b**) and high angles (**c**–**d**).

The high-angle XRD patterns (2θ = 20°–80°) of the modified materials are shown in Figure 1c,d. At high angles, a broad peak at 2θ ≈ 23° was observed for all M-xNiyP and Al-M-xNiyP samples, which could be attributed to amorphous silica. Furthermore, the XRD results indicate that various phases of nickel phosphides were present in the modified samples and that the initial Ni/P molar ratio played a significant role in determining the proportion of each nickel phosphide phase in each sample. In particular, when the initial Ni/P molar ratio of the modified samples in each series was decreased from 10:2 to 10:3, a new phase with lower relative contents of Ni was present. For instance, the XRD result of the M-10Ni2P sample showed two sets of peaks ascribable to the Ni^0 phase (JCPDS No. 89-7128; 2θ ≈ 44.3°, 51.7°, and 76.1°, corresponding to (111), (200), and (220) planes, respectively) and the Ni_3P phase (JCPDS No. 34-0501; 2θ ≈ 41.8°, 43.6°, and 46.6°, corresponding to (321), (112), and (141) planes, respectively). On the other hand, when the initial Ni/P molar ratio of the M-xNiyP series samples was decreased to 10:3, the $Ni_{12}P_5$ phase (JCPDS No. 22-1190; 2θ ≈ 41.6°, 47.0°, and 49.0°, corresponding to (400), (240), and (312) planes, respectively) was also observed in addition to the two aforementioned phases.

Further analysis of the XRD results indicates that the type of support also had an effect on the obtained phases of nickel phosphides. Specifically, for each given initial Ni/P molar ratio, the use of Al-MCM-41 as support material was associated with a lesser presence of nickel phosphide phases as compared to when MCM-41 was used as support. For instance, only the crystalline phase of Ni^0 was present in the XRD pattern of the Al-M-10Ni2P sample, while both Ni^0 and Ni_3P phases were present in the XRD pattern of the M-10Ni2P sample. These findings might be attributed to the relatively larger presence of unreduced phosphorus species such as $P_4O_{12}^{4TM}$ that remained on the surface of the Al-MCM-41 support [27], hindering the production of phosphorus-rich nickel phosphide species on this type of support.

The N_2 adsorption desorption isotherms of both pure and modified MCM-41 and Al-MCM-41 materials exhibited a type IV isotherm (Figure 2), which is a characteristic of mesoporous materials [28]. The textural properties of the prepared materials are reported in Table 1. The total BET surface areas of pure MCM-41 and Al-MCM-41 were 900 and 843 $m^2\ g^{-1}$, respectively, which were significantly larger than those of the nickel phosphide doped counterparts. In particular, the BET surface areas of M-xNiyP and Al-M-xNiyP materials were in the ranges of 577–771 and 622–643 $m^2\ g^{-1}$, respectively. This finding is consistent with the successful incorporation of nickel phosphide nanoparticles into both supports.

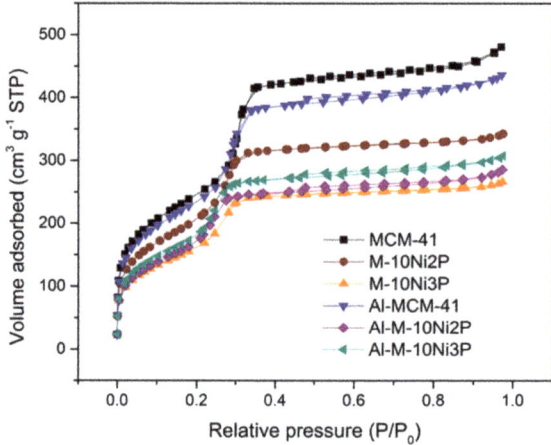

Figure 2. N_2 sorption isotherms of the synthesized materials.

Table 1. Textural properties of the synthesized materials.

Materials	BET Surface Area [a] ($m^2\ g^{-1}$)	Internal Surface Area [b] ($m^2\ g^{-1}$)	External Surface Area [b] ($m^2\ g^{-1}$)	Pore Volume [c] ($cm^3\ g^{-1}$)	Pore Diameter [c] (nm)
MCM-41	900	865	71	0.68	2.43
M-10Ni2P	771	747	27	0.43	2.43
M-10Ni3P	577	552	23	0.33	2.43
Al-MCM-41	843	767	84	0.59	2.43
Al-M-10Ni2P	622	570	48	0.35	2.43
Al-M-10Ni3P	643	616	62	0.39	2.43

[a] Calculated from BET method. [b] Calculated from t-plot method. [c] Calculated from BJH method.

T-plot calculation further revealed that, following the incorporation of nickel phosphide nanoparticles into either supports, the decrease in internal surface area of the materials was more pronounced as compared to the decrease in external surface area. This latter finding indicates that nickel phosphide nanoparticles were relatively well incorporated into the support mesopores. Nevertheless, pore blockage due to agglomerated nickel phosphide nanoparticles and other types of particles, such as unreduced $P_4O_{12}^{4\mathrm{TM}}$ species, could also occur [27]. The change in pore volume of the materials exhibited a similar trend to that of the total BET surface area, where the pore volumes of the nickel phosphide incorporated MCM-41 and Al-MCM-41 samples were significantly smaller than those of the undoped counterparts. On the other hand, all samples had similar average pore diameter of 2.43 nm. The latter finding indicates that the structure of the support materials was stable to the reduction process [29].

The morphology of two representative samples—MCM-41 and M-10Ni3P—was characterized by SEM and TEM, and the results are shown in Figure 3. The SEM results show that both MCM-41 and M-10Ni3P samples exhibited irregular rod shape with smooth surface (Figure 3a,b). Both samples were of similar size, ranging from 0.2–2.4 µm. The TEM image of MCM-41 (Figure 3c) illustrates a uniform array of mesoporous channels with a highly ordered hexagonal structure and a mean pore diameter of ~2.0 nm, which is the characteristic of MCM-41 [30] and is in good agreement with the results from the above N_2 adsorption-desorption analysis. When nickel phosphides were incorporated into the structure of MCM-41, the nanoparticles appeared to be well dispersed over the MCM-41 support with an average particle size of 5.6 ± 3.9 nm (Figure 3d). The fact that the vast majority of these nanoparticles were larger than the average size of the pores indicates that a considerable amount of nickel phosphide nanoparticles was responsible for pore blockage and were otherwise located on the external surface of MCM-41.

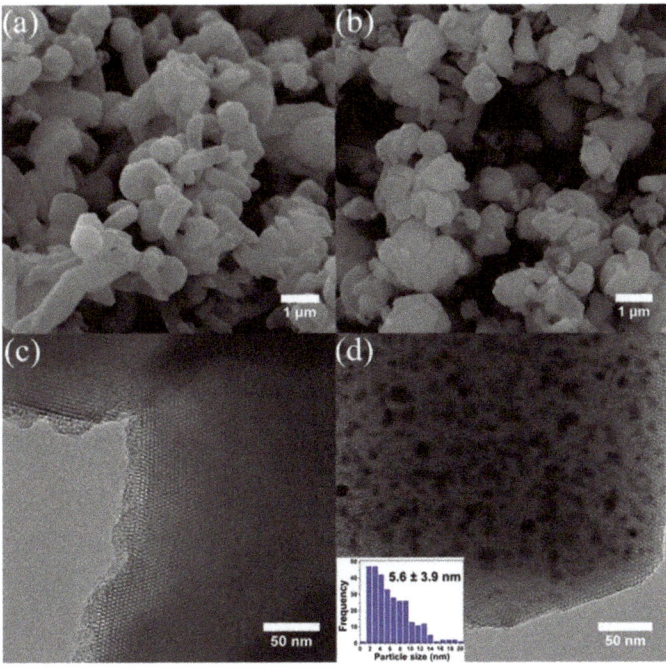

Figure 3. SEM images of MCM-41 (**a**) and M-10Ni3P (**b**); TEM images of MCM-41 (**c**) and M-10Ni3P (**d**).

Two representative samples containing Al: Al-MCM-41 and Al-M-10Ni3P were characterized by ^{27}Al MAS-NMR in order to determine the coordination environment of aluminum in the samples, and the results are shown in Figure 4. The spectrum of Al-MCM-41 exhibited one resonance at 50.7 ppm, indicating that tetrahedral framework aluminum was formed in the mesoporous walls of support [31] and therefore confirming that the post synthesis incorporation of aluminum ions into the MCM-41 structure was successful. The spectrum of the Al-M-10Ni3P showed two strong resonances due to aluminum at 53.2 and 2.1 ppm. These resonances indicate that the Al-M-10Ni3P sample contained both tetrahedral framework and octahedral non-framework aluminum, respectively [31]. In addition to the aforementioned resonances, a weak resonance at about 30–40 ppm was also observed, attributable to penta-coordinated aluminum [32]. The presence of penta- and hexa-coordination of aluminum in Al-M-10Ni3P might have resulted from the reduction process in which the temperature was raised to 750 °C since the use of such a high temperature can potentially cause the formation of these two coordination states of aluminum [33].

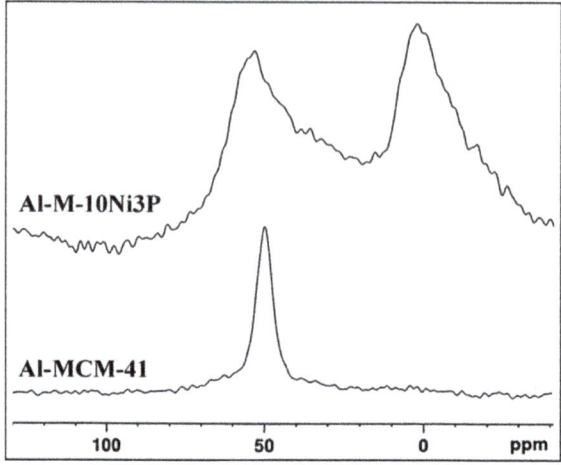

Figure 4. ^{27}Al MAS-NMR spectra of Al-MCM-41 and Al-M-10Ni3P materials.

The contents of nickel and phosphorus in the synthesized samples were determined by ICP-OES, and the results are reported in Table 2. The nickel and phosphorus contents were found to be lower than the corresponding initial amounts used in the synthesis. Specifically, the contents of nickel in all modified samples were in the range of 9.45–9.85 wt.%, a slight decrease (1.5–5.5%) from the initial amount used of 10 wt.%. As for the contents of phosphorus, the decrease was found to be more pronounced (8.5–14.0%). The relatively large decrease observed in the latter case could be attributed to the release of volatile phosphorus species, such as PH$_3$ during the reduction step [34]. Consequently, the resulting Ni/P ratios of all samples were higher than the corresponding initial ratios used in the synthesis.

Table 2. Elemental contents and total acidity of the synthesized materials.

Materials	Ni (wt.%) [a]	P (wt.%) [a]	Total Acidity (μmol g^{-1}) [b]
MCM-41	-	-	2
M-10Ni2P	9.45	1.80	85
M-10Ni3P	9.62	2.58	30
Al-MCM-41	-	-	6
Al-M-10Ni2P	9.59	1.83	125
Al-M-10Ni3P	9.85	2.62	53

[a] measured by ICP-OES [b] estimated from NH$_3$-TPD.

The total acidity of the synthesized samples and the strength of their acid sites were characterized using the NH$_3$-TPD technique. The NH$_3$-TPD profiles are shown in Figure 5. According to the NH$_3$-TPD profiles, the NH$_3$ desorption peaks of the undoped MCM-41 and Al-MCM-41 samples were not clearly observed. Nonetheless, the profiles indicate that the total acidity of the undoped Al-MCM-41 sample was higher than that of the undoped MCM-41 sample (Table 2). This finding is likely attributable to the formation of higher concentration of Brönsted acid sites during the substitution of Al for Si in the MCM-41 framework [33]. Each NH$_3$-TPD profile of the doped MCM-41 and Al-MCM-41 samples, on the other hand, exhibited at least one distinct desorption peak, indicating that nickel phosphide doping led to the increase of acidity. Specifically, the total acidity of the doped samples was in the range of 30–125 μmol g^{-1}, which is significantly higher than that of the undoped samples (2–6 μmol g^{-1}) (Table 2). The high acidity of the doped samples would lead to a higher degree of cellobiose hydrolysis, the details of which will later be discussed.

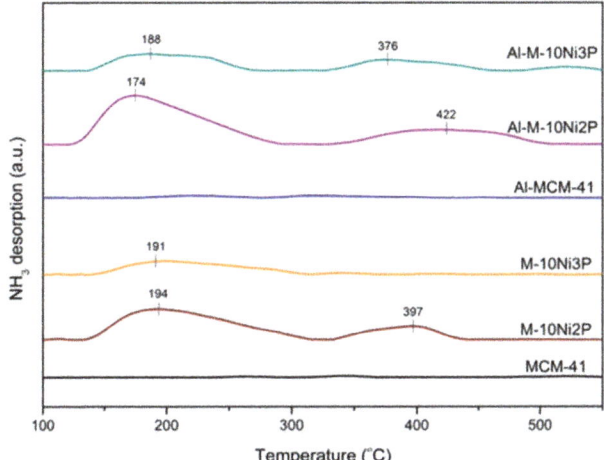

Figure 5. NH$_3$-TPD profiles of the synthesized materials.

As for the strength of the acid sites, the NH$_3$-TPD profiles present two distinct desorption peaks for the M-10Ni2P, Al-M-10Ni2P and Al-M-10N3P samples, suggesting that these doped materials contained acid sites of two different strengths. The lower temperature desorption peaks were observed in the temperature range of 174–194 °C, which can be assigned to the weaker Brönsted acid sites of P-OH groups in the unreduced phosphate species [20,35]. It can be observed that the lower temperature desorption peaks for the M-10Ni2P and Al-M-10Ni2P samples were more intense than those for the M-10Ni3P and Al-M-10Ni3P samples, respectively, even though a higher phosphorus content was initially used for the synthesis of the latter samples. This finding is possibly attributable to the larger presence of the remaining unreduced phosphate species in the M-10Ni2P and Al-M-10Ni2P samples as compared to the M-10Ni3P and Al-M-10Ni3P samples, respectively, resulting in the higher availability of Brönsted acid of the P-OH groups in the former samples. The higher temperature desorption peaks were observed in the temperature range of 376–422 °C and can be assigned to the stronger Lewis acid sites of the electron-deficient Ni$^{\delta+}$ ($0 < \delta < 1$) species [20,35].

The reducibility of two representative materials, MCM-41 and calcined M-10Ni3P precursor, was investigated by H$_2$-TPR technique. According to the H$_2$-TPR profiles (Figure 6), no obvious reduction peak was observed in the temperature range of 100–800 °C for the MCM-41 sample. This finding is consistent with the fact that MCM-41 is a nonreducible oxide [36]. On the other hand, several reduction peaks were observed for the calcined M-10Ni3P precursor sample. First, a minor broad peak at ~300–500 °C was observed, which can be assigned to the reduction of nickel oxide to nickel (0) [20]. Second, a number of overlapping reduction peaks were also observed over a temperature range of ~500–800 °C. In particular, peaks in the range of ~500–684 °C can be ascribed to the reduction of the highly stable P-O bonds in nickel phosphate. Also, within this temperature range, the nickel (0) species previously generated at lower temperatures can adsorb and dissociate hydrogen. The dissociated hydrogen species can subsequently spill over to the nickel phosphate and promote the formation of nickel phosphide species [34]. While it is to be noted that the reduction was not complete at the highest temperature of this study (800 °C), the peak observed in the temperature range above 720 °C can be attributed to the reduction of the excess of phosphate non-associated with nickel [34].

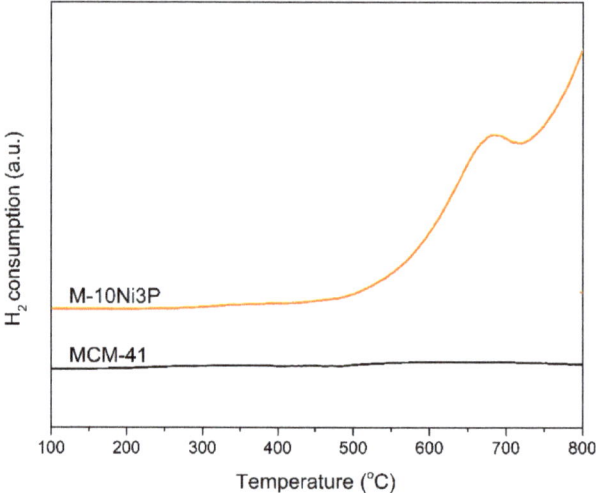

Figure 6. H_2-TPR profiles of MCM-41 and calcined M-10Ni3P precursor.

2.2. Catalytic Conversion of Cellobiose to Sorbitol

The MCM-41- and Al-MCM-41-supported nickel phosphide catalysts were tested for their catalytic performance in the one-pot conversion of cellobiose to sorbitol. In this catalytic reaction, cellobiose molecules had easy access to the active sites located both inside and outside the mesoporous channels of the catalysts. In particular, the calculation by HyperChem showed that the structural size of cellobiose in the samples (length and width of 1.04 and 0.65 nm, respectively) was smaller than the average pore size of the catalysts (2.43 nm).

To investigate the catalytic performance of the synthesized catalysts, the reaction was performed under the following conditions: 150 °C, 3 h, 4 MPa of H_2, 10 mL of 1% cellobiose solution, and 0.08 g of catalyst loading. The control experiments were also performed. First, a blank experiment where no catalyst was used gave 24.6% conversion of cellobiose, 5.8% yield of glucose, and no hexitol products. Second, experiments where the undoped MCM-41 and Al-MCM-41 materials were used as catalysts were also performed. The updoped MCM-41 and Al-MCM-41 materials gave, respectively, 54.8% and 71.9% conversion of cellobiose and 16.7% and 28.0% yield of glucose. The latter, the only product detected in these reactions, was produced from the hydrolysis of cellobiose catalyzed by acid sites on the surface of the undoped supports. The fact that higher conversion was achieved by Al-MCM-41 than by MCM-41 can primarily be attributed to the higher acid density of the former, which was confirmed by NH_3-TPD.

The catalytic performance of the nickel phosphide incorporated catalysts is shown in Figure 7. As compared to the undoped catalysts, the conversion of cellobiose increased to 89.5–95.0%. The fact that the conversion achieved by the M-xNiyP and Al-M-xNiyP catalysts was similarly high indicates that doping the supports with nickel phosphide nanoparticles significantly improved the efficiency of the acid-catalyzed hydrolysis of cellobiose under the reaction conditions used and that the density of acid sites in these catalysts (30–125 µmol g^{-1} from NH_3-TPD) was sufficiently high to convert most of the cellobiose in the reaction into glucose. The distribution of products obtained with the M-xNiyP and Al-M-xNiyP catalysts was also different from that obtained with the undoped counterparts. In particular, in addition to glucose, hexitols (i.e., sorbitol and mannitol) were also obtained.

Despite the above similarity in catalytic performance, the modified catalysts performed differently in terms of sorbitol yield, which can primarily be attributed to the difference in types and amounts of Ni-containing phases present in these catalysts. Specifically, the results indicate that nickel phosphide phases were more active for the hydrogenation of glucose to produce sorbitol than the Ni^0 phase.

For instance, Al-M-10Ni2P, which contained only the Ni^0 phase, gave a 18.6% yield of sorbitol while Al-M-10Ni3P, which contained both the Ni^0 and Ni_3P phases, gave a 25.8% yield. Further investigation confirmed that the Ni_3P species had higher catalytic performance in terms of sorbitol yield than the Ni^0 species, which is in agreement with a previous report [24]. In particular, M-10Ni2P, which had a higher Ni_3P/Ni^0 ratio than Al-M-10Ni3P, gave a higher yield of sorbitol than the latter (33.8% versus 25.8%) even though both catalysts contained the same types of Ni-containing species: Ni^0 and Ni_3P.

The Ni_3P phase, on the other hand, was outperformed by the $Ni_{12}P_5$ phase for the catalytic production of sorbitol from cellobiose. In particular, by comparing the catalytic performance of the M-10Ni2P and M-10Ni3P catalysts, where the latter contained the $Ni_{12}P_5$ phase in addition to the phases that were present in both catalysts: Ni^0 and Ni_3P, it was found that M-10Ni3P gave a 43.5% yield of sorbitol while M-10Ni2P gave only a 33.8% yield.

From the above experimental results, the following order of catalytic performance of the Ni-containing phases for the hydrolytic hydrogenation of cellobiose to sorbitol under the above reaction conditions can be established: $Ni_{12}P_5 > Ni_3P > Ni$. Specifically, sorbitol yield was positively related to the relative amount of phosphorus present in the Ni-containing phases. Additionally, it can be established that the higher Brønsted acidity provided by Al in the Al-MCM-41 supported catalysts did not significantly contribute to the catalytic performance in the conversion of cellobiose to sorbitol. A possible explanation is that, under the reaction conditions studied, the density of Brønsted acid in each catalyst of the M-xNiyP and Al-M-xNiyP series was already sufficiently high for the hydrolysis of cellobiose to glucose. The main factor determining the catalytic performance was therefore the composition of phases present in the catalysts, which affected the performance in terms of sorbitol yield. Moreover, while mannitol was also obtained from this reaction system, the sorbitol/mannitol ratio was high (≥ 10). This latter finding suggests that nickel/nickel phosphide-based catalysts are suitable for the selective one-pot conversion of cellobiose to sorbitol.

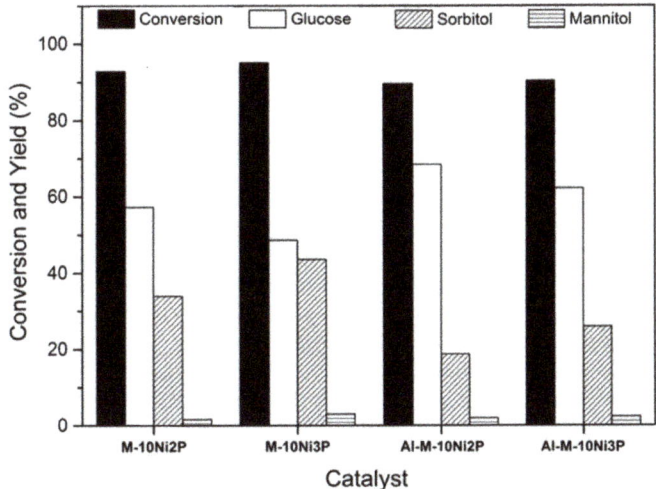

Figure 7. Catalytic performance of M-xNiyP and Al-M-xNiyP catalysts for the hydrolytic hydrogenation of cellobiose at 150 °C, 3 h, 4 MPa of H_2, 10 mL of 1% cellobiose solution, and 0.08 g of catalyst.

The effect of increasing the reaction temperature on the catalytic performance of the M-xNiyP and Al-M-xNiyP catalysts was also investigated. In particular, the reaction temperature was raised to 180 °C, while keeping other reaction conditions constant, and the results are reported in Figure 8. At 180 °C, all catalysts exhibited high cellobiose conversion (93.3–94.0%), which was close to that obtained at 150 °C (89.5–95.0%). Nonetheless, a marked improvement in catalytic performance was observed for the subsequent conversion of glucose to sorbitol via hydrogenation. Specifically, at 180 °C, the amount of glucose remained in the solution and the sorbitol yield significantly decreased and increased, respectively: less than 7% glucose remained (versus 48.6–68.4% at 150 °C) and 81.2–87.8% sorbitol yield was obtained (versus 18.6–43.5% at 150 °C). These findings indicate that temperature also played an important role in the hydrogenation of glucose to sorbitol over the nickel-based catalysts and therefore the performance of both Ni^0 and nickel phosphide phases in catalyzing the hydrolytic hydrogenation under the reaction conditions studied can be effectively optimized by controlling the reaction temperature.

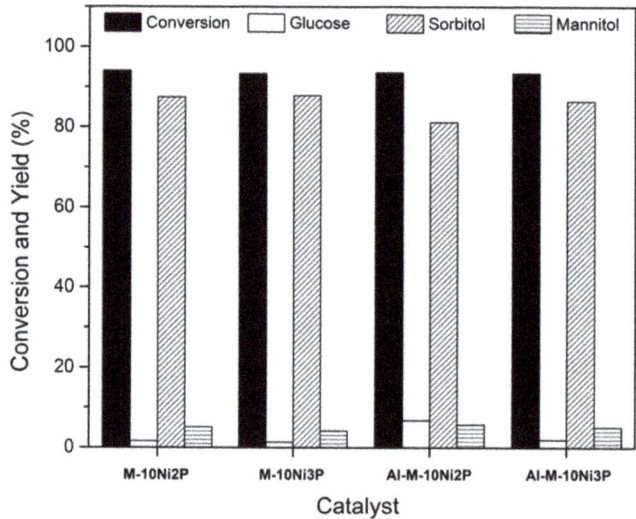

Figure 8. Catalytic performance of M-xNiyP and Al-M-xNiyP catalysts for the hydrolytic hydrogenation of cellobiose at 180 °C, 3 h, 4 MPa of H_2, 10 mL of 1% cellobiose solution, and 0.08 g of catalyst.

In order to identify the suitable temperature range for the two-step hydrolytic hydrogenation of cellobiose, the conversion was carried out under the aforementioned conditions but with reaction temperature ranging from 140–180 °C. M-10Ni3P, the best performing catalyst in this study, was used as a representative catalyst. The results are shown in Figure 9. It was found that the conversion of cellobiose, which was in the high range of 93.3–96.7%, was not significantly affected by reaction temperature. Nevertheless, reaction temperature had significant effect on product distribution. Notably, increasing the temperature from 140 °C to 170 °C led to a large decrease and increase in the contents of glucose and sorbitol, respectively. The effect of reaction temperature on product yields was, however, not linear. In particular, further increasing the temperature from 170 °C to 180 °C resulted only in small changes in glucose and sorbitol yields, which were 1.4% and 87.8% at 180 °C, respectively. Mannitol yield, like the conversion of cellobiose, was relatively unaffected by reaction temperature. Particularly, the yield was found to be about 4% or less across the range of temperature studied. Overall, the suitable reaction temperature for the two-step catalytic reaction was 170–180 °C. Under all reaction conditions employed in this study, the carbon balances for the hydrolytic hydrogenation of cellobiose using the M-xNiyP and Al-M-xNiyP catalysts were above 98.5%. This finding suggests that the occurrence of side reactions where the formation of degradation byproducts and/or that of coke took place were relatively small. It is to be noted that increasing the temperature beyond 180 °C might otherwise deteriorate the yield of sorbitol since its decomposition to form lower alcohols such as glycol and 1,2-propanediol can occur under H_2 atmosphere [14].

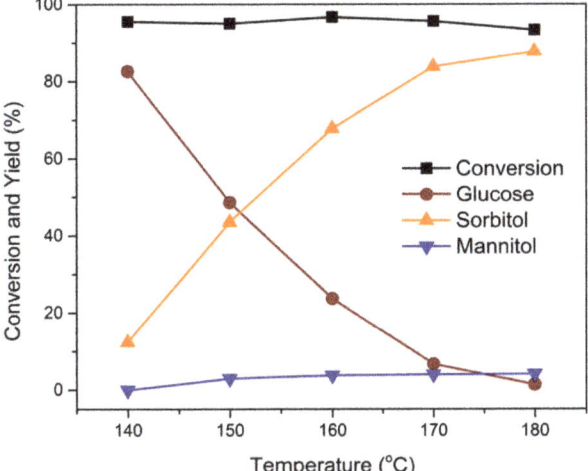

Figure 9. Effect of reaction temperature (140–180 °C) on cellobiose conversion and product distribution in the hydrolytic hydrogenation of cellobiose for 3 h using 4 MPa of H_2, 10 mL of 1% cellobiose solution, and 0.08 g of catalyst.

To investigate the catalyst reusability in the hydrolytic hydrogenation of celobiose, M-10Ni3P was used as the representative catalyst for the reaction under the following conditions: 180 °C, 3 h, 4 MPa of H_2, 10 mL of 1% cellobiose solution, and 0.08 g of catalyst loading. The result indicates that, after the first recycling run, the sorbitol yield significantly decreased from 87.8% to 64.8% despite the fact that the conversion of cellobiose only decreased from 93.3% to 91.7%. The overall decrease in catalyst performance can be attributed to the partial leaching of amorphous nickel phosphides at high temperature, which can occur following the transformation of crystalline nickel phosphides to their amorphous forms. An example of this type of leaching can be represented by the following equation [24]:

$$2Ni_{12}P_5 + 40H_2O \rightarrow 24Ni + 10H_3PO_4 + 25H_2$$

According to the equation, the leaching process leads to the change of a nickel phosphide phase to the less catalytically active Ni^0 phase. Consequently, the sorbitol yield decreased. It can therefore be expected that improving the high-temperature stability in aqueous reactions of the supported nickel phosphide-based catalysts will result in the improvement of overall catalytic performance.

2.3. Catalytic Mechanism

The mechanism that provided the bifunctionality of the nickel phosphide catalysts in the hydrolytic hydrogenation of cellobiose comprised primarily of i) the Brönsted acid sites on the surface of the supports (MCM-41 and Al-MCM-41) and those from the P–OH groups of the unreduced phosphate species and ii) the nickel sites of nickel phosphides. In particular, the Brönsted acid sites helped drive the hydrolysis of cellobiose by cleaving its β-1,4 glycosidic bonds (C-O-C) to produce glucose, while the nickel sites helped in the adsorption and dissociation of hydrogen through catalyzing the hydrogenation of glucose to hexitols (i.e., sorbitol and mannitol) [20,37] (Scheme 1). Additionally, the presence of phosphorus in nickel phosphides also induced the change in electron density of the nickel cations and therefore helped facilitate the hydrogen dissociation in the hydrogenation step [38].

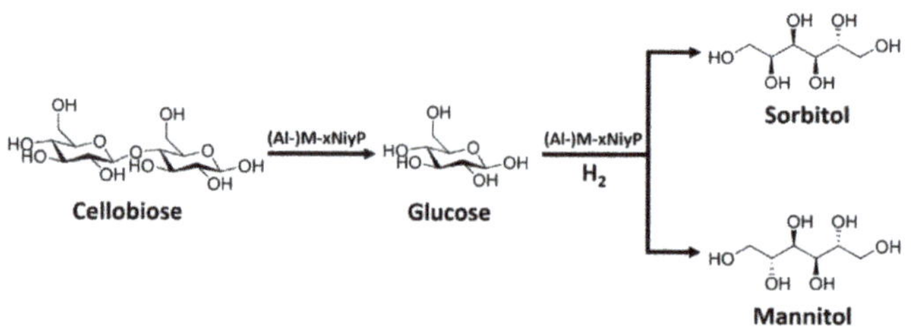

Scheme 1. Hydrolytic hydrogenation of cellobiose to produce hexitols.

In addition to the twoconsecutive reaction pathways mentioned above, the hydrolytic hydrogenation of cellobiose may also involve the hydrogenation of cellobiose to cellobitol followed by the hydrolysis of cellobitol to sorbitol [10,39]. However, this alternative pathway was not present under the reaction conditions used in this study, owing to the fact that cellobitol was not detected as an intermediate. As a matter of fact, the catalytic pathways for the hydrolytic hydrogenation of cellbiose can be affected by a number of factors and conditions including reaction temperature, catalyst loading, as well as the presence of mineral acids [39].

2.4. Comparison of Catalytic Performance of Various Metal-Based Catalysts

The catalytic performance in the one-pot conversion of cellobiose to sorbitol of the M-10Ni3P catalyst—the most efficient catalyst in this study—was compared to that of several other metal-based catalysts and the results are reported in Table 3. It was found that the M-10Ni3P catalyst showed similar or higher sorbitol yield as compared to other catalysts based on noble metals such as Pd, Pt, and Ru. Moreover, unlike the system described in this work, a number of other catalytic systems require the addition of a homogeneous acid solution in order to allow the conversion of cellobiose to sorbitol to proceed in one pot as well as to improve the yield of sorbitol.

Table 3. Comparison of catalytic performance of various catalysts for cellobiose conversion to sorbitol.

Catalyst	Time (h)	Temp. (°C)	Pressure of H_2 (MPa)	Cellobiose Conversion (%)	Sorbitol Yield (%)	Reference
M-10Ni3P	3	150	4	95.0	43.5	This work
M-10Ni3P	3	180	4	93.3	87.8	This work
Pd with pH 2	12	120	4	100	0	[13]
Pt with pH 2	12	120	4	100	18.5	[13]
Ru with pH 2	12	120	4	100	100	[13]
Ru/C	12	120	4	100	<1	[13]
Ru/Cs$_3$PW$_{12}$O$_{40}$	6	140	2	100	86	[40]
Ru/Cs$_2$HPW$_{12}$O$_{40}$	6	140	2	100	93	[40]
Ru/C + 0.05 wt.% H$_3$PO$_4$	1	185	3	n.r. [a]	87.1	[14]
3%RuNPs/Amberlyst 15	5	150	4	100	81.6	[15]

[a] n.r. = not reported.

3. Materials and Methods

3.1. Materials

Hexadecyltrimethylammonium bromide (CTAB), tetraethyl orthosilicate (TEOS), ammonia solution (25%), aluminium isopropoxide, hydrochloric acid and ethanol were purchased from Merck (Darmstadt, Germany). Toluene was purchased from Carlo Erba Reagents (Val de Reuil, France). Ammonium dihydrogen phosphate ($NH_4H_2PO_4$) and nitric acid were purchased from Shantou Xilong Chemical Co., Ltd. (Guangdong, China). Nickel (II) nitrate hexahydrate was purchased from Beijing Yili Chemical Co., Ltd. (Beijing, China). Cellobiose was purchased from J&K Chemical Co., Ltd. (Shanghai, China). Glucose, sorbitol, and mannitol were purchased from Alfa Aesar (Ward Hill, MA, USA). All chemicals were analytical grade reagents and were used without further purification.

3.2. Synthesis of Nickel Phosphide Nanoparticles Supported on MCM-41 and Al-MCM-41

Mesoporous MCM-41 was synthesized by the following procedure: First, 2.4 g of CTAB was dissolved in 120 g of deionized water. Then, 10.24 mL of ammonia solution was added to the above solution and the mixture was stirred for 5 min at room temperature. After that, 10 mL of TEOS was added dropwise to the resulting solution under stirring. The mixture was further stirred at room temperature for 24 h and the resulting product was filtered and washed with ethanol and deionized water until the pH of the wash solution was 7. The solid product was then dried at 60 °C for 12 h followed by calcination at 550 °C for 5 h to obtain MCM-41.

Al-MCM-41 was synthesized by adding 1.0 g of MCM-41 and a required amount of aluminum isopropoxide in 100 mL of toluene such that the Si/Al molar ratio was 14.1. The mixture was then stirred at room temperature for 12 h. The solid product was filtered and subsequently dried at 100 °C for 12 h, followed by calcination at 550 °C for 5 h to obtain Al-MCM-41.

The syntheses of MCM-41 and Al-MCM-41 supported nickel phosphide catalysts were modified from a previous report [24] where two different initial Ni/P ratios were used: a constant nickel loading of 10 wt.% and varied phosphorus loadings of 2 wt.% and 3 wt.%. Briefly, the required amounts of nickel (II) nitrate hexahydrate and dihydrogen phosphate were dissolved in 3 mL of deionized water. A small amount of concentrated HNO_3 was added to the mixture as required to obtain a homogeneous solution. The solution was then sonicated for 5 min, followed by the addition of 0.5 g of MCM-41 or Al-MCM-41. The mixture was subsequently stirred at room temperature for 16 h and dried at 60 °C for 24 h. The resulting solid was then calcined at 550 °C for 4 h prior to being reduced in H_2 atmosphere at 750 °C for 2 h. Finally, the product was cooled to room temperature and passivated in a 0.5% O_2/Ar flow for 3 h. The resulting MCM-41 and Al-MCM-41 supported nickel phosphide catalysts are denoted as M-xNiyP and Al-M-xNiyP, respectively, where x and y represent the initial wt.% loadings of nickel and phosphorus, respectively.

3.3. Materials Characterization

The crystalline phases of the prepared materials were identified by X-ray diffractometer (XRD) (Rigaku D/MaX-2200 Ultima-plus, Tokyo, Japan) with a Cu Kα X-ray source of wavelength 1.5418 Å operated at 40 kV and 30 mA. Low- and high-angle XRD patterns were recorded in the ranges of 1.5° to 6.0° and 20.0° to 80.0°, respectively, with a scanning rate of 5° min^{-1}. The textural properties of the materials were analyzed by N_2 adsorption-desorption with a BELSORP-mini-II apparatus (BEL Inc., Osaka, Japan). The surface area was calculated according to the Brunauer–Emmett–Teller (BET) method and the pore volume and pore size were calculated according to the Barrett–Joyner–Halenda (BJH) method. The position of aluminum ions in the lattice was determined by ^{27}Al-magnetic angle spinning-nuclear magnetic resonance spectrometer (^{27}Al MAS-NMR, Bruker Avance DPX 300 MHz, Karlsruhe, Germany) operated at 78 MHz. The morphology and particle size of the materials were analyzed by transmission electron microscope (TEM, JEOL JEM-2100, Peabody, MA, USA) and scanning electron microscope (SEM, JSM-5410 LV, Peabody, MA, USA). The acidity of the synthesized materials was measured by temperature programmed desorption of ammonia (NH_3-TPD) using a Micromeritics AutoChem II 2920 chemisorption analyzer. Before acidity measurement, each sample was pretreated under helium flow (50 mL min^{-1}) at 500 °C for 5 h. Each sample was then saturated with 10 vol.% NH_3 (10 mL min^{-1}) and the weakly adsorbed NH_3 was removed. After that, each sample was heated from 50 °C to 550 °C at a rate of 10 °C min^{-1}. The reducibility of the catalyst precursors was measured by temperature-programmed reduction (H_2-TPR) using a Micrometrics Chemisorbs 2750 automatic system. In the procedure, 0.1 g of each sample was carried out in a quartz U-tube reactor. The reduction was conducted in 10%H_2 in Ar with a flow rate of 25 mL min^{-1} and a heating rate of 10 °C min^{-1}. The consumption of hydrogen was determined by a thermal conductivity detector (TCD). Nickel and phosphorus contents in the materials were measured by inductively coupled plasma optical emission spectrophotometer (ICP-OES) (Perkin Elmer Optima 2100, Waltham, MA, USA) after acid digestion of the materials.

3.4. Catalytic Cellobiose Conversion and Product Analysis

The conversion of cellobiose was performed in a Teflon-lined stainless-steel autoclave. Unless otherwise stated, the reaction was performed as follows: 10 mL of 1% cellobiose solution and 0.08 g of catalyst were added into the reactor. The reactor was purged four times with H_2 and pressurized with 4 MPa of H_2 at room temperature. The catalytic reaction was then performed at a desired temperature for 3 h with a stirring rate of 750 rpm. The autoclave was subsequently cooled down to room temperature, and the solid catalyst was separated by vacuum filtration. The liquid products were analyzed using a high-performance liquid chromatography (HPLC) (Shimadzu LC-20AD, Kyoto, Japan) connected with refractive index detector using deionized water as mobile phase with the flow rate of 60 mL/min and the oven temperature of 70.0 °C. For the quantification of the remaining cellobiose and the products present in the reaction mixture, external standard calibrations of pure compounds were performed under the same conditions as those used for the analysis of liquid products. The cellobiose conversion and product yield were calculated as follows:

$$\text{Conversion (\%)} = \frac{(\text{mol}_{\text{cellobiose before reaction}} - \text{mol}_{\text{cellobiose after reaction}})}{\text{mol}_{\text{cellobiose before reaction}}} \times 100$$

$$\text{Yield (\%)} = \frac{\text{mol}_{\text{carbon in each product}}}{\text{mol}_{\text{carbon in cellobiose before reaction}}} \times 100$$

4. Conclusions

The synthesized MCM-41- and Al-MCM-4-supported nickel phosphide nanomaterials were highly efficient bifunctional heterogeneous catalysts for the one-pot conversion of cellobiose to sorbitol. The bifunctionality was achieved by combining acid sites originating mainly from P-OH groups of the unreduced phosphate species and metallic sites created by Ni species, which participated in catalyzing the hydrolysis of cellobiose to glucose and the hydrogenation of glucose to sorbitol, respectively. Under the reaction conditions of 150 °C, 3 h, and 4 MPa of H_2, the acidity of the catalysts was sufficient for catalyzing the hydrolysis process where high cellobiose conversion of 89.5–95.0% was obtained. Nonetheless, the catalytic performance in the hydrogenation process varied greatly with the presence of Ni-containing species in the catalysts, where sorbitol yield was positively related to the relative abundance of phosphorus-rich species: $Ni_{12}P_5 > Ni_3P > Ni$. Sorbitol yield was also found to depend on the reaction temperature, where the yield produced by the best catalyst in this study more than doubled (from 43.5 to 87.8%), without significant formation of undesired products, when the temperature was increased from 150 to 180 °C. Due to the environmental friendliness, cost-effectiveness and high efficiency offered by the bifunctional heterogeneous catalytic system described in this work, it has wide potential applications in catalysis and proved to be a promising candidate to replace catalytic systems based on noble metals.

Author Contributions: Conceptualization, W.A., H.L. and D.N.T.; methodology, W.A., K.E., D.J., A.P., and D.N.T.; investigation, W.A., K.E., D.J., and A.P.; writing—original draft preparation, W.A., K.E., and D.J.; writing—review and editing, T.T., H.L., and D.N.T.

Funding: This research was funded by the CU-56-912-AM, International Research Integration: Chula Research Scholar, Ratchadaphiseksomphot Endowment Fund, the Post-doctoral Fellowship, and the Research Grant for New scholar CU Researcher's Project from the Ratchadaphiseksomphot Endowment Fund of Chulalongkorn University.

Conflicts of Interest: The authors declare no conflicts of interest.

References

1. Yabushita, M.; Kobayashi, H.; Fukuoka, A. Catalytic transformation of cellulose into platform chemicals. *Appl. Catal. B* **2014**, *145*, 1–9. [CrossRef]
2. Besson, M.; Gallezot, P.; Pinel, C. Conversion of biomass into chemicals over metal catalysts. *Chem. Rev.* **2014**, *114*, 1827–1870. [CrossRef] [PubMed]

3. Kobayashi, H.; Fukuoka, A. Synthesis and utilisation of sugar compounds derived from lignocellulosic biomass. *Green Chem.* **2013**, *15*, 1740–1763. [CrossRef]
4. Dhepe, P.L.; Fukuoka, A. Cellulose conversion under heterogeneous catalysis. *ChemSusChem* **2008**, *1*, 969–975. [CrossRef] [PubMed]
5. Rinaldi, R.; Palkovits, R.; Schüth, F. Depolymerization of cellulose using solid catalysts in ionic liquids. *Angew. Chem. Int. Ed.* **2008**, *47*, 8047–8050. [CrossRef] [PubMed]
6. Xue, L.; Cheng, K.; Zhang, H.; Deng, W.; Zhang, Q.; Wang, Y. Mesoporous h-zsm-5 as an efficient catalyst for conversions of cellulose and cellobiose into methyl glucosides in methanol. *Catal. Today* **2016**, *274*, 60–66. [CrossRef]
7. Hsu, W.-H.; Lee, Y.-Y.; Peng, W.-H.; Wu, K.C.W. Cellulosic conversion in ionic liquids (ils): Effects of h2o/cellulose molar ratios, temperatures, times, and different ils on the production of monosaccharides and 5-hydroxymethylfurfural (hmf). *Catal. Today* **2011**, *174*, 65–69. [CrossRef]
8. An, D.; Ye, A.; Deng, W.; Zhang, Q.; Wang, Y. Selective conversion of cellobiose and cellulose into gluconic acid in water in the presence of oxygen, catalyzed by polyoxometalate-supported gold nanoparticles. *Chem. Eur. J.* **2012**, *18*, 2938–2947. [CrossRef]
9. Zhang, Y.; Wang, A.; Zhang, T. A new 3d mesoporous carbon replicated from commercial silica as a catalyst support for direct conversion of cellulose into ethylene glycol. *Chem. Commun.* **2010**, *46*, 862–864. [CrossRef]
10. Deng, W.; Liu, M.; Tan, X.; Zhang, Q.; Wang, Y. Conversion of cellobiose into sorbitol in neutral water medium over carbon nanotube-supported ruthenium catalysts. *J. Catal.* **2010**, *271*, 22–32. [CrossRef]
11. Zhang, J.; Li, J.-B.; Wu, S.-B.; Liu, Y. Advances in the catalytic production and utilization of sorbitol. *Ind. Eng. Chem. Res.* **2013**, *52*, 11799–11815. [CrossRef]
12. Fukuoka, A.; Dhepe, P.L. Catalytic conversion of cellulose into sugar alcohols. *Angew. Chem. Int. Ed. Engl.* **2006**, *45*, 5161–5163. [CrossRef] [PubMed]
13. Yan, N.; Zhao, C.; Luo, C.; Dyson, P.J.; Liu, H.; Kou, Y. One-step conversion of cellobiose to c6-alcohols using a ruthenium nanocluster catalyst. *J. Am. Chem. Soc.* **2006**, *128*, 8714–8715. [CrossRef] [PubMed]
14. Zhang, J.; Wu, S.; Li, B.; Zhang, H. Direct conversion of cellobiose into sorbitol and catalyst deactivation mechanism. *Catal. Commun.* **2012**, *29*, 180–184. [CrossRef]
15. Almeida, J.M.A.R.; Da Vià, L.; Demma Carà, P.; Carvalho, Y.; Romano, P.N.; Peña, J.A.O.; Smith, L.; Sousa-Aguiar, E.F.; Lopez-Sanchez, J.A. Screening of mono- and bi-functional catalysts for the one-pot conversion of cellobiose into sorbitol. *Catal. Today* **2017**, *279*, 187–193. [CrossRef]
16. Geboers, J.; Van de Vyver, S.; Carpentier, K.; Jacobs, P.; Sels, B. Efficient hydrolytic hydrogenation of cellulose in the presence of ru-loaded zeolites and trace amounts of mineral acid. *Chem. Commun. (Camb.)* **2011**, *47*, 5590–5592. [CrossRef]
17. Kusserow, B.; Schimpf, S.; Claus, P. Hydrogenation of glucose to sorbitol over nickel and ruthenium catalysts. *Adv. Synth. Catal.* **2003**, *345*, 289–299. [CrossRef]
18. Zhang, B.; Li, X.; Wu, Q.; Zhang, C.; Yu, Y.; Lan, M.; Wei, X.; Ying, Z.; Liu, T.; Liang, G.; et al. Synthesis of ni/mesoporous zsm-5 for direct catalytic conversion of cellulose to hexitols: Modulating the pore structure and acidic sites via a nanocrystalline cellulose template. *Green Chem.* **2016**, *18*, 3315–3323. [CrossRef]
19. Zhang, J.; Wu, S.; Liu, Y.; Li, B. Hydrogenation of glucose over reduced ni/cu/al hydrotalcite precursors. *Catal. Commun.* **2013**, *35*, 23–26. [CrossRef]
20. Yang, Y.; Chen, J.; Shi, H. Deoxygenation of methyl laurate as a model compound to hydrocarbons on ni2p/sio2, ni2p/mcm-41, and ni2p/sba-15 catalysts with different dispersions. *Energy Fuels* **2013**, *27*, 3400–3409. [CrossRef]
21. Koranyi, T.; Vit, Z.; Poduval, D.; Ryoo, R.; Kim, H.; Hensen, E. Sba-15-supported nickel phosphide hydrotreating catalysts. *J. Catal.* **2008**, *253*, 119–131. [CrossRef]
22. Oyama, S.; Lee, Y. The active site of nickel phosphide catalysts for the hydrodesulfurization of 4,6-dmdbt. *J. Catal.* **2008**, *258*, 393–400. [CrossRef]
23. Alexander, A.M.; Hargreaves, J.S. Alternative catalytic materials: Carbides, nitrides, phosphides and amorphous boron alloys. *Chem. Soc. Rev.* **2010**, *39*, 4388–4401. [CrossRef] [PubMed]
24. Yang, P.; Kobayashi, H.; Hara, K.; Fukuoka, A. Phase change of nickel phosphide catalysts in the conversion of cellulose into sorbitol. *ChemSusChem* **2012**, *5*, 920–926. [CrossRef]
25. Mathew, A.; Parambadath, S.; Kim, S.Y.; Ha, H.M.; Ha, C.-S. Diffusion mediated selective adsorption of zn2+ from artificial seawater by mcm-41. *Microporous Mesoporous Mater.* **2016**, *229*, 124–133. [CrossRef]

26. Song, H.; Wang, J.; Wang, Z.; Song, H.; Li, F.; Jin, Z. Effect of titanium content on dibenzothiophene HDS performance over Ni$_2$P/Ti-MCM-41 catalyst. *J. Catal.* **2014**, *311*, 257–265. [CrossRef]
27. Wang, R.; Smith, K.J. The effect of preparation conditions on the properties of high-surface area ni2p catalysts. *Appl. Catal. A* **2010**, *380*, 149–164. [CrossRef]
28. Rouquerol, J.; Avnir, D.; Fairbridge, C.W.; Everett, D.H.; Haynes, J.M.; Pernicone, N.; Ramsay, J.D.F.; Sing, K.S.W.; Unger, K.K. Recommendations for the characterization of porous solids (technical report). *Pure Appl. Chem.* **1994**, *66*, 1739–1758. [CrossRef]
29. Kadi, M.W.; Hameed, A.; Mohamed, R.M.; Ismail, I.M.I.; Alangari, Y.; Cheng, H.-M. The effect of pt nanoparticles distribution on the removal of cyanide by tio2 coated al-mcm-41 in blue light exposure. *Arab. J. Chem.* **2016**. [CrossRef]
30. Kresge, C.T.; Leonowicz, M.E.; Roth, W.J.; Vartuli, J.C.; Beck, J.S. Ordered mesoporous molecular sieves synthesized by a liquid-crystal template mechanism. *Nature* **1992**, *359*, 710. [CrossRef]
31. Gokulakrishnan, N.; Pandurangan, A.; Somanathan, T.; Sinha, P.K. Uptake of decontaminating agent from aqueous solution: A study on adsorption behaviour of oxalic acid over al-mcm-41 adsorbents. *J. Porous Mater.* **2010**, *17*, 763–771. [CrossRef]
32. Chen, S.; Li, J.; Zhang, Y.; Zhao, Y.; Liew, K.; Hong, J. Ru catalysts supported on al–sba-15 with high aluminum content and their bifunctional catalytic performance in fischer–tropsch synthesis. *Catal. Sci. Technol.* **2014**, *4*, 1005–1011. [CrossRef]
33. Mokaya, R. Post-synthesis grafting of al onto mcm-41. *Chem. Commun.* **1997**, *22*, 2185–2186. [CrossRef]
34. Yang, Y.; Ochoa-Hernández, C.; Pizarro, P.; de la Peña O'Shea, V.A.; Coronado, J.M.; Serrano, D.P. Influence of the ni/p ratio and metal loading on the performance of nixpy/sba-15 catalysts for the hydrodeoxygenation of methyl oleate. *Fuel* **2015**, *144*, 60–70. [CrossRef]
35. Wu, S.-K.; Lai, P.-C.; Lin, Y.-C. Atmospheric hydrodeoxygenation of guaiacol over nickel phosphide catalysts: Effect of phosphorus composition. *Catal. Lett.* **2014**, *144*, 878–889. [CrossRef]
36. Prins, R. Hydrogen spillover. Facts and fiction. *Chem. Rev.* **2012**, *112*, 2714–2738. [CrossRef] [PubMed]
37. Ding, L.N.; Wang, A.Q.; Zheng, M.Y.; Zhang, T. Selective transformation of cellulose into sorbitol by using a bifunctional nickel phosphide catalyst. *ChemSusChem* **2010**, *3*, 818–821. [CrossRef] [PubMed]
38. Berenguer, A.; Sankaranarayanan, T.M.; Gómez, G.; Moreno, I.; Coronado, J.M.; Pizarro, P.; Serrano, D.P. Evaluation of transition metal phosphides supported on ordered mesoporous materials as catalysts for phenol hydrodeoxygenation. *Green Chem.* **2016**, *18*, 1938–1951. [CrossRef]
39. Negahdar, L.; Oltmanns, J.U.; Palkovits, S.; Palkovits, R. Kinetic investigation of the catalytic conversion of cellobioseto sorbitol. *Appl. Catal. B* **2014**, *147*, 677–683. [CrossRef]
40. Liu, M.; Deng, W.; Zhang, Q.; Wang, Y.; Wang, Y. Polyoxometalate-supported ruthenium nanoparticles as bifunctional heterogeneous catalysts for the conversions of cellobiose and cellulose into sorbitol under mild conditions. *Chem. Commun. (Camb.)* **2011**, *47*, 9717–9719. [CrossRef]

© 2019 by the authors. Licensee MDPI, Basel, Switzerland. This article is an open access article distributed under the terms and conditions of the Creative Commons Attribution (CC BY) license (http://creativecommons.org/licenses/by/4.0/).

Article

Selective Synthesis of Furfuryl Alcohol from Biomass-Derived Furfural Using Immobilized Yeast Cells

Xue-Ying Zhang [1], Zhong-Hua Xu [1], Min-Hua Zong [1], Chuan-Fu Wang [2] and Ning Li [1,*]

[1] School of Food Science and Engineering, South China University of Technology, 381 Wushan Road, Guangzhou 510640, China; zxy1989318@126.com (X.-Y.Z.); xzh199000@163.com (Z.-H.X.); btmhzong@scut.edu.cn (M.-H.Z.)
[2] National Institute of Clean-and-Low-Carbon Energy, Future Science Park, Beijing 102211, China; wangchuanfu@nicenergy.com
* Correspondence: lining@scut.edu.cn

Received: 19 December 2018; Accepted: 6 January 2019; Published: 10 January 2019

Abstract: Furfuryl alcohol (FA) is an important building block in polymer, food, and pharmaceutical industries. In this work, we reported the biocatalytic reduction of furfural, one of the top value-added bio-based platform chemicals, to FA by immobilized *Meyerozyma guilliermondii* SC1103 cells. The biocatalytic process was optimized, and the tolerance of this yeast strain toward toxic furfural was evaluated. It was found that furfural of 200 mM could be reduced smoothly to the desired product FA with the conversion of 98% and the selectivity of >98%, while the FA yield was only approximately 81%. The gap between the substrate conversion and the product yield might partially be attributed to the substantial adsorption of the immobilization material (calcium alginate) toward the desired product, but microbial metabolism of furans (as carbon sources) made a negligible contribution to it. In addition, FA of approximately 156 mM was produced within 7 h in a scale-up reaction, along with the formation of trace 2-furoic acid (1 mM) as the byproduct. The FA productivity was up to 2.9 g/L/h, the highest value ever reported in the biocatalytic synthesis of FA. The crude FA was simply separated from the reaction mixture by organic solvent extraction, with the recovery of 90% and the purity of 88%. FA as high as 266 mM was produced by using a fed-batch strategy within 15.5 h.

Keywords: biocatalysis; bio-based platform chemicals; furans; reduction; whole cells

---

1. Introduction

Recently, the production of biofuels and bio-based chemicals from renewable biomass has attracted outstanding interest for reducing the dependence on fossil fuel sources, as well as mitigating global warming [1,2]. Like 5-hydroxymethylfurfural (HMF), furfural that can be synthesized readily from xylose is one of the "Top 10 + 4" bio-based platform chemicals [3–5]. There exist two active functional groups, including the aromatic ring and formyl group, in furfural, which are responsible for the high chemical reactivity of this bio-based molecule. Therefore, it can be upgraded readily into a group of value-added products via some typical chemical transformations such as reduction, oxidation, and Diels-Alder reactions [6]. Furfuryl alcohol (FA) is the most important derivative of furfural, since approximately 65% of furfural produced worldwide is utilized for the synthesis of FA [4,6]. FA has been used widely in polymer, food, and pharmaceutical industries [4,6]. FA is primarily used in the production of thermostatic resins, corrosion resistant fiber glass, and polymer concrete [7]. In addition, it is a building block in the manufacture of fragrances and pharmaceuticals, and for the synthesis of useful chemicals, such as tetrahydrofurfuryl alcohol, ethyl furfuryl ether, levulinic acid, and γ-valerolactone [6].

The industrial large-scale production of FA from furfural is performed over Cu-based catalysts, especially Cu-Cr catalysts, in gas or liquid phase [6]. Although the industrial routes have been well-established, they suffer from some problems, such as catalyst deactivation, serious environmental problems (due to the high toxicity of chromium), and overreduction of the desired product (thus leading to the formation of 2-methylfuran and furan) [5,6,8]. To address the drawbacks, chemists have devoted great efforts in the last decades. Significant advances have been achieved in the chemical catalytic hydrogenation of furfural to FA. A variety of new chemical catalysts, as well as efficient reaction engineering strategies, have also been developed to improve the production of this commercially important chemical [4,6,9].

Biotransformation is generally performed under mild conditions and is exquisitely selective, and biocatalysts are environmentally friendly [10]. Hence, biocatalysis represents a promising strategy for upgrading bio-based furans such as HMF and furfural [11], because of the inherent instability of these chemicals. Nevertheless, biocatalytic upgrading of these furans remains a great challenge, since they are proverbial inhibitors against enzymes and microorganisms [12]. Previously, many microorganisms (e.g., bacteria and yeasts) were reported to enable the detoxification of furfural into less toxic FA during the fermentation of biofuels and chemicals from lignocellulosic hydrolysates [13–18]. However, these reported microbes exhibited an unsatisfactory furan tolerance, and the biotransformation efficiencies were very low, especially at moderate to high substrate concentrations [19,20]. High substrate concentrations are highly desired for achieving satisfactory productivities in the biocatalytic synthesis of FA, which is crucial for moving these kinds of green technologies toward and into successful applications. Recently, we isolated *Meyerozyma guilliermondii* SC1103 from soil for the reduction of furans, including HMF and furfural [21]. A chemo-enzymatic method was reported for the synthesis of FA from xylose by He's group [22,23], in which the intermediate furfural derived from xylose was reduced into FA by a group of recombinant *Escherichia coli* strains. Interestingly, some *E. coli* strains exhibited good catalytic performances when the furfural concentrations were up to 200–300 mM [22,23]. Recently, Yan et al. reported the synthesis of FA using *Bacillus coagulans* NL01 [24]; the furfural tolerance of this bacterium (less than 50 mM) was unsatisfactory, although the conversion of 92% and the selectivity of 96% were obtained.

We recently performed preliminary experiments in which furfural of 50 mM was reduced into FA by resting cells of wild-type *M. guilliermondii* SC1103, with the yield of 83% and the selectivity of 96% [21]. To tap the application potential of this yeast in furfural reduction, the biocatalytic process was optimized and reaction engineering strategies were applied. Additionally, adaptively evolved yeast cells entrapped in calcium alginate were exploited for the synthesis of FA, in which NAD(P)H-dependent alcohol dehydrogenases catalyze the reduction of furfural and glucose as a co-substrate is enzymatically oxidized for generating NAD(P)H (Scheme 1). The effect of various key parameters on this reaction was studied. Interestingly, furfural of up to 200 mM could be transformed smoothly into the desired product with the selectivity of approximately 98%. The scale-up synthesis and purification of FA were carried out. In addition, a high concentration of FA was produced in the reaction mixture using a fed-batch strategy.

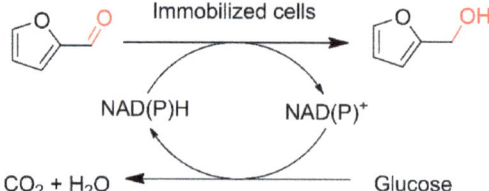

Scheme 1. Biocatalytic synthesis of FA from furfural with immobilized cells.

2. Results and Discussion

2.1. Process Optimization

We recently found that the catalytic activities of microbial cells were enhanced markedly when they were cultivated in the presence of a low concentration of HMF and FA (as the inducers) [25,26], likely because of the improved induction expression of the enzyme(s) responsible for the redox reactions. Therefore, the effect of the furfural addition during cell cultivation on the catalytic performances of whole cells in the synthesis of FA was examined (Figure 1a). It was found that the substrate (approximately 97%) was almost used up within 2 h in the control. FA was afforded with a yield of 89%, along with trace 2-furoic acid as the byproduct. The selectivity was up to 98%. Compared to the results in the control, no improvements in FA yields and selectivities were observed with furfural-induced cells as biocatalysts. It is well-known that furfural is more strongly inhibitory toward microbial cells than HMF and FA [27]. It can exert significantly negative effects on the cells (e.g., membrane, chromatin, and actin damage, and reduced intracellular ATP and NAD(P)H levels) during cultivation in the presence of this substance [15,27], which might offset the positive effect caused by the improved expression of the related enzymes. Therefore, no apparent improvements in the catalytic performances of these induced cells were observed compared to those of the control.

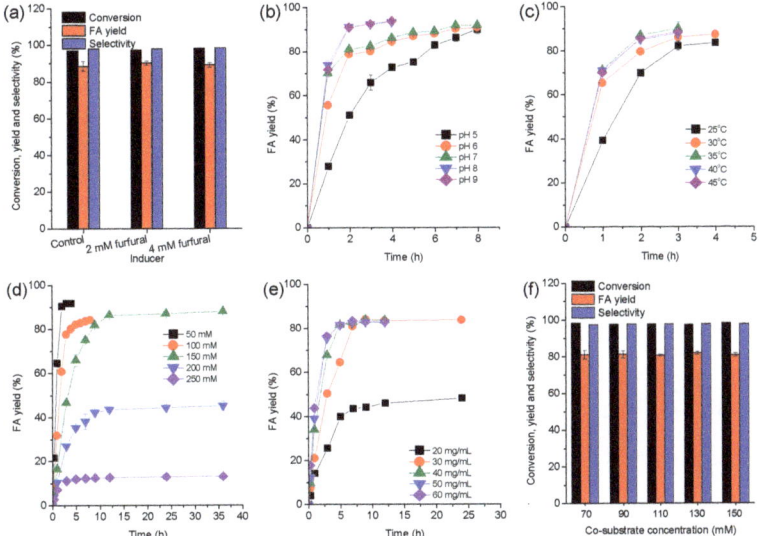

Figure 1. Effect of process parameters on the synthesis of FA: (**a**) effect of inducer; (**b**) effect of pH; (**c**) effect of reaction temperature; (**d**) effect of substrate concentrations; (**e**) effect of cell concentrations; (**f**) effect of co-substrate concentrations. General conditions unless otherwise stated: 50 mM furfural, glucose (Mol$_{glucose}$:Mol$_{furfural}$ = 3:5), immobilized cells (containing 30 mg/mL cells), 4 mL Tris-HCl buffer (100 mM, pH 8.0), 35 °C, 200 r/min; (**a**) free cells of 30 mg/mL, 2 h; (**b**) pH 5–9; (**c**) 25–45 °C; (**d**) 50–250 mM furfural; (**e**) 150 mM furfural, immobilized cells (containing 20–60 mg/mL cells); (**f**) 150 mM furfural, glucose (Mol$_{glucose}$:Mol$_{furfural}$ = 7:15–1:1), immobilized cells (containing 40 mg/mL cells), 5–7 h.

Figure 1b shows the effect of pH on the biocatalytic synthesis of FA. It was found that the catalytic activities of yeast cells were pH-dependent, and increased with the increment of buffer pH values from 5.0 to 9.0. Interestingly, FA was produced with high yields (more than 90%) and good selectivities (more than 95%) in all cases, which indicates the excellent tolerance of this strain toward acidic and alkaline environments. This is in good agreement with our recent results [21]. The subsequent studies were performed under pH 8.0. In addition, the effect of reaction temperature on this reaction was also examined (Figure 1c). Reaction temperature exerted a significant effect on the synthetic rates of FA, whereas its influence on the maximal yields and selectivities was slight. The temperature of 35 °C was considered optimal for the biotransformation.

Then, the tolerance of the immobilized cells toward furfural was evaluated, within the concentration range of 50–250 mM (Figure 1d). Almost complete exhaustion of the substrate (>97%) occurred when its concentrations were less than 150 mM. Furthermore, the desired product FA was obtained with yields of 84–92% and selectivities of 97–99%. Nonetheless, the substrate conversions decreased remarkably to 12–49% when its concentrations were more than 200 mM. Unsatisfied FA yields (12–45%) were found, although excellent selectivities (approximately 98%) were retained. To evaluate substrate inhibition, the initial reaction rates were compared (Figure S1). The initial reaction rates increased gradually from 33 to 44 mM/h when the substrate concentrations increased from 50 to 200 mM. Additionally, a further increase in the substrate concentrations (250 mM) led to a reduced initial reaction rate (32 mM/h). This suggests that substrate inhibition toward this whole-cell biocatalyst occurs when its concentration is up to 250 mM. As mentioned above, furfural does not only severely inhibit the activities of a variety of dehydrogenases, but is also highly toxic to microorganisms. It causes cell damage, and low intracellular ATP and NAD(P)H levels [27]. In addition, substantial cell decay readily occurs in the presence of high concentrations of furfural. Hence, the moderate conversion and yield at the substrate concentration of 200 mM may be explained by these toxic effects of furfural, in spite of the absence of substrate inhibition. Great substrate toxicity plus strong substrate inhibition resulted in poor results when the furfural concentration was 250 mM.

Biocatalyst concentrations were optimized when the substrate concentration was 150 mM (Figure 1e). It was found that the FA synthesis was significantly accelerated with the increment of the cell concentrations. The reaction periods required for achieving the maximal yields decreased to 5 h when the cell concentrations were more than 40 mg/mL. Moreover, the yields and selectivities were approximately 83% and 98%, respectively. It is well-known that efficient recycling of the reduced cofactors (e.g., NAD(P)H) is crucial for efficient biocatalytic reduction, which is closely relevant to co-substrates and their concentrations. In fact, we also previously found that glucose was important for HMF reduction [21]. Consequently, the effect of the glucose concentrations on the reduction of furfural was studied (Figure 1f). As shown in Figure 1f, co-substrate concentrations had a slight impact on the biocatalytic reduction of furfural within the concentration range tested. This suggests that NAD(P)H can be effectively regenerated via oxidation of the co-substrate for furfural reduction. Therefore, the molar ratio of furfural to glucose (15:7) will be used in subsequent studies.

2.2. Improved Synthesis of FA

With optimal conditions in hand, improved synthesis of FA was conducted. Figure 2 shows time courses of the biocatalytic synthesis of FA. The maximal substrate conversion reached approximately 98% when the substrate concentration was 200 mM, which was much higher than the corresponding value (49%) at the same substrate concentration before optimization (Figure 1d). Also, a significant improvement in the substrate conversions (64% vs. 12%) was observed upon optimization when the substrate concentration was 250 mM. Unfortunately, the substrate conversions (less than 22%) remained very poor when its concentrations were more than 300 mM, indicating its significant inhibitory and toxic effects on the biocatalyst.

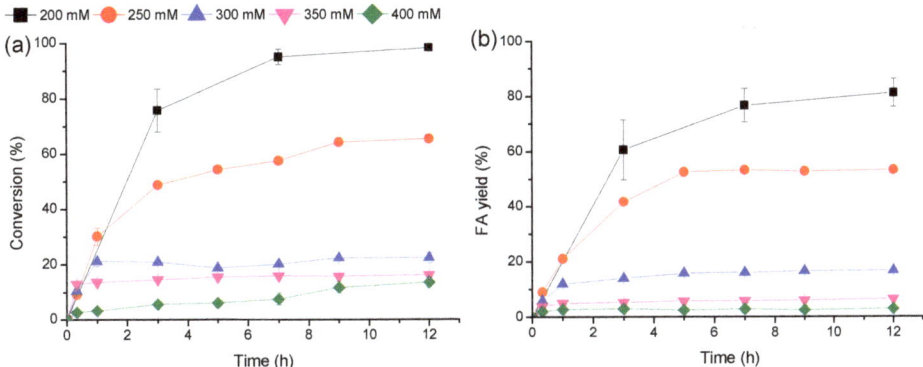

Figure 2. Time courses of the synthesis of FA from furfural: (**a**) furfural reduction; (**b**) FA synthesis. Reaction conditions: 200–400 mM furfural, glucose (Mol$_{glucose}$:Mol$_{furfural}$ = 7:15), immobilized cells (around 350 mg, containing 50 mg/mL cells), 4 mL Tris-HCl buffer (100 mM, pH 8.0), 35 °C, 200 r/min.

As shown in Figure 2b, FA was afforded with the yield of 81% after 12 h when the substrate concentration was 200 mM. In addition, only 2 mM of 2-furoic acid was produced as the byproduct. Therefore, there exists a considerable gap between the substrate conversion and the total yield of FA and byproduct (98% vs. 82%). We postulated two possible reasons for these results: (1) furans (furfural and FA) as carbon sources are metabolized by microbial cells, where 2-furoic acid as the intermediate enters into the tricarboxylic acid cycle [11]; (2) the desired product is adsorbed by the immobilized material.

To verify our assumption, biotransformation of FA and 2-furoic acid using free cells (to avoid the adsorption of the immobilized material) was carried out in the presence and absence of glucose (Figure 3). As shown in Figure 3a, FA was totally transformed in the absence of glucose after 9 h, affording 2-furoic acid with the yield of approximately 90%. A slight decrease in 2-furoic acid yields (3%) was observed with the elongation of the reaction period to 24 h. However, the oxidation of FA was greatly inhibited in the presence of glucose, since the latter was the preferred substrate for yeast cells. As shown in Figure 3a, only 20% of FA was oxidized in the presence of glucose in 12 h. Then FA conversion greatly increased to approximately 50% at 24 h, likely due to the exhaustion of glucose in the reaction mixture. Indeed, we previously found that glucose of 30 mM was quickly used up within 3 h by this yeast strain [21]. As described above, 2-furoic acid is the key intermediate in the metabolism of furfural and FA; thus, the biodegradation of this compound was examined (Figure 3b). It was found that 2-furoic acid of around 7% and 3% was degraded in the presence and absence of glucose, respectively. It suggests that this yeast strain has an extremely weak ability to biodegrade 2-furoic acid. Therefore, cellar metabolism of 2-furoic acid may make a negligible contribution to the above results.

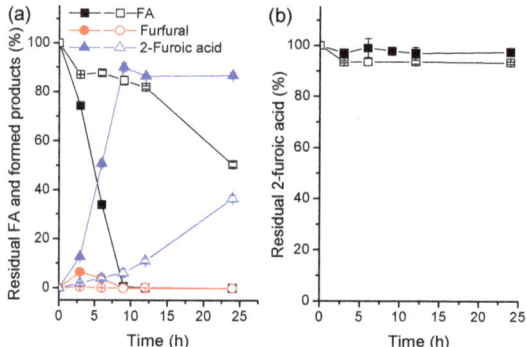

Figure 3. Transformation of FA (**a**) and 2-furoic acid (**b**) with free cells. Reaction conditions: 50 mM FA or 20 mM 2-furoic acid, in the presence (blank symbols) or absence (full symbols) of glucose (93 mM), 50 mg/mL free cells, 4 mL Tris-HCl buffer (100 mM, pH 8.0), 35 °C, 200 r/min.

Table 1 summarizes the adsorption of various immobilization materials toward the substrate and products. It was found that furfural of approximately 4% and FA of 3% were lost after incubation of 12 h in the control, likely due to evaporation, because their boiling points are approximately 170 °C. Regardless of the substances tested, much higher losses were observed in the presence of immobilization matrices compared to in the control (Table 1). In addition, losses showed a close dependence on the amounts of materials. For example, furfural of 7% and FA of 10% were reduced in the presence of 150 mg calcium alginate, whereas 350 mg of alginate beads resulted in losses of approximately 14% of furfural and FA. Compared to furfural and FA, the losses of 2-furoic acid were much lower (2–3%) in the presence of alginate beads. Therefore, the considerable adsorption of alginate beads toward the desired product FA may partially account for the lower product yields compared to the conversions.

Table 1. Adsorption of immobilization materials toward substrate and products.

Polymers (w/v)	Polymer Amount (mg)	Losses (%)		
		Furfural	FA	2-Furoic Acid
Control	none	4 ± 2	3 ± 1	n.d. [1]
Calcium alginate (2.5%)	150	7 ± 2	10 ± 1	2 ± 1
	350	14 ± 1	15 ± 3	3 ± 3
Agar (2%)	150	10 ± 1	10 ± 0	9 ± 1
	350	16 ± 1	15 ± 1	14 ± 0
Gelatin (15%)	350	26 ± 1	14 ± 0	9 ± 2
Carrageenan (3%)	350	10 ± 3	11 ± 0	15 ± 0
Poly(vinyl alcohol) (10%)	350	21 ± 1	18 ± 2	9 ± 2

Conditions: 150 or 350 mg of gel beads without the cells were incubated in 4 mL Tris-HCl buffer (100 mM, pH 8.0) at 35 °C and 200 r/min for 12 h in the presence of 200 mM furfural, 200 mM FA, or 20 mM 2-furoic acid; the changes in the concentrations of furfural, FA, and 2-furoic acid were determined, respectively. [1] none detection.

To identify promising materials that slightly adsorb the desired product for cell immobilization, a variety of commonly used polymer beads were prepared and their adsorption toward substrate and products was examined (Table 1). Unfortunately, these materials tested exhibited great adsorption capacities toward furans. The losses were dependent on the nature of the materials, as well as on the chemicals tested. In addition, most of the materials showed higher adsorption capacities toward furfural than calcium alginate (16–26% vs. 14%), with the exception of carrageenan (10%). Furthermore, considerable losses of FA (11–18%) were observed in the presence of other polymer beads. Among the polymers tested, calcium alginate showed the poorest adsorption toward 2-furoic acid, which may be an attractive material for immobilizing cells for the production of 2-furoic acid.

2.3. Scale-Up Synthesis and Separation of FA

To envisage the practicality of this bioprocess, the synthesis of FA was scaled up when the substrate concentration was 200 mM (Figure 4). As shown in Figure 4, furfural was almost used up within 5 h. The desired product FA was quickly formed in the initial stage of 5 h. FA of approximately 157 mM was finally produced, together with the formation of 2-furoic acid of around 1 mM. The yield and selectivity of FA were around 79% and 99%, respectively. The FA productivities of approximately 2.9 and 2.2 g/L/h were obtained in the scale-up synthesis, respectively, based on the FA concentrations at 5 and 7 h. After the reaction, the immobilized cells were filtered off, and the desired product was extracted three times from the reaction mixture by ethyl acetate. The crude product was furnished with the recovery of approximately 90% and the purity of 88% after evaporating the organic solvent.

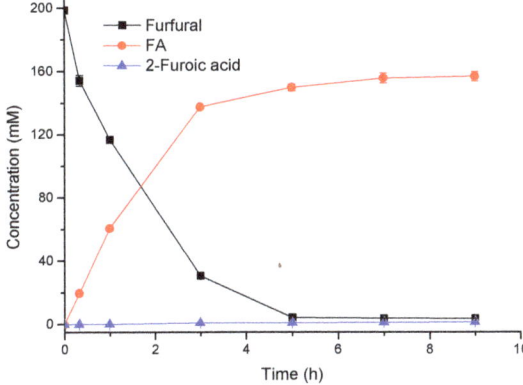

Figure 4. Scale-up synthesis of FA. Reaction conditions: 200 mM furfural, glucose ($Mol_{glucose}$:$Mol_{furfural}$ = 7:15), immobilized cells (containing 50 mg/mL cells), 20 mL Tris-HCl buffer (100 mM, pH 8.0), 35 °C, 200 r/min.

2.4. FA Synthesis by a Fed-Batch Strategy

In addition to the yield and productivity, the titer is an important index for evaluating the economic viability of a process [28], because an appropriate titer can not only significantly improve the reactor productivity, but also facilitate the product purification. Considering the high toxicity of the substrate, a fed-batch strategy was applied for accumulating the desired product of high concentrations in the reaction mixture (Figure 5). As shown in Figure 5, FA of approximately 190 mM was produced after 5.5 h. In a preliminary study, we found that the cells significantly deactivated at 5.5 h (data not shown), likely due to the great cell damage caused by toxic furfural. Accordingly, the immobilized cells were isolated from the reaction mixture, and reactivated by cultivation in fresh culture medium. It was observed that the reactivated cells were capable of transforming 90% furfural into FA within 5 h. Unfortunately, significant deactivation of the biocatalysts remained after the third substrate feeding. After 15.5 h, substrate of up to 62 mM was found in the mixture, and its concentration reduced slightly with the elongation of the reaction period. This might be ascribed to the biocatalyst deactivation caused by the toxic substrate. On the other hand, the product inhibition toward microbial cells occurred readily when its concentration was as high as 266 mM. Overall, the desired product of up to 266 mM was produced within 15.5 h, along with the formation of 2-furoic acid of 6 mM.

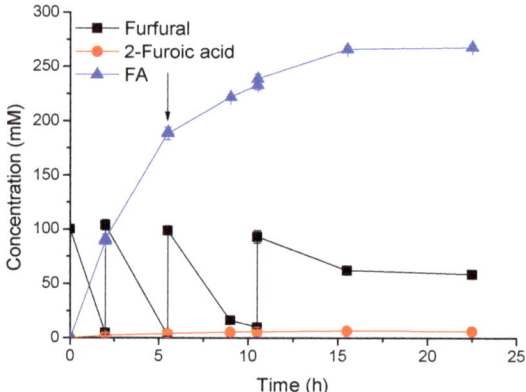

Figure 5. FA synthesis by a fed-batch strategy. Reaction conditions: 100 mM furfural, glucose (Mol$_{glucose}$:Mol$_{furfural}$ = 7:15), immobilized cells (containing 50 mg/mL cells), 8 mL Tris-HCl buffer (100 mM, pH 8.0), 35 °C, 200 r/min; after furfural was almost used up, substrate and co-substrate were supplemented. Arrows indicate the reactivation of the immobilized cells by cultivation.

To directly evidence the adsorption of immobilized material toward furans, the immobilized cells were isolated and deconstructed, followed by extraction by ethyl acetate. The furan concentrations were measured. It was found that FA of approximately 25 mM was adsorbed by the immobilized material, together with 3 mM furfural. However, 2-furoic acid was not observed. The results suggest that the desired product of at least 10% was adsorbed by the immobilized material, leading to the underestimation of the real FA yield.

Table 2 shows biocatalytic FA synthesis using a variety of microorganisms. In general, the FA productivities (<1 g/L/h) were low in the biodetoxification of lignocellulosic hydrolysates. This is due to the fact that the primary aims of these studies are not the production of FA, but the production of biofuels and chemicals from hydrolysates through fermentation. Therefore, the furfural concentrations tested were low in biodetoxification, resulting in low FA productivities. Compared to *B. coagulans* NL01 reported recently [24], *M. guilliermondii* SC1103 exhibited a much higher furfural tolerant level (42 vs. 200 mM). He et al. reported that FA was obtained in 12 h using recombinant *E. coli* CCZU-A13 cells, with yields of 94% and 74% at the furfural concentrations of 200 and 300 mM, respectively [23]. The concentrations of both the biocatalyst and co-substrate glucose were two-fold higher in He's report than those in this work, although better results (higher substrate tolerance and higher FA yields) were obtained in the former. More importantly, the productivity of FA was much higher in this work than in He's report (2.9 vs. 1.8 g/L/h).

Table 2. Comparison of FA synthesis using various biocatalysts.

Biocatalyst	Reaction Conditions	Substrate Concentration (mM)	Time (h)	C/Y [1] (%)	Selectivity (%)	Productivity [2] (g/L/h)	Ref.
Enterobacter sp. FDS8	4.6 mg (dry weight)/mL cells, pH 7, 30 °C, in lignocellulosic hydrolysate	19	3	n.a. [3]	n.a.	0.5 [4]	[20]
Clostridium acetobutylicum ATCC 824	1% of the inoculate culture, pH 5.6, 30 °C, in a sugar cane molasses medium	17	12	100 (C)	100	0.1	[15]
S. cerevisiae 307-12-F40	1% of the inoculate culture, 30 °C, in a synthetic complete medium containing 20 mM glucose	30	30	70 (Y)	n.a.	<0.1 [4]	[14]
S. cerevisiae 354	Cell concentration is not available, pH 6.7, 35 °C, in a P2 medium containing approximately 333 mM glucose	62.5	6	100 (Y)	100	1.0	[29]
B. coagulans NL01	Mol$_{glucose}$:Mol$_{furfural}$ = 5:2, 9 mg (dry weight)/mL cells, pH 7, 50 °C, in phosphate buffer	42	3	96 (C)	87	1.3 [4]	[24]
E. coli CCZU-A13	Mol$_{glucose}$:Mol$_{furfural}$ = 1:1, 100 mg (wet weight)/mL cells, pH 6.5, 30 °C, in KH$_2$PO$_4$-K$_2$HPO$_4$ buffer	200 300	12 12	94 (Y) 74 (Y)	n.a. n.a.	1.5 1.8	[23] [23]
E. coli CCZU-K14	Mol$_{glucose}$:Mol$_{furfural}$ = 3:2, 400 mM xylose, 100 mg (wet weight)/mL cells, pH 6.5, 30 °C, in KH$_2$PO$_4$-K$_2$HPO$_4$ buffer	200	24	100 (Y)	100	0.8	[22]
M. guilliermondii SC1103	Mol$_{glucose}$:Mol$_{furfural}$ = 7:15, 50 mg (wet weight)/mL cells, pH 8, 35 °C, in Tris-HCl buffer	200	7	79 (Y)	99	2.2/2.9 (5 h)	This work

[1] C, conversion; Y, yield. [2] based on the formed FA. [3] not available. [4] based on the consumed substrate.

3. Materials and Methods

3.1. Materials

M. guilliermondii SC1103 maintained in the China Center for Type Culture Collection (CCTCC, Wuhan, China; with CCTCC No. M2016144) was isolated by our laboratory [21]. Furfural (99%) was purchased from Macklin Biochemical Co., Ltd. (Shanghai, China). FA (98%) was obtained from J&K Scientific Ltd. (Guangzhou, China). 2-Furoic acid (98%) was obtained from TCI (Japan). Other chemicals were of the highest purity commercially available.

3.2. Cultivation and Immobilization of Microbial Cells

Acclimatized *M. guilliermondii* SC1103 cells described in a recent report [26] were used in this work, which were cultivated prior to immobilization. Briefly, the cells were pre-cultivated at 30 °C and 200 r/min for 12 h in the yeast extract peptone dextrose medium (YPD, 1% yeast extract, 2% peptone and 2% glucose) in the presence of 0–4 mM furfural. Then, the 2% seed culture was inoculated to the fresh YPD medium in the presence of 0–4 mM furfural. After incubation at 30 °C and 200 r/min for 12 h, the cells were harvested by centrifugation (6000 r/min, 10 min, 4 °C) and washed twice with distilled water.

Cell immobilization was performed according to a recent method [26]. Typically, 2.0–6.0 g cells (cell wet weight) were mixed with 10 mL 2.5% (w/v) sodium alginate. Then, the mixture was added drop-wise from a syringe to 0.2 M $CaCl_2$ solution. The resulting gel beads (with the diameters of approximately 1.4 mm) were hardened at room temperature for 4 h. Then, the beads were washed three times with Tris-HCl buffer (100 mM, pH 7.2) and stored at 4 °C in this buffer until use. Other polymer gel beads without cells were prepared according to previous methods [30–33], with slight modifications.

3.3. General Procedure for the Synthesis of FA

Typically, 4 mL Tris-HCl buffer (100 mM, pH 8.0) containing 150 mM furfural, 70 mM glucose, and immobilized cells (containing 50 mg cells (cell wet weight) per mL buffer) was incubated at 35 °C and 200 r/min. Aliquots were withdrawn from the reaction mixtures at specified time intervals and diluted with the corresponding mobile phase prior to HPLC analysis. The initial reaction rate was calculated based on the changes in the substrate concentrations in the initial stage (usually 20 min). The conversion was defined as the ratio of the consumed substrate to the initial substrate amount (in mol). The yield was defined as the ratio of the formed FA to the initial substrate amount (in mol). The selectivity was defined as the ratio of the formed FA to the total amount of all products (in mol). All the experiments were conducted at least in duplicate, and the values were expressed as the means ± standard deviations.

3.4. Scale-Up Production and Extraction of FA

The reaction mixture was composed of 20 mL Tris-HCl buffer (100 mM, pH 8.0), 200 mM furfural, and 93 mM glucose. After the immobilized cells (containing 50 mg cells (cell wet weight) per mL buffer) were added to the reaction mixture, the reaction was conducted at 35 °C and 200 r/min. The immobilized cells were removed by filtration upon reaction. The reaction solution was saturated with NaCl by adding this substance in excess, followed by extraction three times with ethyl acetate. The organic phases were combined and dried over anhydrous Na_2SO_4 overnight. The crude product was obtained upon evaporation of the organic solvent.

3.5. FA Synthesis via Fed-Batch Feeding of Substrate

The reaction mixture containing 8 mL Tris-HCl buffer (100 mM, pH 8.0), 100 mM furfural, 47 mM glucose, and the immobilized cells (containing 50 mg cells (cell wet weight) per mL buffer) was incubated at 35 °C and 200 r/min. When furfural was almost used up, 0.8 mmol furfural and

0.38 mmol glucose were supplemented into the reaction mixture. After 5.5 h, the immobilized cells were isolated from the mixture and re-activated at 30 °C and 200 r/min for 10 h in fresh YPD medium. Upon re-activation, the immobilized cells accompanied by 0.8 mmol furfural and 0.38 mmol glucose were added into the reaction mixture. When furfural was almost used up in the reaction mixture, supplementation of furfural and glucose was repeated. The changes in the concentrations of various compounds were monitored by HPLC. After the reaction, the immobilized cells were isolated from the reaction mixture, followed by rinsing by Tris-HCl buffer (100 mM, pH 8.0) to remove furans on the bead surfaces. Then, the beads were treated ultrasonically in 3 mL sodium phosphate buffer (200 mM, pH 7.0) for 1 h, resulting in complete deconstruction of the beads. After removal of the particles by centrifugation (12,000 r/min, 10 min), the supernatant was saturated with NaCl by adding this substance in excess, followed by extraction by 6 mL ethyl acetate. The furan concentrations were determined by HPLC.

3.6. HPLC Analysis

The reaction mixtures were analyzed on an Eclipse XDB-C18 column (4.6 mm × 250 mm, 5 μm, Agilent Technologies, Santa Clara, CA, USA) by reversed phase HPLC equipped with a Waters 996 photodiode array detector (Waters Corporation, Milford, MA, USA). The mobile phase was the mixture of acetonitrile/0.4% $(NH_4)_2SO_4$ solution (10:90, v/v) with the flow rate of 0.6 mL/min. The retention times of 2-furoic acid, FA, and furfural were 9.1, 13.8, and 15.7 min, respectively.

4. Conclusions

In summary, an efficient biocatalytic process for FA synthesis was successfully developed using immobilized *M. guilliermondii* SC1103 cells. The yeast cells were tolerant to furfural of up to 200 mM. A high conversion (98%) and an excellent selectivity (>98%) were achieved. The desired product was significantly adsorbed by immobilization material, resulting in a relatively low yield (81%). In the scale-up synthesis, FA productivity as high as 2.9 g/L/h was obtained, which is the highest productivity of FA ever reported. FA of up to 266 mM was produced within 15.5 h by a fed-batch strategy. Although satisfactory results were obtained by this bioprocess, some problems, such as long-term stability of the biocatalyst and longer reaction periods compared to chemical methods, should be addressed in the future to move this clean technology forward to successful applications.

Supplementary Materials: The following are available online at http://www.mdpi.com/2073-4344/9/1/70/s1, Figure S1: Time courses of substrate conversion in the reduction of furfural. The detailed methods for the preparation of other gel beads.

Author Contributions: Conceptualization, N.L.; investigation, X.-Y.Z. and Z.-H.X.; funding acquisition, N.L.; resources, N.L.; writing—original draft preparation, N.L.; writing—review and editing, M.-H.Z. and C.-F.W.

Funding: This research was financially supported by the Natural Science Foundation of Guangdong Province (No. 2017A030313056), the Science and Technology Project of Guangzhou City (201804010179), the National Natural Science Foundation of China (No. 21676103), and the Fundamental Research Funds for the Central Universities (No. 2017ZD065).

Conflicts of Interest: The authors declare no conflict of interest.

References

1. Tuck, C.O.; Pérez, E.; Horváth, I.T.; Sheldon, R.A.; Poliakoff, M. Valorization of biomass: Deriving more value from waste. *Science* **2012**, *337*, 695–699. [CrossRef] [PubMed]
2. Sheldon, R.A. Green and sustainable manufacture of chemicals from biomass: State of the art. *Green Chem.* **2014**, *16*, 950–963. [CrossRef]
3. Bozell, J.J.; Petersen, G.R. Technology development for the production of biobased products from biorefinery carbohydrates-the US department of energy's "top 10" revisited. *Green Chem.* **2010**, *12*, 539–554. [CrossRef]
4. Li, X.; Jia, P.; Wang, T. Furfural: A promising platform compound for sustainable production of C_4 and C_5 chemicals. *ACS Catal.* **2016**, *6*, 7621–7640. [CrossRef]

5. Lange, J.-P.; van der Heide, E.; van Buijtenen, J.; Price, R. Furfural—A promising platform for lignocellulosic biofuels. *ChemSusChem* **2012**, *5*, 150–166. [CrossRef] [PubMed]
6. Mariscal, R.; Maireles-Torres, P.; Ojeda, M.; Sadaba, I.; Lopez Granados, M. Furfural: A renewable and versatile platform molecule for the synthesis of chemicals and fuels. *Energy Environ. Sci.* **2016**, *9*, 1144–1189. [CrossRef]
7. Sharma, R.V.; Das, U.; Sammynaiken, R.; Dalai, A.K. Liquid phase chemo-selective catalytic hydrogenation of furfural to furfuryl alcohol. *Appl. Catal. A Gen.* **2013**, *454*, 127–136. [CrossRef]
8. Liu, D.; Zemlyanov, D.; Wu, T.; Lobo-Lapidus, R.J.; Dumesic, J.A.; Miller, J.T.; Marshall, C.L. Deactivation mechanistic studies of copper chromite catalyst for selective hydrogenation of 2-furfuraldehyde. *J. Catal.* **2013**, *299*, 336–345. [CrossRef]
9. Gupta, K.; Rai, R.K.; Singh, S.K. Metal catalysts for the efficient transformation of biomass-derived HMF and furfural to value added chemicals. *ChemCatChem* **2018**, *10*, 2326–2349. [CrossRef]
10. Sheldon, R.A.; Pereira, P.C. Biocatalysis engineering: The big picture. *Chem. Soc. Rev.* **2017**, *46*, 2678–2691. [CrossRef]
11. Domínguez de María, P.; Guajardo, N. Biocatalytic valorization of furans: Opportunities for inherently unstable substrates. *ChemSusChem* **2017**, *10*, 4123–4134. [CrossRef] [PubMed]
12. Palmqvist, E.; Hahn-Hägerdal, B. Fermentation of lignocellulosic hydrolysates. II: Inhibitors and mechanisms of inhibition. *Bioresour. Technol.* **2000**, *74*, 25–33. [CrossRef]
13. Liu, Z.L.; Slininger, P.J.; Dien, B.S.; Berhow, M.A.; Kurtzman, C.P.; Gorsich, S.W. Adaptive response of yeasts to furfural and 5-hydroxymethylfurfural and new chemical evidence for HMF conversion to 2,5-bis-hydroxymethylfuran. *J. Ind. Microbiol. Biotechnol.* **2004**, *31*, 345–352. [CrossRef] [PubMed]
14. Liu, Z.L.; Slininger, P.J.; Gorsich, S.W. Enhanced biotransformation of furfural and hydroxymethylfurfural by newly developed ethanologenic yeast strains. *Appl. Biochem. Biotechnol.* **2005**, *121*, 451–460. [CrossRef]
15. Zhang, Y.; Han, B.; Ezeji, T.C. Biotransformation of furfural and 5-hydroxymethyl furfural (HMF) by *Clostridium acetobutylicum* ATCC 824 during butanol fermentation. *New Biotechnol.* **2012**, *29*, 345–351. [CrossRef] [PubMed]
16. Boopathy, R.; Bokang, H.; Daniels, L. Biotransformation of furfural and 5-hydroxymethyl furfural by enteric bacteria. *J. Ind. Microbiol.* **1993**, *11*, 147–150. [CrossRef]
17. Liu, Z.L.; Moon, J.; Andersh, B.J.; Slininger, P.J.; Weber, S. Multiple gene-mediated NAD(P)H-dependent aldehyde reduction is a mechanism of in situ detoxification of furfural and 5-hydroxymethylfurfural by saccharomyces cerevisiae. *Appl. Microbiol. Biotechnol.* **2008**, *81*, 743–753. [PubMed]
18. Wang, X.; Yomano, L.P.; Lee, J.Y.; York, S.W.; Zheng, H.; Mullinnix, M.T.; Shanmugam, K.T.; Ingram, L.O. Engineering furfural tolerance in *Escherichia coli* improves the fermentation of lignocellulosic sugars into renewable chemicals. *Proc. Natl. Acad. Sci. USA* **2013**, *110*, 4021–4026. [CrossRef]
19. Boopathy, R. Anaerobic biotransformation of furfural to furfuryl alcohol by a methanogenic archaebacterium. *Int. Biodeterior. Biodegrad.* **2009**, *63*, 1070–1072. [CrossRef]
20. Zhang, D.; Ong, Y.L.; Li, Z.; Wu, J.C. Biological detoxification of furfural and 5-hydroxyl methyl furfural in hydrolysate of oil palm empty fruit bunch by *Enterobacter* sp. FDS8. *Biochem. Eng. J.* **2013**, *72*, 77–82. [CrossRef]
21. Li, Y.M.; Zhang, X.Y.; Li, N.; Xu, P.; Lou, W.Y.; Zong, M.H. Biocatalytic reduction of HMF to 2,5-bis(hydroxymethyl)furan by HMF-tolerant whole cells. *ChemSusChem* **2017**, *10*, 372–378. [CrossRef] [PubMed]
22. He, Y.-C.; Jiang, C.-X.; Jiang, J.-W.; Di, J.-H.; Liu, F.; Ding, Y.; Qing, Q.; Ma, C.-L. One-pot chemo-enzymatic synthesis of furfuralcohol from xylose. *Bioresour. Technol.* **2017**, *238*, 698–705. [CrossRef] [PubMed]
23. He, Y.; Ding, Y.; Ma, C.; Di, J.; Jiang, C.; Li, A. One-pot conversion of biomass-derived xylose to furfuralcohol by a chemo-enzymatic sequential acid-catalyzed dehydration and bioreduction. *Green Chem.* **2017**, *19*, 3844–3850. [CrossRef]
24. Yan, Y.; Bu, C.; He, Q.; Zheng, Z.; Ouyang, J. Efficient bioconversion of furfural to furfuryl alcohol by *Bacillus coagulans* NL01. *RSC Adv.* **2018**, *8*, 26720–26727. [CrossRef]
25. Zhang, X.Y.; Zong, M.H.; Li, N. Whole-cell biocatalytic selective oxidation of 5-hydroxymethylfurfural to 5-hydroxymethyl-2-furancarboxylic acid. *Green Chem.* **2017**, *19*, 4544–4551. [CrossRef]

26. Xu, Z.H.; Cheng, A.D.; Xing, X.P.; Zong, M.H.; Bai, Y.P.; Li, N. Improved synthesis of 2,5-bis(hydroxymethyl)furan from 5-hydroxymethylfurfural using acclimatized whole cells entrapped in calcium alginate. *Bioresour. Technol.* **2018**, *262*, 177–183. [CrossRef]
27. Almeida, J.R.M.; Modig, T.; Petersson, A.; Hähn-Hägerdal, B.; Lidén, G.; Gorwa-Grauslund, M.F. Increased tolerance and conversion of inhibitors in lignocellulosic hydrolysates by *Saccharomyces cerevisiae*. *J. Chem. Technol. Biotechnol.* **2007**, *82*, 340–349. [CrossRef]
28. Meynial-Salles, I.; Dorotyn, S.; Soucaille, P. A new process for the continuous production of succinic acid from glucose at high yield, titer, and productivity. *Biotechnol. Bioeng.* **2008**, *99*, 129–135. [CrossRef]
29. Villa, G.P.; Bartroli, R.; López, R.; Guerra, M.; Enrique, M.; Peñas, M.; Rodríquez, E.; Redondo, D.; Jglesias, I.; Díaz, M. Microbial transformation of furfural to furfuryl alcohol by *Saccharomyces cerevisiae*. *Acta Biotechnol.* **1992**, *12*, 509–512. [CrossRef]
30. Vassileva, A.; Burhan, N.; Beschkov, V.; Spasova, D.; Radoevska, S.; Ivanova, V.; Tonkova, A. Cyclodextrin glucanotransferase production by free and agar gel immobilized cells of *Bacillus circulans* ATCC 21783. *Process Biochem.* **2003**, *38*, 1585–1591. [CrossRef]
31. He, Y.-C.; Xu, J.-H.; Su, J.-H.; Zhou, L. Bioproduction of glycolic acid from glycolonitrile with a new bacterial isolate of *Alcaligenes* sp. ECU0401. *Appl. Biochem. Biotechnol.* **2010**, *160*, 1428–1440. [CrossRef] [PubMed]
32. Takei, T.; Ikeda, K.; Ijima, H.; Kawakami, K. Fabrication of poly(vinyl alcohol) hydrogel beads crosslinked using sodium sulfate for microorganism immobilization. *Process Biochem.* **2011**, *46*, 566–571. [CrossRef]
33. Aparecida de Assis, S.; Ferreira, B.S.; Fernandes, P.; Guaglianoni, D.G.; Cabral, J.M.S.; Oliveira, O.M.M.F. Gelatin-immobilized pectinmethylesterase for production of low methoxyl pectin. *Food Chem.* **2004**, *86*, 333–337. [CrossRef]

© 2019 by the authors. Licensee MDPI, Basel, Switzerland. This article is an open access article distributed under the terms and conditions of the Creative Commons Attribution (CC BY) license (http://creativecommons.org/licenses/by/4.0/).

Article

A Comparative Study of MFI Zeolite Derived from Different Silica Sources: Synthesis, Characterization and Catalytic Performance

Jianguang Zhang [1], Xiangping Li [2,*], Juping Liu [2] and Chuanbin Wang [2]

1. School of Petroleum engineering, China University of Petroleum (East China), Qingdao 266580, China; eduzjg@163.com
2. School of Environmental Science and Engineering/China-Australia Centre for Sustainable Urban Development, Tianjin University, Tianjin 300072, China; liujuping1234@163.com (J.L.); wangchuanbin6@163.com (C.W.)
* Correspondence: xiangping.li@tju.edu.cn; Tel.: +86-022-8740-1929

Received: 13 November 2018; Accepted: 25 December 2018; Published: 26 December 2018

Abstract: In this paper, a comparative study of MFI zeolite derived from different silica sources is presented. Dry gel conversion (DGC) method is used to synthesize silicalite-1 and ZSM-5 with MFI structure. Two kinds of silica sources with different particle sizes are used during the synthesis of MFI zeolite. The as-prepared samples were characterized by X-ray diffraction (XRD), N_2-sorption, Fourier transform infrared spectroscopy (FTIR), Scanning electron microscopy (SEM) and X-ray fluorescence spectrometer (XRF). From the characterization results, it could be seen that the high-quality coffin-like silicalite-1 was synthesized using silica sphere with particle size of 300 nm as silica source, with crystallization time being shortened to 2 h. The schematic diagram of silicalite-1 formation using silica sources with different particle sizes is summarized. ZSM-5 was obtained by adding Al atoms to raw materials during the synthesis of MFI zeolite. The performance of aqueous phase eugenol hydrodeoxygenation over Pd/C-ZSM-5 catalyst is evaluated.

Keywords: dry gel conversion; MFI zeolite; particle sizes; silica sources; hydrodeoxygenation

1. Introduction

The structure of crystalline aluminosilicates is three dimensional and always contains cages or pores, making them very favorable. Meanwhile, due to their extremely high thermal stability and chemical resistance, they have been widely used in industrial production [1–6]. Owing to tunable porosity and molecule shape selectivity, zeolites can be used as adsorbent, ion-exchange material and catalysts, and so on [7–11]. A large variety of reactions such as cracking, isomerization, dewaxing, dehydration, hydrodeoxygenation and alkylation can be accomplished with microporous zeolites [12–18]. Moreover, the applications of zeolites in the fields of separation, chemical sensors, anticorrosive coating, low-k materials, and hydrophilic antimicrobial coatings are a research hotspot at present [19–22]. As a well-known microporous aluminosilicate [23,24] firstly synthesized by scientists in 1969, ZSM-5 zeolite has a 3D host framework of intersecting 10-membered rings, with a pore size of (0.51 × 0.55 nm) in the [100] direction.

Various methods can be used to synthesize MFI zeolite, including the hydrothermal synthesis method [25–27], the microwave irradiation method [28], and so on. Conventional methods for synthesis of MFI zeolite are often associated with long synthesis times and large quantities of template agents, which leads to increasing cost. In addition, MFI zeolite with low degrees of crystallinity is obtained from synthesis processes using conventional methods, and massive waste materials will be generated, resulting in environment pollution [29,30]. With the deepening of consciousness throughout the whole society of environmental protection, economical and environmentally friendly methods are increasingly favored. The method of dry gel conversion (DGC) has advantages including higher zeolite yield, lower template agent usage, rapid crystallization, environment-friendliness and economic efficiency for zeolite synthesis [29].

The key factors influencing the synthesis of zeolite include the overall chemical composition of the reactant mixture and thermodynamic variables. Among these, silica source plays an important role in the synthesis process and also determines the morphology of the synthesized product [31]. Different silica are used as silica sources during the synthesis process of zeolite [32–34]. The effect of silica sources (natural silica nanoparticles derived from rice husk and commercial Ludox) on zeolite of NaY synthesis was been studied by Najat and coworkers [35]. NaY is the sodium type of Y zeolite. It has been found that adding natural silica from rice husk in both feedstock gel and seed gel for the preparation of Y zeolite gel can produce a nanosilica catalyst and enhance the catalytic performance of catalytic cracking [35]. However, the effect of silica sources on the synthesis process of MFI zeolite has not yet been deeply investigated.

Silicalite-1, with an MFI topology structure, is an all-silica zeolite that only contains Si, O and H in the framework. Silicalite-1 has a pore diameter of about 5–6 Å and possesses a 3D channel structure, with sinusoidal channels in the x-direction and straight channels in the y-direction. The two types of channels intersect with each other, forming a 3D porous structure [36,37]. Silicalite-1 has high-temperature resistance, as well as strong hydrophobic and oil-wet properties, due to the absence of aluminum in the structure. Organic molecules such as arenes, short-chain alkanes and polyhydric alcohols can be absorbed by silicalite-1 [38]. Therefore, it has been extensively applied in catalysis and separation [39–41]. Since the supply of aluminum and other inorganic ions is avoided during the synthesis process of silicalite-1, high-quality crystals, rather than ZSM-5, are easier to synthesize. Silicalite-1 is the best model for studying the synthesis mechanism, crystalline regulation, and control of MFI zeolite, as well as the dispersion and size of the particles. Meanwhile, research on silicalite-1 synthesis provides references for the synthesis of a series of ZSM-5 zeolite.

Hydrodeoxygenation (HDO) is a catalytic upgrading process, which has been considered to be the most effective method for bio-oil upgrading [42]. ZSM-5 with moderate acid intensity is a suitable acidic supplier for bio-oil upgrading, especially for lignin derived phenolic compounds upgrading.

In this work, in order to investigate the effect of silica sources on the synthesis of MFI zeolite, silicalite-1 was synthesized using tetrapropylammonium hydroxide (TPAOH) as a template agent and silica of different particle sizes as silica sources. The dry gel conversion method was selected as the synthesis method of MFI zeolite in this study. The synthesis mechanisms of silicalite-1 derived from different silica sources were investigated. After careful adjustment of the silica sources, silicalite-1 with nanoparticle size, smooth surface and coffin-like structure was synthesized. The synthesis mechanisms of silicalite-1 with different silica sources were discussed. ZSM-5 was synthesized by adding Al atoms to the raw materials in the dry gel conversion synthesis process. The physical and chemical characters of as-prepared samples were analyzed by X-ray diffraction (XRD), scanning electron microscope (SEM), N_2-sorption, Fourier transform infrared spectroscopy (FTIR) and X-ray fluorescence spectrometer (XRF). The reaction activity of eugenol hydrodeoxygenation over ZSM-5 combined with Pd/C catalysts was obtained.

2. Results and Discussion

2.1. The Effect of Silica Sources and Synthesis Time on the Preparation of Silicalite-1

To study the effect of silica sources on preparation of silicalite-1, fume silica with different primary particle sizes and spherical silica with a particle size of 300 nm were applied as silica sources in the synthesis process.

In the process of synthesis, fume silica (AEROSIL200) with an average particle size of 12 nm was used as silica source, and crystallization time was changed from 1 h to 12 h. The XRD spectra were obtained, as shown in Figure 1a. It can be seen that the characteristic diffraction peaks of silicalite-1 framework at around 7.97°, 8.87°, 23.17°, 24.02° and 24.46° corresponded to the [101], [020], [501], [151] and [303] reflections, respectively [43]. After placing the mixture in the autoclave at 453 K for 1 h, the characteristic peaks of MFI disappeared from the XRD spectrum. With the extending of crystallization time, characteristic peaks of MFI appeared in the XRD spectra, but with broad peaks existing between 23° and 25°, which demonstrates that amorphous substances also existed in the sample. After crystallization at 453 K for 4 h, the characteristic peaks of MFI were apparent and there was no significant difference among the XRD spectra of samples after crystallizing for 8 h.

The FTIR spectra of materials synthesized using fume silica (AEROSIL200) as silica source with different crystallization times are shown in Figure 1b. The vibrational modes near 1100, 800 and 450 cm^{-1} are assigned to internal vibrations of SiO_4. These kinds of vibration can also be observed in silica, quartz and cristobalite. Meanwhile, the vibrational modes near 1210 and 550 cm^{-1} were due to the asymmetric stretching of SiO_4 and the double-ring tetrahedra vibration in the zeolite framework, respectively [44]. After crystallization for 2 h, a very weak band at 550 cm^{-1} showed. It has been reported that a weak band at 550 cm^{-1} implies a low ordering of the material [44]. This result is in good agreement with the low crystallinity detected from XRD studies. With prolongation of the crystallization time, the strength of the band at 550 cm^{-1} increased significantly. After crystallization for 4 h, the band near 1109 cm^{-1} was significant, which suggests that the internal linked antisymmetric stretching of Si-O-Si increased. Moreover, with the extending of crystallization time, the band near 1109 cm^{-1} became increasingly sharper, which indicates larger numbers of Si-O-Si bonds were formed.

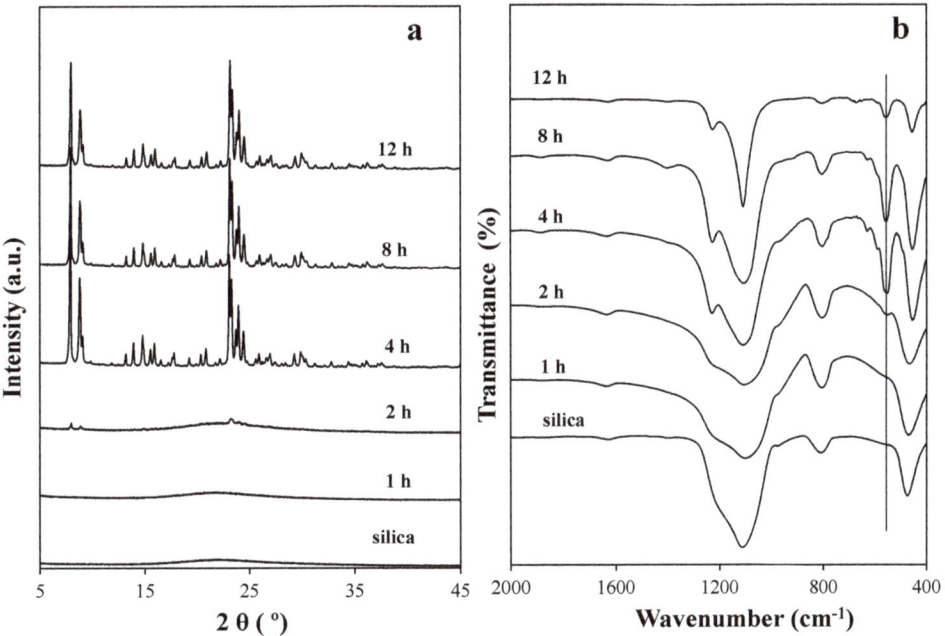

Figure 1. The XRD pattern (**a**) and FT-IR spectra (**b**) of silicalite-1 synthesized from silica with particle size of 12 nm.

When the crystallization time was 1 h, the synthesized sample had no significant difference from the fume silica, according to the SEM images shown in Figure 2. When the crystallization time was prolonged to 2 h, the morphology of sample still had no significant changes, and the sample was composed by aggregated particles in irregular shapes. However, it can be seen from the XRD spectra of the sample in Figure 1a that the characteristic peaks appeared after 2 h, which reveals that silicalite-1 particles were generated. When the crystallization time was prolonged to 4 h, silicalite-1 in regular morphology could be obtained. Furthermore, when the crystallization time was prolonged to 8 h, silicalite-1 with uniform particles was obtained. When the crystallization time was extended to 12 h, there was no significant change in the morphology of silicalite-1 samples.

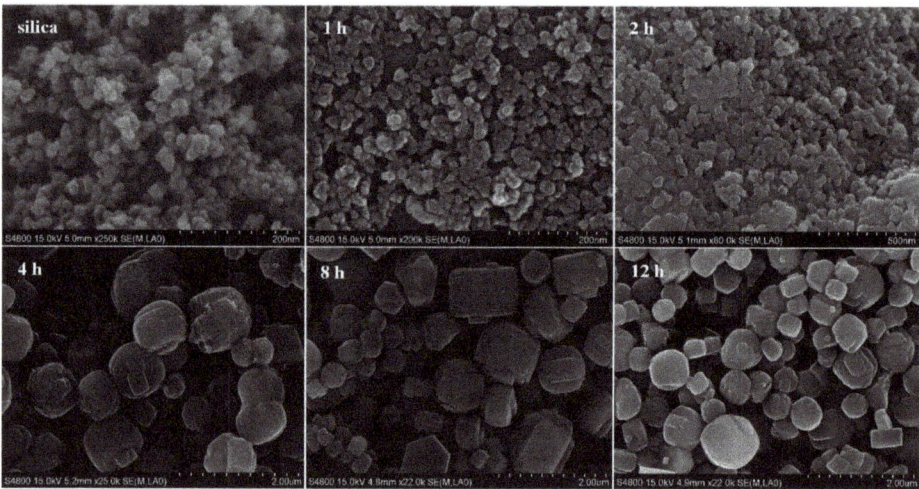

Figure 2. The SEM images of silicalite-1 synthesized from silica with particle size of 12 nm.

Figure 3a shows the XRD patterns of samples synthesized with fume silica (AEROSIL380). After crystallization at 453 K for 2 h, characterization peaks of MFI structure appeared, and the peak intensity was higher than that when using fume silica (AEROSIL200) as the silica source. This is possibly due to the fact that the fume silica of AEROSIL380 has a smaller particle size than fume silica of AEROSIL200, so that it is easier to combine the directing agent with silica source, and thus the MFI structure can be formed more easily.

Figure 3b shows the FTIR spectra of silicalite-1 synthesized by using fume silica AEROSIL380 with different synthesis time. After crystallization for 2 h, a very weak band at 550 cm^{-1} showed the low ordering of the material, which is in good agreement with the low crystallinity detected from XRD studies. With the extension of the crystallization time, the strength of the band at 550 cm^{-1} increased significantly. After crystallization for 4 h, a band near 1109 cm^{-1} was observed, which suggests that the internally linked antisymmetric stretching of Si-O-Si had increased. With the further increase of crystallization time, the band near 1109 cm^{-1} became sharper and sharper, which indicates that more and more Si-O-Si bonds had been formed.

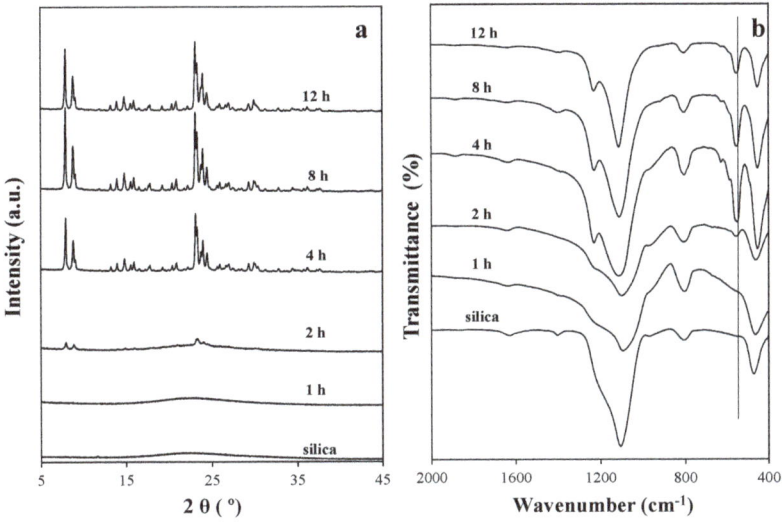

Figure 3. The XRD pattern (**a**) and FTIR (**b**) of silicalite-1 synthesized from silica with particle size of 7 nm.

Unlike the synthesis of silicalite-1 using fume silica of AEROSIL200 as the silica source, the synthesis of silicalite-1 using AEROSIL380 as silica source involved crystallization at 453 K for 2 h, particles aggregated into bigger particles with the size larger than 200 nm (Figure 4). According to FTIR spectra and XRD patterns, it can be concluded that crystal of silicalie-1 appeared after crystallization at 453 K for 2 h, with small grains aggregated into big particles.

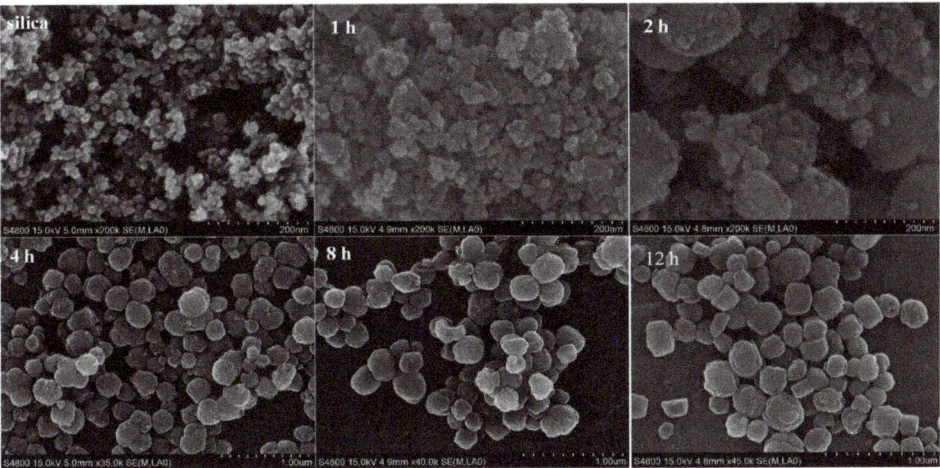

Figure 4. The SEM images of silicalite-1 synthesized from silica with particle size of 7 nm.

Figure 5a shows the XRD patterns of samples synthesized with spherical silica (particle size of 300 nm) as silica source. After crystallization at 453 K for 2 h, the characterization peaks of MFI structure appeared. Moreover, the silicalite-1 had 100% crystallization calculated based on the XRD pattern, and 71% crystallization calculated based on the FTIR spectrum.

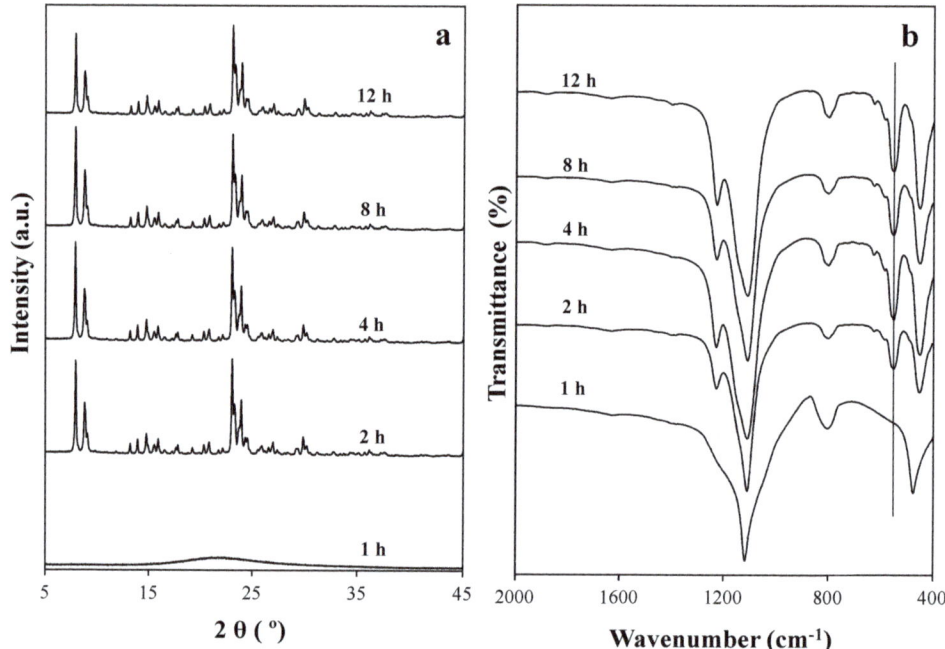

Figure 5. The XRD pattern (**a**) and FTIR (**b**) of silicalite-1 synthesized from silica spheres of 300 nm.

As shown in Figure 5b, a very strong band at 550 cm^{-1} could be observed in the FTIR spectra of silicalite-1 when using spherical silica (particle size of 300 nm) as the silica source and crystallizing for a time of 2 h or longer. At the same time, the band near 1109 cm^{-1} also appeared, which suggests that the internal linked antisymmetric stretching of Si-O-Si was strong. These results also demonstrate that the speed of crystallization using spherical silica with particle size of 300 nm as silica source was faster than that using the other two kinds of silica as silica source, which is well consistent with the result of the XRD studies.

According to Figure 6, after crystallization for 1 h, the surface of the straw material presented a certain degree of damage when synthesizing silicalite-1 using spherical silica as silica source. This indicates that there exist solutions of spherical silica at certain extent. Nevertheless, the overall spherical morphology still remained unchanged. High-quality silicalite-1 was synthesized after crystallization for 2 h at 453 K, which is in concordance with the results of the XRD pattern. When extending the crystallization time to 12 h, some cracks appeared on the crystal particles, which is possibly due to the desilication of silicalite-1 under an alkali system.

The crystallization degree of samples synthesized with different silica sources for different crystallization times was calculated by XRD patterns and FTIR spectra. The results are shown in Table 1. The crystallization degrees were calculated according to the three strongest peaks in the scope of 22–25° of XRD patterns with reflections of [501], [151] and [303] and the bands of 550 and 450 cm^{-1} in the FTIR spectra [44,45]. The ZSM-5 purchased from the catalyst Plant of Nankai University (commercial code: NKF-5) was chosen as reference.

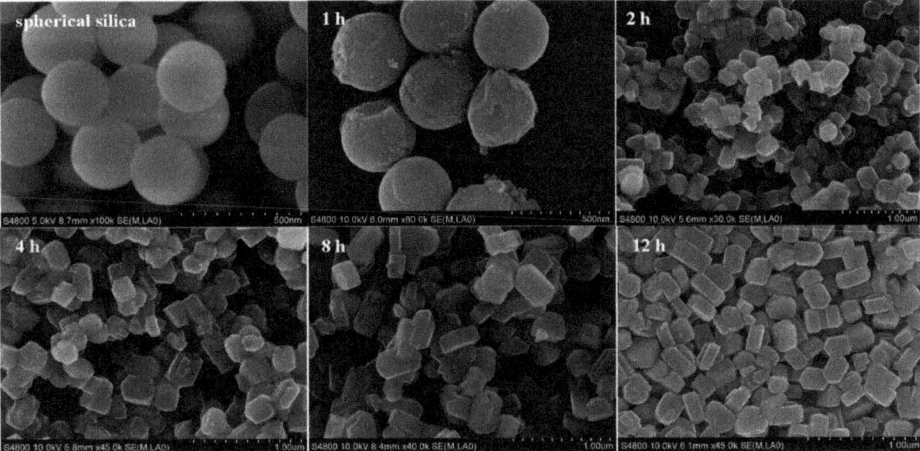

Figure 6. The SEM images of silicalite-1 synthesized from silica spheres of 300 nm.

Table 1. Percent of crystallinities for silicalite-1 samples synthesized using different silica sources.

Time	200 m^2/g SiO$_2$		380 m^2/g SiO$_2$		300 nm SiO$_2$ Spherical Silica	
	IR	XRD	IR	XRD	IR	XRD
1 h	n.d.	n.d.	n.d.	n.d.	n.d.	n.d.
2 h	7	16	47	17	71	100
4 h	83	100	90	74	73	99

When using fume silica as silica source, with the extending of crystallization time, the crystallization degree of samples gradually increased and reached over 74% after crystallization for 4 h at 453 K. In comparison, when using spherical silica as silica source, with the extending of crystallization time, the crystallization degree of silicalite-1 reached to over 71% after 2 h of crystallization. Based on the results of SEM, silicalite-1 samples synthesized with spherical silica as source had better morphology compared with those synthesized with fume silica (with average particle sizes of 12 and 7 nm) as silica source.

2.2. The Effect of Spherical Silica Particle Size on Synthesis of Silicalite-1

To study the effect of spherical silica size on the synthesis of silicalite-1, spherical silica with different particles sizes were synthesized and used as silica sources in the crystallization process.

The amounts of agents needed for synthesizing spherical silica with different sizes are showed in Table S1. The SEM images of spherical silica with particles size of 50, 100, 300 and 500 nm are as shown in Figure S1. All spherical silicas obtained have narrow particle size distributions.

Based on the results mentioned above, the synthesis time was set to 8 h. The samples synthesized with spherical silica of different sizes as silica sources were characterized by XRD, as shown in Figure S2. As can be seen, the XRD patterns show straight basic line with no broad peaks, which indicates that silicalite-1 with a high degree of crystallization was obtained.

According to the SEM image in Figure 7, the morphology of samples can be analyzed. The particle size of silicalite-1 was around 200 nm, which was significantly bigger than the size of spherical silica. This was due to the fact that a large quantity of spherical silica with small particle sizes were thoroughly decomposed, leaving small amounts of spherical silica in the solution. Therefore, there was a mass of nuclear phase present in the solution, while there was little spherical silica serving as a silica source for the growth of silicalite-1. With the growth of silicalite-1, nuclear phases joined with others, resulting in bigger silicalite-1 crystals. The aggregation of silicalite-1 synthesized with spherical silica (particle size of 50 nm) as the silica source was basically a consequence of the aggregation of spherical silica. With the particle size of spherical silica increasing from 50 to 300 nm, the morphology of silicalite-1 became more and more regular, and finally coffin-like silicalite-1 was obtained. As the particle size of spherical silica further increased to 500 nm, there were no significant changes in the morphology of silicalite-1.

Figure 7. SEM images of silicalite-1 synthesized from silica spheres with various particle sizes of (**a**), 50 nm; (**b**), 100 nm; (**c**), 300 nm; (**d**), 500 nm.

As is depicted above, the particle size of the silica source can impact the synthesis time of silicalite-1. Using spherical silica as the silica source can result in samples with better properties than using fume silica as the silica source, which is probably due to the larger particle size of spherical silica. When the silicalite-1 was synthesized using fume silica (7–12 nm) as the silica source, the silica nucleated automatically and dispersed in the solution, and then the smaller crystals dispersed in the solution jointed with each other, resulting in bigger silicalite-1 crystals due to the high surface energy of the small crystals. In contrast, when the silicalite-1 was synthesized using spherical silica as the silica source, the silica nucleated automatically on the surface of spherical silica due to the large particle size of the spherical silica, supplying silica source continuously. As a result, the formation of smaller crystals was reduced, and the joint time during the silicalite-1 synthesis was shortened, and thus the total time required for silicalite-1 growth was shortened as well.

The schematic diagram of silicalite-1 formation using different silicon sources is speculated to be as shown in Figure 8. As for fume silica as silica source for MFI zeolite synthesis, due to the small particle sizes of fume silica sources (7–12 nm), the fume silica is quickly hydrolyzed into nuclear phases and homogeneously dispersed in the alkaline solution. Then these nuclear phases are aggregated, and crystals of MFI zeolite are formed and grow gradually. However, when spherical silica with particle sizes higher than 50 nm are used as the silica source for MFI zeolite synthesis, the larger particle sizes of spherical silica cause crystallization of the nuclear phases and hydrolyzation of the silica source to co-exist on the surface of the spherical silica. As time goes by, the volume of nuclear phases gradually increases and the spherical silica disappears, finally leading to the formation of MFI crystal. It is indicated that the dispersion of nuclear phases in the alkaline solution is the most decisive step for the formation of MFI. Compared with the nuclear phases being dispersed independently in the solution when using fume silica as silicon source, the nuclear phases when using spherical silica as silicon source were close to each other and polymerized with each other faster, leading to a shortened time of crystallization for MFI zeolite. Meanwhile, the regular shape of the silicon source also had a significant impact on the morphology of as-prepared MFI zeolite.

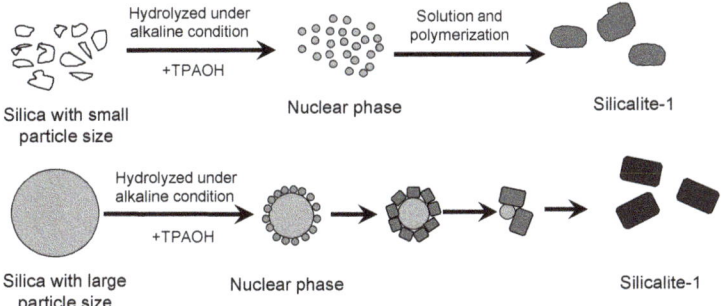

Figure 8. Schematic diagram of silicalite-1 growth using silica sources with different particle sizes during the synthesis process.

2.3. Synthesis of ZSM-5 and Catalytic Performance

Based on the above experiments, Al atoms were added to obtain ZSM-5 zeolite. Silica with a particle size of 300 nm was selected as the silica source, and then thermal treatment was conducted at 453 K for 1 day to obtain ZSM-5 samples. The XRD spectrum and SEM image of ZSM-5 are shown in Figure 9. The obtained ZSM-5 samples had small particle sizes, and pure and high crystallization of MFI structure was obtained based on the SEM image and XRD spectrum of ZSM-5. The XRF result showed that the actual Si to Al ratio of ZSM-5 was 38. The main products after eugenol hydrodeoxygenation in the aqueous phase over Pt/C-based HZSM-5 catalysts were hydrocarbon, 2-methoxy-4-propyl-cyclohexanol, propyl-cyclohexanone in liquid phase and methanol in gas phase. The carbon balance of the product was 91%. High hydrocarbon selectivity (73.8%) and eugenol conversion (96.5%) were obtained from the HDO reaction when using the synthesized ZSM-5 as supporter and acidic site supplier. This result is higher than those reported in the references [46,47]. The selectivity of hydrocarbon is lower than 50%, when using Pd/C and HZSM-5 as the catalysts for eugenol hydrodeoxygenation in aqueous phase [47]. However, the morphology of ZSM-5 was irregular, and needs to be improved in the future research.

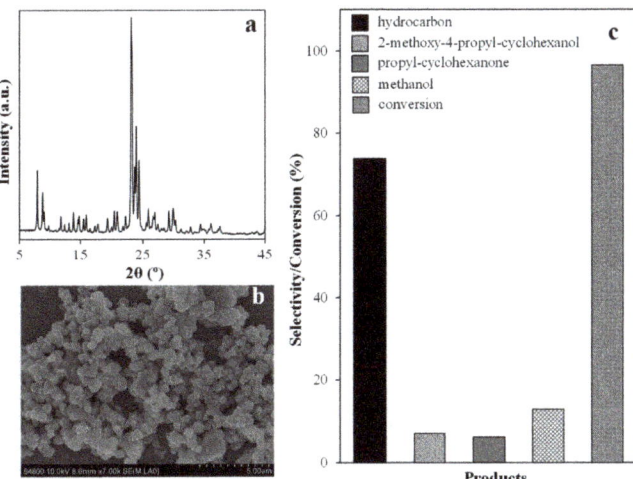

Figure 9. The XRD spectrum (**a**) and SEM image (**b**) of ZSM-5 synthesized from silica spheres, and the catalytic performance of APR hydrodeoxygenation over Pd/C-based HZSM-5 catalysts (**c**).

3. Experimental

3.1. Chemicals

The chemical products used in this experiment include ammonium hydroxide solution (25–28% $NH_3 \cdot H_2O$, Sinopharm Chemical Reagent Shanghai Co., Ltd., Shanghai, China), tetraethoxysilane (TEOS, Sinopharm Chemical Reagent Shanghai Co., Ltd., Shanghai, China), ethanol (Sinopharm Chemical Reagent Shanghai Co., Ltd., Shanghai, China), n-propanol (Sinopharm Chemical Reagent Shanghai Co., Ltd., Shanghai, China), fumed silica of AEROSIL200 (99.8%, average particle size of 7 nm) and AEROSIL380 (99.8%, average particle size of 12 nm) from Evonik Degussa Specialty Chemicals Shanghai Co., Ltd., (Shanghai, China), sodium aluminate ($NaAlO_2$, AR, from Aladdin, Shanghai, China) and TPAOH (25% in water, Tianjin Guangfu Fine Chemical Research Institute, Tianjin, China).

3.2. Synthesis Process

The general synthesis method for spherical silica can be described as follows: firstly, a certain volume of 28% $NH_3 \cdot H_2O$ and alcohol or n-propanol were mixed with a certain volume of deionized water, resulting in a mixed solution named as solution A. Then, a quantity of TEOS and alcohol, methyl alcohol or n-propanol were mixed, resulting in mixed solution is named as solution B. Then, the obtained solution B was poured into solution A before magnetic stirring at high speed for 2 h at room temperature. Finally, the mixture was centrifuged, washed with ethanol three times and dried at 333 K overnight.

The synthesis process of silicalite-1 can be described as follows. Firstly, 20 mL of 25% TPAOH solution and 10 g of silica source were put into a 50 mL beaker, and then mixed evenly with a glass rod. The beaker was sealed with para film and was placed on the table at room temperature for 1 h. Then, the excess water was absorbed using filter paper, and the obtained mixture was put in the oven at 313 K for 1.5 h. Subsequently, the mixture was transferred to an autoclave, sealed and heated at 453 K for a certain period of time under static conditions. After cooling to room temperature, the mixture was filtered and washed with distilled water three times. The solid precipitates were collected and dried at 353 K overnight. Finally, the template was removed by calcining the solid precipitates in a muffle furnace up to 823 K at a rate of 3 K/min, and then kept at this temperature for 5 h. The synthesis procedure for ZSM-5 was the same as shown above for silicalite-1, except that TPAOH was added to the solution. HZSM-5 was synthesized by adding 0.054 g (0.66 mmol) of sodium aluminate into the synthesis process of MFI and silica with a particle size of 300 nm was chosen as the silica source. The thermal treatment of HZSM-5 was conducted at 453 K for 1 day.

3.3. Characterization

XRD measurements were carried out on a Bruker D8 Advance powder diffractometer, using Cu Kα radiation (wavelength λ = 1.5147 Å, 40 kV, 40 mA), with a step size of 0.02°(2θ) and 2 s per step over the 2θ ranging from 5° to 45°. The XRD crystallinities of the silicatlite-1 samples were determined by comparison of the intensities of the four major reflections in region of 22.5° to 24° relative to those of a reference NKF-5 with Si/Al ratio as 50 [48]. Si and Al contents of ZSM-5 were determined by XRF (Axios PW4400, Panalytical, The Netherlands). SEM analysis was conducted on a Hitachi S-4800 electronic microscope at 200 kV. Nitrogen adsorption-desorption isotherms were measured at 77 K on a micromeritics ASAP 2020 sorptometer. FTIR spectra were recorded on a Nicolet 6700 with a resolution less than 0.4 cm^{-1} and signal-to-noise ratio of 50000:1. The samples were mixed with KBr and pressed into flakes before testing. The intensity ratio of the 550 and 450 cm^{-1} band, namely the I_{550}/I_{450} ratio, is used to assess the crystallinity of silicalite-1 samples.

3.4. HDO of Eugenol

Eugenol (1.2 μmol), Pd/C (0.03 g) and HZSM-5 (0.5 g) were loaded into a stainless-steel autoclave (100 mL) with distilled water (20 mL). The reactor was flushed with H_2 three times, and the pressure was adjusted to 2 Mpa with H_2. Then, the temperature was increased to 513 K while maintaining (700 rpm), and the reaction was held at 513 K for 3 h. After the reactor was quenched with ice to room temperature, the aqueous and gas phase were collected directly, and the organic mixture was extracted by ethyl acetate. The organic and aqueous phases were both quantitatively analyzed by gas chromatography-mass spectrometry (GC-MS; Agilent 7890A-Agilent 5975C, Santa Clara, CA, USA) equipped with a capillary column (HP-5; 30 m × 250 μm).

4. Conclusions

High-quality coffin-like silicalite-1 was synthesized via dry gel conversion method by using silica sphere with a particle size of 300 nm as the silica source, with the crystallization time being decreased to 2 h. The time for crystallization was curtailed by using silica spheres instead of fume silica (AEROSIL200 and AEROSIL380) as the silica source during the synthesis process of MFI. Silica spheres with a particle size of 300 nm is much better for using as silica source during the synthesis process of MFI than silica spheres with larger particle size or fume silica. The formation mechanisms of silicalite-1 using silica sources with different particle sizes were concluded. The particle size of silica sources and the polymerization speed of silica sources were two key factors impacting the sizes of the final product of silicalite-1. ZSM-5 samples with Si to Al ratio of 38 were obtained by the DGC method. The catalytic activity of ZSM-5 with Pd/C as catalyst for eugenol hydrodeoxygenation in aqueous phase was investigated. High hydrocarbon selectivity (73.8%) and eugenol conversion (96.5%) were obtained from the HDO reaction when using the synthesized ZSM-5 as supporter and acidic sites supplier.

Supplementary Materials: The following are available online at http://www.mdpi.com/2073-4344/9/1/13/s1, Table S1. Amount of agents needed for spherical silica during the synthesis process, Figure S1. SEM images of silica sphere with different particle sizes, Figure S2. XRD patterns of silicalite-1 synthesized with silica spheres of different particle sizes.

Author Contributions: Data Curation, J.L.; Methodology, C.W.; Writing—Original Draft Preparation, X.L.; Writing—review & Editing, J.Z.

Acknowledgments: This work is financially supported by National Natural Science Foundation of China through project (grant number: 51602215 and 41502131), the Fundamental Research Funds for the Central Universities (grant number: 18CX02101A) and National Science and Technology Major Project (grant number: 2016ZX05014-0004-07).

Conflicts of Interest: The authors declare no conflict of interest. The funders had no role in the design of the study; in the collection, analyses, or interpretation of data; in the writing of the manuscript, or in the decision to publish the results.

References

1. Zhou, W.; Zhang, S.Y.; Hao, X.Y.; Guo, H.; Zhang, C.; Zhang, Y.Q.; Liu, S.X. MFI-type boroaluminosilicate: A comparative study between the direct synthesis and the templating method. *J. Solid State Chem.* **2006**, *179*, 855–865. [CrossRef]
2. Sarmah, B.; Satpati, B.; Srivastava, R. Highly efficient and recyclable basic mesoporous zeolite catalyzed condensation, hydroxylation, and cycloaddition reactions. *J. Colloid. Interface Sci.* **2017**, *493*, 307–316. [CrossRef] [PubMed]
3. Akhmetzyanova, U.; Opanasenko, M.; Horacek, J.; Montanari, E.; Cejka, J.; Kikhtyanin, O. Zeolite supported palladium catalysts for hydroalkylation of phenolic model compounds. *Micropous Mesoporous Mater.* **2017**, *252*, 116–124. [CrossRef]
4. Silva, A.F.; Fernandes, A.; Antunes, M.M.; Neves, P.; Rocha, S.M.; Ribeiro, M.F.; Pillinger, M.; Ribeiro, J.; Silva, C.M.; Valente, A.A. TUD-1 type aluminosilicate acid catalysts for 1-butene oligomerisation. *Fuel* **2017**, *209*, 371–382. [CrossRef]

5. Grenev, I.V.; Gavrilov, V.Y. Calculation of adsorption properties of aluminophosphate and aluminosilicate zeolites. *Adsorption* **2017**, *23*, 903–915. [CrossRef]
6. Moreno-Recio, M.; Jimenez-Morales, I.; Arias, P.L.; Santamaria-Gonzalez, J.; Maireles-Torres, P. The Key Role of Textural Properties of Aluminosilicates in the Acid-Catalysed Dehydration of Glucose into 5-Hydroxymethylfurfural. *Chemistryselect* **2017**, *2*, 2444–2451. [CrossRef]
7. Rillig, M.C.; Wagner, M.; Salem, M.; Antunes, P.M.; George, C.; Ramke, H.G.; Titirici, M.M.; Antonietti, M. Material derived from hydrothermal carbonization: Effects on plant growth and arbuscular mycorrhiza. *Appl. Soil Ecolog.* **2010**, *45*, 238–242. [CrossRef]
8. Al-Jubouri, S.M.; Holmes, S.M. Hierarchically porous zeolite X composites for manganese ion-exchange and solidification: Equilibrium isotherms, kinetic and thermodynamic studies. *Chem. Eng. J.* **2017**, *308*, 476–491. [CrossRef]
9. Losch, P.; Pinar, A.B.; Willinger, M.G.; Soukup, K.; Chavan, S.; Vincent, B.; Pale, P.; Louis, B. H-ZSM-5 zeolite model crystals: Structure-diffusion-activity relationship in methanol-to-olefins catalysis. *J. Catal.* **2017**, *345*, 11–23. [CrossRef]
10. Hernandez-Tamargo, C.E.; Roldan, A.; de Leeuw, N.H. A density functional theory study of the structure of pure-silica and aluminium-substituted MFI nanosheets. *J. Solid State Chem.* **2016**, *237*, 192–203. [CrossRef]
11. Barakov, R.; Shcherban, N.; Yaremov, P.; Solomakha, V.; Vyshnevskyy, A.; Ilyin, V. Low-temperature synthesis, structure, sorption properties and acidity of zeolite ZSM-5. *J. Porous Mater.* **2016**, *23*, 517–528. [CrossRef]
12. Mi, S.; Wei, T.; Sun, J.; Liu, P.; Li, X.; Zheng, Q.; Gong, K.; Liu, X.; Gao, X.; Wang, B.; Zhao, H.; Liu, H.; Shen, B. Catalytic function of boron to creating interconnected mesoporosity in microporous Y zeolites and its high performance in hydrocarbon cracking. *J. Catal.* **2017**, *347*, 116–126. [CrossRef]
13. Chung, K.H.; Park, H.; Jeon, K.J.; Park, Y.K.; Jung, S.C. Microporous Zeolites as Catalysts for the Preparation of Decyl Glucoside from Glucose with 1-Decanol by Direct Glucosidation. *Catalysts* **2016**, *6*, 216. [CrossRef]
14. Menoufy, M.F.; Nadia, A.E.; Ahmed, H.S. Catalytic Dewaxing for Lube Oil Production. *Petrol. Sci. Technol.* **2009**, *27*, 568–574. [CrossRef]
15. Xin, H.; Li, X.; Fang, Y.; Yi, X.; Hu, W.; Chu, Y.; Zhang, F.; Zheng, A.; Zhang, H.; Li, X. Catalytic dehydration of ethanol over post-treated ZSM-5 zeolites. *J. Catal.* **2014**, *312*, 204–215. [CrossRef]
16. Li, X.; Xing, J.; Zhou, M.; Zhang, H.; Huang, H.; Zhang, C.; Song, L.; Li, X. Influence of crystal size of HZSM-5 on hydrodeoxygenation of eugenol in aqueous phase. *Catal. Commun.* **2014**, *56*, 123–127. [CrossRef]
17. Xie, J.; Zhuang, W.; Wei, Z.; Ning, Y.; Yu, Z.; Ju, W. Construction of Acid-Base Synergetic Sites on Mg-bearing BEA Zeolites Triggers the Unexpected Low-Temperature Alkylation of Phenol. *Chemcatchem* **2017**, *9*, 1076–1083. [CrossRef]
18. Wang, Y.; Wu, J.; Wang, S. Hydrodeoxygenation of bio-oil over Pt-based supported catalysts: Importance of mesopores and acidity of the support to compounds with different oxygen contents. *Rsc Adv.* **2013**, *3*, 12635–12640. [CrossRef]
19. Beiragh, H.H.; Omidkhah, M.; Abedini, R.; Khosravi, T.; Pakseresht, S. Synthesis and characterization of poly (ether-block-amide) mixed matrix membranes incorporated by nanoporous ZSM-5 particles for CO_2/CH_4 separation. *Asia-Pac. J. Chem. Eng.* **2016**, *11*, 522–532. [CrossRef]
20. Lakhane, M.; Khairnar, R.; Mahabole, M. Metal oxide blended ZSM-5 nanocomposites as ethanol sensors. *Bull. Mater. Sci.* **2016**, *39*, 1483–1492. [CrossRef]
21. Pande, H.B.; Parikh, P.A. Novel Application of ZSM-5 Zeolite: Corrosion-Resistant Coating in Chemical Process Industry. *J. Mater. Eng. Perform.* **2013**, *22*, 190–199. [CrossRef]
22. McDonnell, A.M.P.; Beving, D.; Wang, A.J.; Chen, W.; Yan, Y.S. Hydrophilic and antimicrobial zeolite coatings for gravity-independent water separation. *Adv. Funct. Mater.* **2005**, *15*, 336–340. [CrossRef]
23. Jiao, K.; Xu, X.; Lv, Z.; Song, J.; He, M.; Gies, H. Synthesis of nanosized Silicalite-1 in F- media. *Micropous Mesoporous Mater.* **2016**, *225*, 98–104. [CrossRef]
24. Mohammed, M.A.A.; Salmiaton, A.; Azlina, W.A.K.G.W.; Amran, M.S.M.; Fakhru'l-Razi, A.; Taufiq-Yap, Y.H. Hydrogen rich gas from oil palm biomass as a potential source of renewable energy in Malaysia. *Renew. Sustain. Energy Rev.* **2011**, *15*, 1258–1270. [CrossRef]
25. Cimenler, U.; Joseph, B.; Kuhn, J.N. Hydrocarbon steam reforming using Silicalite-1 zeolite encapsulated Ni-based catalyst. *Aiche J.* **2017**, *63*, 200–207. [CrossRef]

26. Ge, C.; Li, Z.; Chen, G.; Qin, Z.; Li, X.; Dou, T.; Dong, M.; Chen, J.; Wang, J.; Fan, W. Kinetic study of vapor-phase Beckmann rearrangement of cyclohexanone oxime over silicalite-1. *Chem. Eng. Sci.* **2016**, *153*, 246–254. [CrossRef]
27. Kabalan, I.; Rioland, G.; Nouali, H.; Lebeau, B.; Rigolet, S.; Fadlallah, M.B.; Toufaily, J.; Hamiyeh, T.; Daou, T.J. Synthesis of purely silica MFI-type nanosheets for molecular decontamination. *Rsc Adv.* **2014**, *4*, 37353–37358. [CrossRef]
28. Sanhoob, M.A.; Muraza, O. Synthesis of silicalite-1 using fluoride media under microwave irradiation. *Microporous Mesoporous Mater.* **2016**, *233*, 140–147. [CrossRef]
29. Han, S.W.; Kim, J.; Ryoo, R. Dry-gel synthesis of mesoporous MFI zeolite nanosponges using a structure-directing surfactant. *Microporous Mesoporous Mater.* **2017**, *240*, 123–129. [CrossRef]
30. Meng, X.; Xiao, F.S. Green Routes for Synthesis of Zeolites. *Chem. Rev.* **2014**, *114*, 1521–1543. [CrossRef]
31. Mintova, S.; Valtchev, V. Effect of the silica source on the formation of nanosized silicalite-1: An in situ dynamic light scattering study. *Micropous Mesoporous Mater.* **2002**, *55*, 171–179. [CrossRef]
32. Pan, F.; Lu, X.; Zhu, Q.; Zhang, Z.; Yan, Y.; Wang, T.; Chen, S. Direct synthesis of HZSM-5 from natural clay. *J. Mater. Chem. A* **2015**, *3*, 4058–4066. [CrossRef]
33. Jesudoss, S.K.; Vijaya, J.J.; Kaviyarasu, K.; Rajan, P.I.; Narayanan, S.; Kennedy, L.J. In vitro anti-cancer activity of organic template-free hierarchical M (Cu, Ni)-modified ZSM-5 zeolites synthesized using silica source waste material. *J. Photochem. Photobiol. B* **2018**, *186*, 178–188. [CrossRef] [PubMed]
34. Khoshbin, R.; Oruji, S.; Karimzadeh, R. Catalytic cracking of light naphtha over hierarchical ZSM-5 using rice husk ash as silica source in presence of ultrasound energy: Effect of carbon nanotube content. *Adv. Powder Technol.* **2018**, *29*, 2176–2187. [CrossRef]
35. Saleh, N.J.; Al-Zaidi, B.Y.S.; Sabbar, Z.M. A Comparative Study of Y Zeolite Catalysts Derived from Natural and Commercial Silica: Synthesis, Characterization, and Catalytic Performance. *Arab. J. Sci. Eng.* **2018**, *43*, 5819–5836. [CrossRef]
36. Bai, P.; Tsapatsis, M.; Siepmann, J.I. Multicomponent Adsorption of Alcohols onto Silicalite-1 from Aqueous Solution: Isotherms, Structural Analysis, and Assessment of Ideal Adsorbed Solution Theory. *Langmuir* **2012**, *28*, 15566–15576. [CrossRef] [PubMed]
37. Qi, J.; Zhao, T.; Xu, X.; Li, F.; Sun, G. Hydrothermal synthesis of size-controlled silicalite-1 crystals. *J. Porous Mater.* **2011**, *18*, 509–515. [CrossRef]
38. Sanchez-Flores, N.A.; Solache, M.; Olguin, M.T.; Legaspe, J.; Pacheco-Malagon, G.; Saniger, J.M.; Martinez, E.; Bulbulian, S.; Fripiat, J.J. Silicalite-1, an adsorbent for 2-, 3-, and 4-chlorophenols. *Water Sci. Technol.* **2012**, *66*, 247–253. [CrossRef]
39. Shi, L.; Song, X.; Liu, G.; Guo, H. Effect of Catalyst Preparation on Hydroisomerization of n-Heptane over Pt/Silicalite-1. *Catal. Lett.* **2017**, *147*, 2549–2557. [CrossRef]
40. Amiri, H.; Charkhi, A.; Moosavian, M.A.; Ahmadi, S.J.; Nourian, H. Performance improvement of PDMS/PES membrane by adding silicalite-1 nanoparticles: Separation of xenon and krypton. *Chem. Pap.* **2017**, *71*, 1587–1596. [CrossRef]
41. Ding, X.; Chen, F.; Ju, Y.; Lu, S. Simulation and Thermodynamic Analysis of the Adsorption of Mixed CH_4 and N-2 on Silicalite-1 Molecular Sieve. *J. Nanosci. Nanotechnol.* **2017**, *17*, 6732–6737. [CrossRef]
42. Moreira, R.; Ochoa, E.; Pinilla, J.L.; Portugal, A.; Suelves, I. Liquid-Phase Hydrodeoxygenation of Guaiacol over Mo2C Supported on Commercial CNF. Effects of Operating Conditions on Conversion and Product Selectivity. *Catalysts* **2018**, *8*, 127. [CrossRef]
43. Barot, S.; Nawab, M.; Bandyopadhyay, R. Alkali metal modified nano-silicalite-1: An efficient catalyst for transesterification of triacetin. *J. Porous Mater.* **2016**, *23*, 1197–1205. [CrossRef]
44. Xue, T.; Wang, Y.M.; He, M.Y. Facile synthesis of nano-sized NH_4-ZSM-5 zeolites. *Micropous Mesoporous Mater.* **2012**, *156*, 29–35. [CrossRef]
45. Xin, H.; Koekkoek, A.; Yang, Q.; van Santen, R.; Li, C.; Hensen, E.J.M. A hierarchical Fe/ZSM-5 zeolite with superior catalytic performance for benzene hydroxylation to phenol. *Chem. Commun.* **2009**, 7590–7592. [CrossRef] [PubMed]
46. Zhang, C.; Xing, J.; Song, L.; Xin, H.; Lin, S.; Xing, L.; Li, X. Aqueous-phase hydrodeoxygenation of lignin monomer eugenol: Influence of Si/Al ratio of HZSM-5 on catalytic performances. *Catal. Today* **2014**, *234*, 145–152. [CrossRef]

47. Xing, J.; Song, L.; Zhang, C.; Zhou, M.; Yue, L.; Li, X. Effect of acidity and porosity of alkali-treated ZSM-5 zeolite on eugenol hydrodeoxygenation. *Catal. Today* **2015**, *258*, 90–95. [CrossRef]
48. Koekkoek, A.J.J.; Xin, H.; Yang, Q.; Li, C.; Hensen, E.J.M. Hierarchically structured Fe/ZSM-5 as catalysts for the oxidation of benzene to phenol. *Micropous Mesoporous Mater.* **2011**, *145*, 172–181. [CrossRef]

© 2018 by the authors. Licensee MDPI, Basel, Switzerland. This article is an open access article distributed under the terms and conditions of the Creative Commons Attribution (CC BY) license (http://creativecommons.org/licenses/by/4.0/).

Article

The Influence of Texture on Co/SBA–15 Catalyst Performance for Fischer–Tropsch Synthesis

Jun Han [1,2], Zijiang Xiong [1], Zelin Zhang [1], Hongjie Zhang [1,*], Peng Zhou [3] and Fei Yu [3,*]

[1] Hubei Key Laboratory for Efficient Utilization and Agglomeration of Metallurgic Mineral Resources, Wuhan University of Science and Technology, Wuhan 430081, China; hanjun@wust.edu.cn (J.H.); xzj0507@outlook.com (Z.X.); zhangzelin@wust.edu.cn (Z.Z.)
[2] Industrial Safety Engineering Technology Research Center of Hubei Province, Wuhan University of Science and Technology, Wuhan 430081, China
[3] Department of Agricultural and Biological Engineering, Mississippi State University, Mississippi State, MS 39762, USA; zhoupengwust@hotmail.com
* Correspondence: zhanghongjie@wust.edu.cn (H.Z.); fyu@abe.msstate.edu (F.Y.); Tel.: +86-27-6886-2880 (H.Z.)

Received: 28 November 2018; Accepted: 14 December 2018; Published: 16 December 2018

Abstract: The influence of the Co/SBA–15 catalyst texture, such as pore size and pore length on Fischer–Tropsch (FT) Synthesis, was investigated in this paper. The morphology, structure, and microstructures of Co/SBA–15 catalysts were characterized by SEM, Brunauer–Emmett–Teller (BET), TPR, HRTEM, and XRD. The experimental results indicated that the increase of pore size could improve the activity of the Co/SBA–15 catalyst, and the further increase of pore size could not significantly promote the activity. Moreover, it was also found that the pore length of the Co/SBA–15 catalyst played a key role in the catalytic activity. CO_2 and C4+ selectivity were 2.0% and 74% during the simulated syngas (64% H_2: 32% CO: balanced N_2) FT over the Co/SBA–15 catalysts, and CO conversion rate and CH_4 selectivity were 10.8% and 15.7% after 100 h time on stream.

Keywords: Fischer–Tropsch synthesis; Co/SBA–15; pore size; pore length

1. Introduction

The world's increasing energy demand and the depletion of crude oil has stimulated great interest in Fischer–Tropsch synthesis (FTS), which is a key step in transforming various non-petroleum carbon resources, such as natural gas, coal, and biomass into liquid fuels or valuable chemicals [1–4]. It is well known that Ni [5], Fe [6], Co [7], and Ru [8] are the most active metals for the hydrogenation of carbon monoxide. However, only cobalt-based catalysts and iron-based catalysts have been successfully applied in the industrial FTS application [9]. Cobalt based catalysts are preferred for FTS in a hydrogen rich syngas due to their characteristics of low activity for water-gas shift reaction, high selectivity for linear hydrocarbons, and better resistance to the deactivation by water (a byproduct of FTS reaction) [10,11]. The catalytic properties of the cobalt catalysts are usually affected by the chemical interactions between the supports and cobalt metal, the texture of the catalysts, and crystal morphology. Weak interactions suppress the dispersion of cobalt metal, while strong interactions promote the difficulty of reducing cobalt species. Therefore, a balance of the interactions between the supports and cobalt is significantly important for FTS cobalt-based catalyst [12]. The dispersion of cobalt metal and the surface area of this active ingredient are significantly dependent on the support, and larger particle sizes result in a lower dispersion of metallic Co metallic [13,14].

SBA–15 is a highly ordered mesoporous molecular sieve, which has the characteristics of a narrow pore size distribution (3–30) nm and large surface area (600–1000 m^2/g) [15]. Moreover, SBA–15 has uniform hexagonal channels and high thermal stability. The above features of SBA–15 lead to it being regarded as a suitable support for FTS catalyst [16]. Cai et al. and Wang et al. reported that the CO conversion was 6.51%–20.51%, and C^{5+} selectivity was 42.43%–77.14% when Ru promoted Co/SBA–15 catalysts was used in FTS [17,18]. García et al. explored the effects of the Co–SiO_2 interaction, and thought that SBA could suppress the aggregation of metal particles and promote the metal dispersion [19].

In this paper, the effect of the texture of Co/SBA–15 on FTS performance was investigated. Moreover, the reaction temperature and Co loading were also discussed.

2. Results and Discussion

SEM images of SBA–15 are presented in Figure 1. SBA–15 consisted of aligned rod-like particles with a diameter of 1 μm and a length of 2–3 μm longwise, which coincided with the results of Prieto et al. [20]. It could be clearly observed from Figure 2 that SBA–15 had a well-ordered hexagonal crystal structure composed of one-dimensional channels, which was highly ordered and stable [21].

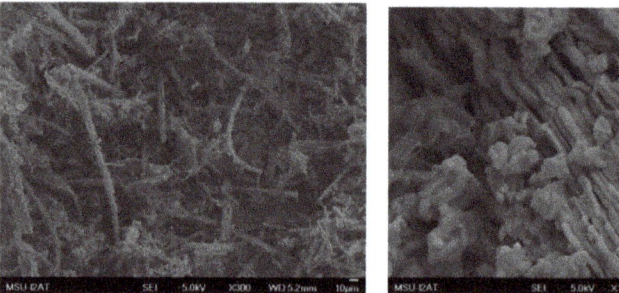

Figure 1. SEM image of SBA–15.

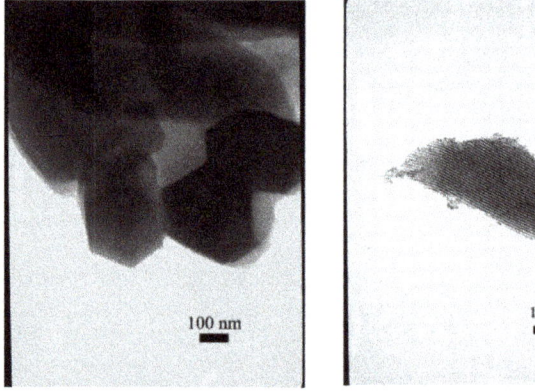

Figure 2. TEM image of SBA–15.

The N_2 adsorption-desorption isotherms and the pore size distributions of Co/SBA–15 catalysts are shown in Figure 3. A type IV adsorption isotherm with a H1 hysteresis loop is observed. The pore size distribution of SBA–15 catalysts Figure 4 demonstrated that Co/SBA–15 catalysts had a narrow pore size distribution. The catalysts that were obtained at 120 °C aging temperature had narrower pore size distributions than the catalysts that were obtained at 100 °C. As shown in Table 1, the average diameters of the SBA–15 catalyst varied from 5 to 10 nm. The pore volume was increased with an increasing average pore diameter, while the surface area of catalyst was decreased with increasing the average pore diameter.

Table 1. Texture properties of 20% Co/SBA–15 catalysts.

Sample	Surface Area (m²/g)	Pore Diameter (nm)	Pore Volume (cm³/g)	Average Pore Length [1] (μm)
S1	457.5	4.9	0.719	1.6
S2	455.6	6.5	0.802	1.90
S3	395.9	7.9	0.853	1.80
S4	353.7	9.7	0.900	1.70

[1] Estimated by TEM measurements along the pore direction and correction for curved pores.

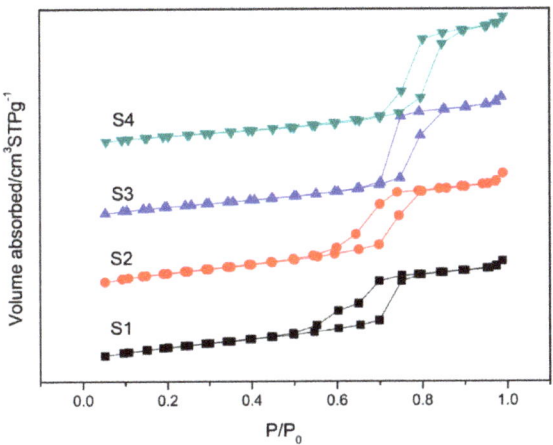

Figure 3. N_2 adsorption-desorption isotherms of Co supported on SBA–15 with different pore sizes.

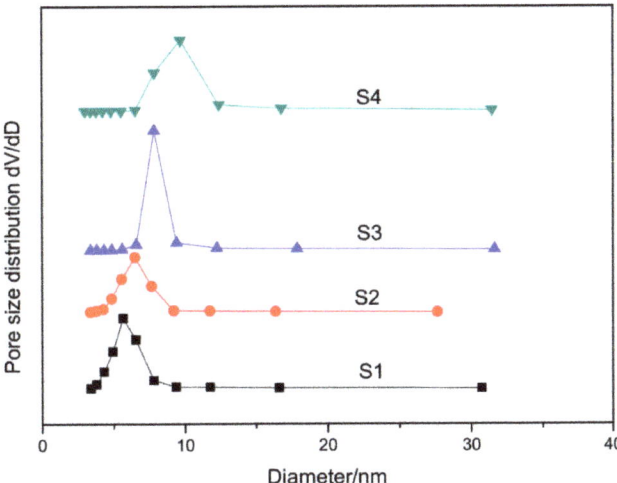

Figure 4. Pore size distribution of the catalysts.

The TPR profiles of 20% Co/SBA–15 catalysts with different pore sized are shown Figure 5. The main peak at about 360~390 °C can be attributed to the reduction of Co_3O_4 to CoO, and subsequently to metallic Co. Besides the main peak, two broad reduction peaks at 480~600 °C and 600~850 °C are also observed, which meant that there was the interaction between the surface Co species with the support [22]. As the pore size decreases, the broad peak shifted to the higher temperature. At the same time, the variation of the main peak was limited. The above results indicated that the catalyst with a smaller pore size was more difficulty reduced. The average crystallite Co_3O_4 particle size was calculated from the Scherrer Equation based on the most intense reflection peak at 36.9 °C, as Equation (1). The crystallite size of Co after the reduction could be calculated from the crystallite size of Co_3O_4, according to Equation (2).

$$D_{Co3O4} = 0.89\lambda/(\beta \cos\theta) \tag{1}$$

$$D_{Co} = 0.75 D_{Co3O4} \tag{2}$$

Table 2 presented that the average crystallite size of cobalt particles was increased as the pore size of SBA–15 increases. Xiong et al. thought that the smaller CoO clusters could interact more strongly with the support than the larger ones [22]. Hence, the smaller Co_3O_4 crystallites were reduced with more difficulty than the larger ones [23].

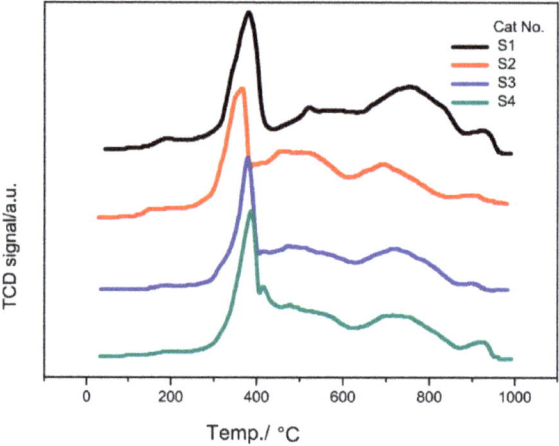

Figure 5. H_2-TPR profiles of Co/SBA–15 samples with different pore sizes.

Table 2. Average crystalline size of Co_3O_4 estimated by the XRD method.

Sample	Pore Size (nm)	D_{Co3O4} (nm)	D_{Co} (nm)
S1	4.9	10.5	7.9
S2	6.5	13.8	10.4
S3	7.9	15	11.3
S4	9.7	16	12.0

The XRD patterns in Figure 6 showed there was the diffraction peaks at $2\theta = 0.7°$, which was characteristic of the hexagonal mesoporous structure (p6m) [24,25]. This reflected the ordered structure of SBA–15 was kept after the impregnation of cobalt. The high-angle XRD patterns of catalysts were displayed in Figure 7. Diffraction peaks 31.3, 36.9, 45.1, 59.4, and 65.48 corresponds to the planes (220), (311), (400), (511), and (440) of Co_3O_4 spinel phase (JCPDS 42-1467) [26].

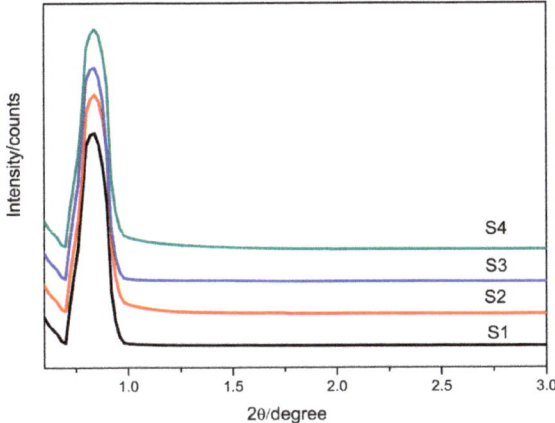

Figure 6. Small-angle XRD pattern of Co/SBA–15 catalyst with different pore sizes.

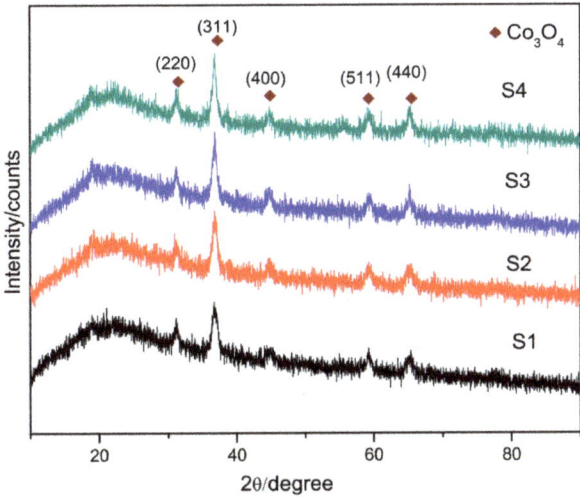

Figure 7. High-angle XRD patterns of the catalysts.

After the comparison of Figures 2 and 8, it was found that the structure of SBA–15 was kept after the cobalt impregnation and calcination. TEM images showed that there were many crystals in the pores, which caused the estimated $D_{Co_3O_4}$ in Table 2 to be 10.5–16 nm and to be larger than the pore size of SBA–15. Moreover, Co dispersion in S2 catalyst was more uniform than other catalysts in Figure 8. HRTEM image further confirmed that cobalt was presented in the form of Co_3O_4 crystalline phase after the calcination; (220) planes with a lattice space of 0.287 nm and (400) planes with a lattice space of 0.208 nm were observed for Co_3O_4.

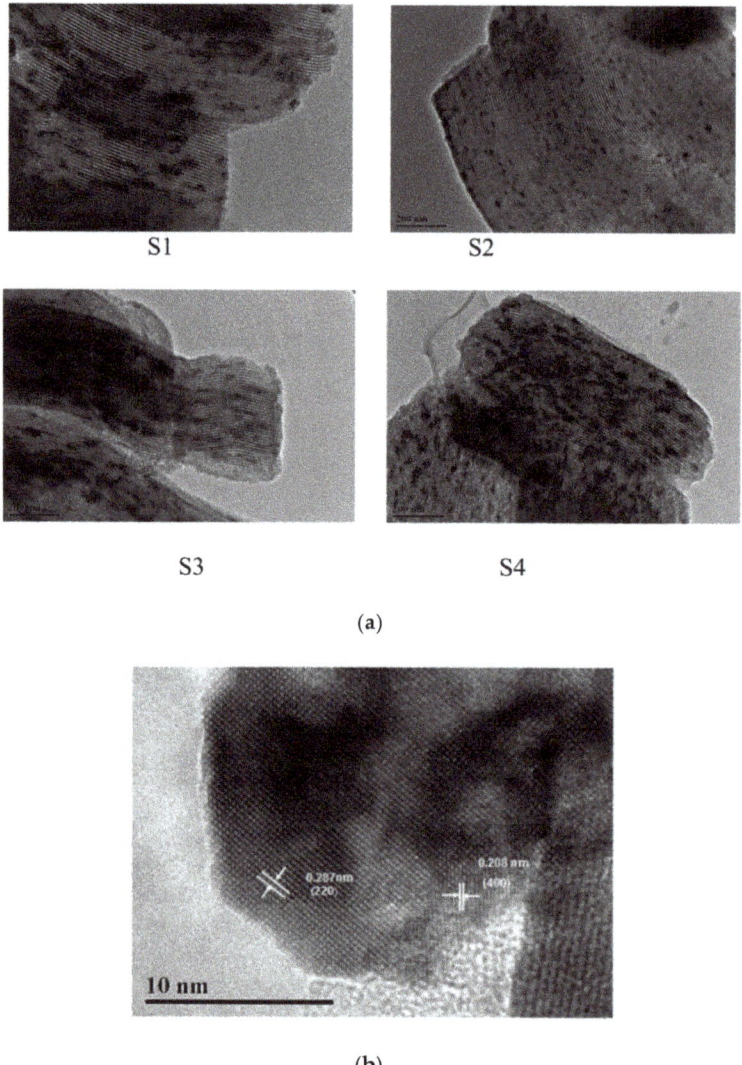

Figure 8. (a) TEM images of calcined 20% Co/SBA–15 catalysts and (b) HRTEM image of Co_3O_4.

The catalytic performance of Co/SBA–15 catalysts during FTS was evaluated at 215–265 °C and 2 MPa for 15 h. The gas hourly space velocity (GHSV) in all tests was 3600 h^{-1}. Figure 9 presented the influence of the reaction temperature on CO conversion and product selectivity for Co/SBA–15 catalysts with different pore sizes. All Co/SBA–15 catalysts exhibited low CO_2 selectivity and high C4+ hydrocarbon selectivity (~80%), which can be attributed to the well pore arrangement of SBA–15. The variation of the pore size in 4.9–10 nm had no obvious effect on C4+ selectivity, except for S1 catalyst. The above result was agreed with the previous literatures [13,27]. The larger cobalt particles were formed on the larger pore size of the support, which was easier to form a long chain of hydrocarbon [13]. However, the further increasing the pore size of catalyst had no significant effect on C4+ selectivity. Prieto et al. stated the length of the pore in catalyst was more important than the pore size for the catalyst activity [20]. In this test, S2 has the higher CO conversion and C4+ selectivity.

The length of S2 catalyst is the longest among the four catalysts. Moreover, the Co particle was more uniformly distributed on SBA–15, as shown in Figure 8.

Figure 9 also presented the increase of reaction temperature that led to a decrease in C4+ selectivity and an increase in methane selectivity. According to the literatures, the destroyed of mesoporous structure of silica and the formation of $CoSiO_3$ spinel accounted for the decrease of catalytic activity and selectivity. The better performance of the catalysts for C4+ selectivity was thought to be 215 °C.

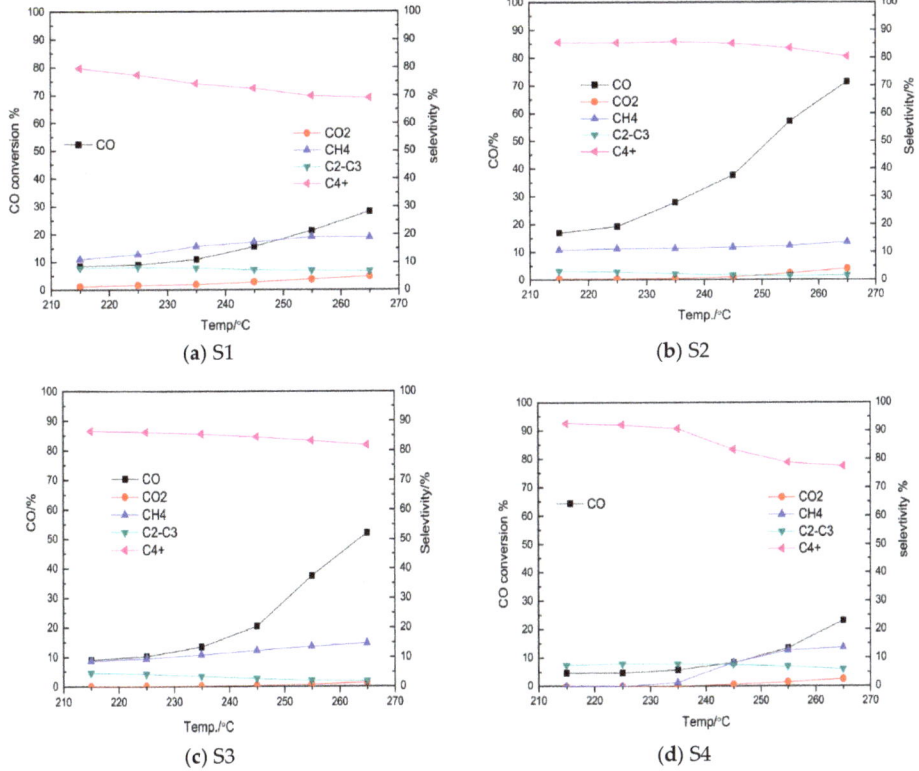

Figure 9. Effect of temperature on CO conversion and product selectivity for the catalysts (**a**) S1, (**b**) S2, (**c**) S3, and (**d**) S4 with different pore size and length.

S2 catalyst was tested over 100 h to investigate the lifetime activity of the catalyst. Figure 10 illustrated the effects of TOS on the syngas conversion rate and product selectivity at 215 °C. Initially, 12 h after introducing the feed to the reactor, the total conversion of CO and H_2 was about 20%. After 20 h on stream, a constant value was achieved. The overall conversion after 20 h was ~10 mol%. Selectivity towards C4+ was 72% after 20 h and it showed a little increase to 75% after 60 h. Selectivity towards CO2 and C2−C3 were very low and kept slowly decrease during 100 h. It has reported various mechanisms for the deactivation in cobalt-based FTS, such as the sintering, oxidation of the active Co species, destruction of porous structure, the coke formation, the aggregation of Co particles, poisoning, and cobalt reconstruction [28,29]. Overall, the S2 catalyst displayed good stability and product selectivity over 100 h, which was expected be kept active, even for longer time period.

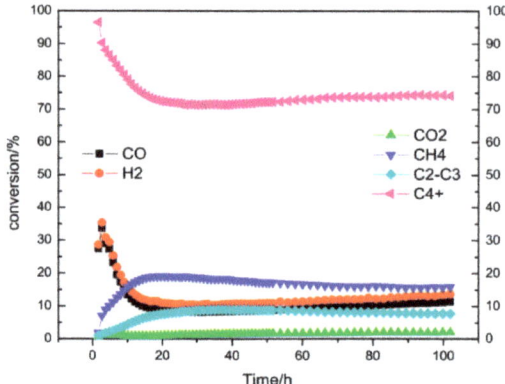

Figure 10. CO conversion rate and CO_2/C4+ selectivity with time on stream for S2 catalyst.

3. Experimental

3.1. Catalyst Preparation

First, the triblock copolymer (EO20PO70EO20, P123, Sigma Aldrich, 99%, Dorset, UK) was dissolved in the deionized water at room temperature and the solution was stirred for 2 h. Subsequently, HCl (Jiangcheng Chemical Co., Ltd., Wuhan, China) solution was added and stirred from 30 min. and then the required amount of tetraethyl orthosilicate (TEOS, Sigma Aldrich, 98%, Dorset, UK) was added to the above preparation solution at 36 °C and was kept under the stirring for 20 h. Afterwards, the gel mixture was moved into polypropylene bottles and heated at the predetermined aging temperature for the predetermined aging time (As in Table 3). After the synthesis, the solid obtained was filtered, exhaustively washed with the distilled water until neutral pH, dried at 80 °C, and lastly calcined in a flow of air at 500 °C for 6 h to remove the organic template. Co/SBA–15 catalysts were obtained by the incipient wetness impregnation using $Co(NO_3)_2 \cdot 6H_2O$/ethanol solution (Sinopharm Chemical Reagent Co., Ltd., Shanghai, China). All of the samples were calcined in air at 450 °C for 6h (heating rate 1 °C/min) and reduced with H_2 at 400 °C (heating rate 1 °C/min) for 4 h before reaction.

Table 3. The parameters of catalyst preparation.

Sample	Aging Time/h	Aging Temp./°C
S1	24	100
S2	48	100
S3	24	120
S4	48	120

3.2. Catalyst Characterization

The N_2 adsorption–desorption measurements were used to analyze the texture parameters of catalysts by a Micromerictics ASAP 2020 apparatus (Norcross, GA, USA) at −196 °C. The specific surface area was calculated by the Brunauer–Emmett–Teller (BET) method, which N_2-adsorption data was obtained in the relative pressure (P/P_0) range of 0.05–0.30. The total pore volume was determined from the amount of nitrogen at P/P_0 of 0.995. The average pore diameter and pore sizes distribution were obtained by the Barrett–Joyner–Halenda (BJH) method (Micromeritics, Norcross, GA, USA). Before the analysis, the catalysts were degassed at 120 °C for 4 h to remove the impurities from the samples. The reduction characteristic of the oxidized cobalt was studied by the temperature-programed reduction in a ChemBET Pulsar TPR/TPD (Quantachrome Instruments, Boynton Beach, FL, USA) equipment.

X-ray diffraction patterns (Rigaku, Tokyo, Japan) were investigated at room temperature though monochromatized Cu-Kα (λ = 0.15418 nm) radiation. The average diameters of Co_3O_4 in the catalysts were estimated by the Scherrer Equation. The morphology of supports and catalyst was observed using a field emission scanning electron microscope (FE-SEM) equipped with an energy dispersive X-ray analysis (EDX) detector. A JEOL instrument (Tokyo, Japan) operating at 80 keV or 200 keV was used in this paper.

3.3. Fischer–Tropsch Test

The Fischer–Tropsch synthesis tests were performed in a fixed-bed reactor (diameter is 1/2 inch), which was made with stainless-steel, as shown in Figure 11. Before each experiment, 0.5 g of catalyst was uniformly mixed with quartz sand and was loaded in the reactor. The catalyst was reduced by 50% H_2/N_2 at an atmospheric pressure. During the reduction, the temperature of catalyst was increased to 400 °C with a heating rate of 1 °C min^{-1} and was maintained at 400 °C for 6 h. After the reduction, the heating was stopped and the temperature was naturally decreased to 120 °C under 50% H_2/N_2. Subsequently, the pressure of the reactor was slowly increased up to 2 MPa by feeding the reactant gases (64% H_2: 32% CO: balanced N_2). Lastly, the temperature of the catalyst was heated to 220 °C (or other designed temperature) at a controlled heating rate of 1 °C/min. In each run, the time of the Fischer–Tropsch reaction was about 10–15 h to evaluate the stabilization and activity of the catalyst. During the reaction, the effluent gases were successively passed two traps to collect waxes, water and the residual products. The off-gas after the cold trap was analyzed by an Agilent 8990 chromatograph.

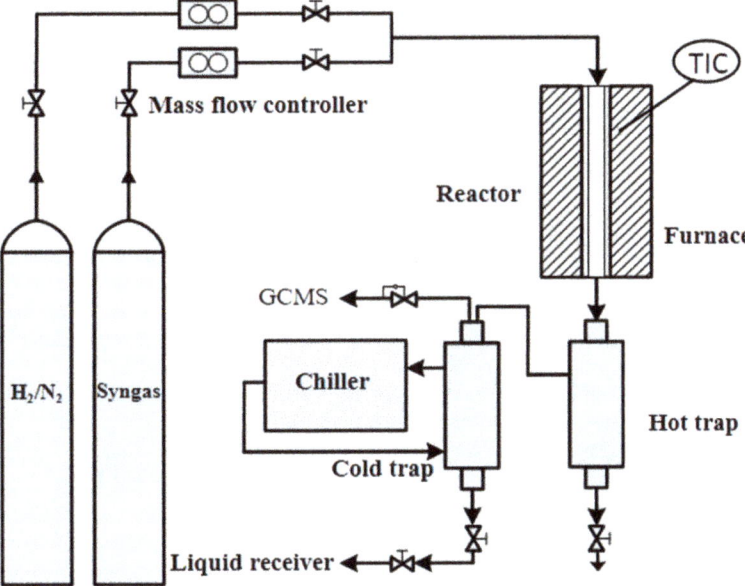

Figure 11. Schematic of experimental setup for FTS.

The CO conversion was defined as Equation (3) and the product selectivity was calculated according to Equations (4). Here, F^0 and F were the flow rates of the feeding syngas and effluent gas, respectively, C_i^0 and C_i were i component concentrations of the feeding syngas and effluent gas, which can be quantitatively analyzed by GC. n was the carbon number in the i species.

$$\text{Conversion of CO (\%)} = \frac{F^0 C_{CO}^0 - F C_{CO}}{F^0 C_{CO}^0} = \frac{C_{CO}^0 - C_{N_2}^0 C_{CO}/C_{N_2}}{C_{CO}^0} \times 100 \tag{3}$$

$$\text{Selectivity of product } i\ (\%) = \frac{FC_i n}{F^0 C_{CO}^0 - FC_{CO}} = \frac{C_{N_2}^0 C_i n}{C_{N_2} C_{CO}^0 - C_{N_2}^0 C_{CO}} \times 100 \tag{4}$$

The selectivity of C4+ (S_{C4+}) was determined, as follows:

$$S_{C4+}\ (\%) = 100 - S_{C1} - S_{C2} - S_{C3} \tag{5}$$

4. Conclusions

The influence of texture of the Co/SBA–15 catalyst such as pore size and pore length on Fischer–Tropsch (FT) Synthesis was investigated. The morphology, structure, and microstructures of Co/SBA–15 catalysts were analyzed by SEM, BET, TPR, HRTEM, and XRD. The experimental results demonstrated that the improvement of pore size of the Co/SBA–15 catalyst could improve the activity of catalyst, the further increase of pore size had a limited influence on the activity. However, the pore length was more important to the activity of the Co/SBA–15 catalyst. Moreover, the Co/SBA–15 catalyst had well C4+ selectivity and CO conversion rate during FTS. CO_2 and C4+ selectivity were 2.0% and 74% during the simulated syngas (64% H_2: 32% CO: balanced N_2) FT over the Co/SBA–15 catalysts, and the CO conversion rate and CH_4 selectivity were 10.8% and 15.7% after 100 h time on stream.

Author Contributions: Conceptualization, F.Y. and H.Z.; methodology, J.H.; software, Z.X.; validation, P.Z., J.H. and Z.Z.; formal analysis, P.Z.; investigation, J.H.; writing—original draft preparation, J.H.; writing—review and editing, F.Y.; supervision, F.Y.; project administration, H.Z.; funding acquisition, F.Y., H.Z. and Z.Z.

Funding: This research was funded by Department of Energy under Awards (grant number DE-FG3606GO86025, DE-FC2608NT01923), US Department of Agriculture under Award (grant number AB567370MSU), National Natural Science Foundation of China (grant number 51706160), Fundation of State Key Laboratory of Mineral Processing (grant number BGRIMM-KJSKL-2016-06) and China Postdoctoral Science Foundation (grant number 2017M612522). The APC was funded by National Natural Science Foundation of China (grant number 51706160).

Conflicts of Interest: The authors declare no conflict of interest.

References

1. Han, J.; Liang, Y.; Hu, J.; Qin, L.; Street, J.; Lu, Y.; Yu, F. Modeling downdraft biomass gasification process by restricting chemical reaction equilibrium with aspen plus. *Energy Convers. Manag.* **2017**, *153*, 641–648. [CrossRef]
2. Han, J.; Zhang, L.; Lu, Y.; Hu, J.; Cao, B.; Yu, F. The effect of syngas composition on the fischer tropsch synthesis over three-dimensionally ordered macro-porous iron based catalyst. *Mol. Catal.* **2017**, *440*, 175–183. [CrossRef]
3. Han, J.; Zhang, L.; Kim, H.J.; Kasadani, Y.; Li, L.; Shimizu, T. Fast pyrolysis and combustion characteristic of three different brown coals. *Fuel Process. Technol.* **2018**, *176*, 15–20. [CrossRef]
4. Han, J.; Li, W.; Liu, D.; Qin, L.; Chen, W.; Xing, F. Pyrolysis characteristic and mechanism of waste tyre: A thermogravimetry-mass spectrometry analysis. *J. Anal. Appl. Pyrolysis* **2018**, *129*, 1–5. [CrossRef]
5. Saheli, S.; Rezvani, A.R.; Malekzadeh, A. Study of structural and catalytic properties of Ni catalysts prepared from inorganic complex precursor for Fischer-Tropsch synthesis. *J. Mol. Struct.* **2017**, *1144*, 166–172. [CrossRef]
6. Roe, D.P.; Xu, R.; Roberts, C.B. Influence of a carbon nanotube support and supercritical fluid reaction medium on Fe-catalyzed Fischer-Tropsch synthesis. *Appl. Catal. A Gen.* **2017**, *543*, 141–149. [CrossRef]
7. Huang, J.; Qian, W.; Zhang, H.; Ying, W. Influences of ordered mesoporous silica on product distribution over Nb-promoted Cobalt catalyst for Fischer-Tropsch synthesis. *Fuel* **2018**, *216*, 843–851. [CrossRef]
8. Yang, X.; Wang, W.; Wu, L.; Li, X.; Wang, T.; Liao, S. Effect of confinement of TiO_2 nanotubes over the Ru nanoparticles on Fischer-Tropsch synthesis. *Appl. Catal. A Gen.* **2016**, *526*, 45–52. [CrossRef]
9. Luque, R.; Osa, A.R.D.L.; Campelo, J.M.; Romero, A.A.; Valverde, J.L.; Sanchez, P. Design and development of catalysts for biomass-to-liquid-fischer-tropsch (BTL-FT) processes for biofuels production. *Energy Environ. Sci.* **2012**, *5*, 5186–5202. [CrossRef]

10. Phaahlamohlaka, T.N.; Dlamini, M.W.; Mogodi, M.W.; Kumi, D.O.; Jewell, L.L.; Billing, D.G.; Coville, N.J. A sinter resistant CO Fischer-Tropsch catalyst promoted with Ru and supported on titania encapsulated by mesoporous silica. *Appl. Catal. A Gen.* **2018**, *552*, 129–137. [CrossRef]
11. Li, Z.; Wu, J.; Yu, J.; Han, D.; Wu, L.; Li, J. Effect of incorporation manner of zr on the Co/SBA-15 catalyst for the Fischer–Tropsch synthesis. *J. Mol. Catal. A Chem.* **2016**, *424*, 384–392. [CrossRef]
12. Liu, Y.; Jia, L.; Hou, B.; Sun, D.; Li, D. Cobalt aluminate-modified alumina as a carrier for cobalt in fischer–tropsch synthesis. *Appl. Catal. A Gen.* **2017**, *530*, 30–36. [CrossRef]
13. Intarasiri, S.; Ratana, T.; Sornchamni, T.; Phongaksorn, M.; Tungkamani, S. Effect of pore size diameter of Cobalt supported catalyst on gasoline-diesel selectivity. *Energy Procedia* **2017**, *138*, 1035–1040. [CrossRef]
14. Rytter, E.; Holmen, A. On the support in Cobalt Fischer–Tropsch synthesis—Emphasis on alumina and aluminates. *Catal. Today* **2016**, *275*, 11–19. [CrossRef]
15. Liu, J.; Yu, L.; Zhao, Z.; Chen, Y.; Zhu, P.; Wang, C.; Luo, Y.; Xu, C.; Duan, A.; Jiang, G. Potassium-modified molybdenum-containing sba-15 catalysts for highly efficient production of acetaldehyde and ethylene by the selective oxidation of ethane. *J. Catal.* **2012**, *285*, 134–144. [CrossRef]
16. Osakoo, N.; Henkel, R.; Loiha, S.; Roessner, F.; Wittayakun, J. Effect of support morphology and pd promoter on Co/SBA-15 for fischer–tropsch synthesis. *Catal. Commun.* **2014**, *56*, 168–173. [CrossRef]
17. Cai, Q.; Li, J. Catalytic properties of the ru promoted Co/SBA-15 catalysts for fischer–tropsch synthesis. *Catal. Commun.* **2008**, *9*, 2003–2006. [CrossRef]
18. Wang, Y.; Jiang, Y.; Huang, J.; Liang, J.; Wang, H.; Li, Z.; Wu, J.; Li, M.; Zhao, Y.; Niu, J. Effect of hierarchical crystal structures on the properties of cobalt catalysts for Fischer–Tropsch synthesis. *Fuel* **2016**, *174*, 17–24. [CrossRef]
19. Wu, L.; Li, Z.; Han, D.; Wu, J.; Zhang, D. A preliminary evaluation of ZSM-5/SBA-15 composite supported Co catalysts for Fischer–Tropsch synthesis. *Fuel Process. Technol.* **2015**, *134*, 449–455. [CrossRef]
20. Prieto, G.; Martínez, A.; Murciano, R.; Arribas, M.A. Cobalt supported on morphologically tailored SBA-15 mesostructures: The impact of pore length on metal dispersion and catalytic activity in the Fischer–Tropsch synthesis. *Appl. Catal. A Gen.* **2009**, *367*, 146–156. [CrossRef]
21. Sevimli, F.; Yılmaz, A. Surface functionalization of SBA-15 particles for amoxicillin delivery. *Microporous Mesoporous Mater.* **2012**, *158*, 281–291. [CrossRef]
22. Xiong, H.; Zhang, Y.; Liew, K.; Li, J. Ruthenium promotion of Co/SBA-15 catalysts with high cobalt loading for Fischer–Tropsch synthesis. *Fuel Process. Technol.* **2009**, *90*, 237–246. [CrossRef]
23. Carrero, A.; Vizcaíno, A.J.; Calles, J.A.; García-Moreno, L. Hydrogen production through glycerol steam reforming using Co catalysts supported on SBA-15 doped with Zr, Ce and La. *J. Energy Chem.* **2017**, *26*, 42–48. [CrossRef]
24. Zhu, J.; Yang, J.; Miao, R.; Yao, Z.; Zhuang, X.; Feng, X. Nitrogen-enriched, ordered mesoporous carbons for potential electrochemical energy storage. *J. Mater. Chem. A* **2016**, *4*, 2286–2292. [CrossRef]
25. Han, J.; Zhang, L.; Zhao, B.; Qin, L.; Wang, Y.; Xing, F. The N-doped activated carbon derived from sugarcane bagasse for CO_2 adsorption. *Ind. Crop. Prod.* **2019**, *128*, 290–297. [CrossRef]
26. Vizcaíno, A.J.; Carrero, A.; Calles, J.A. Comparison of ethanol steam reforming using co and ni catalysts supported on SBA-15 modified by Ca and Mg. *Fuel Process. Technol.* **2016**, *146*, 99–109. [CrossRef]
27. Xiong, K.; Zhang, Y.; Li, J.; Liew, K. Catalytic properties of Ru nanoparticles embedded on ordered mesoporous carbon with different pore size in Fischer-Tropsch synthesis. *J. Energy Chem.* **2013**, *22*, 560–566. [CrossRef]
28. Zhuo, L.; Wu, L.; Han, D.; Wu, J. Characterizations and product distribution of Co-based Fischer-Tropsch catalysts: A comparison of the incorporation manner. *Fuel* **2018**, *220*, 257–263.
29. Kang, J.; Wang, X.; Peng, X.; Yang, Y.; Cheng, K.; Zhang, Q. Mesoporous zeolite Y-supported Co nanoparticles as efficient Fischer–Tropsch catalysts for selective synthesis of diesel fuel. *Ind. Eng. Chem. Res.* **2016**, *55*, 13008–13019. [CrossRef]

© 2018 by the authors. Licensee MDPI, Basel, Switzerland. This article is an open access article distributed under the terms and conditions of the Creative Commons Attribution (CC BY) license (http://creativecommons.org/licenses/by/4.0/).

Article

MPV Reduction of Furfural to Furfuryl Alcohol on Mg, Zr, Ti, Zr–Ti, and Mg–Ti Solids: Influence of Acid–Base Properties

Jesús Hidalgo-Carrillo, Almudena Parejas, Manuel Jorge Cuesta-Rioboo, Alberto Marinas * and Francisco José Urbano

Departamento de Química Orgánica, Instituto Universitario de Investigación en Química Fina y Nanoquímica IUIQFN, Universidad de Córdoba, Campus de Rabanales, Edificio Marie Curie, E-14071 Córdoba, Spain; yimo@hotmail.com (J.H.-C.); q12pabaa@uco.es (A.P.); qo2maara@uco.es (M.J.C.-R.); qo1urnaf@uco.es (F.J.U.)
* Correspondence: alberto.marinas@uco.es; Tel.: +34-957-218-622

Received: 1 October 2018; Accepted: 9 November 2018; Published: 13 November 2018

Abstract: The Meerwein–Ponndorf–Verley (MPV) reaction is an environmentally-friendly process consisting of the reduction of a carbonyl compound through hydrogen transfer from a secondary alcohol. This work deals with MPV reduction of furfural to furfuryl alcohol on different ZrO_x, MgO_x, TiO_x, and Mg–Ti, as well as Zr–Ti mixed systems. The solids were synthesized through the sol–gel process and subsequently calcined at 200 °C. Characterization was performed using a wide range of techniques: ICP-MS, N_2 adsorption-desorption isotherms, EDX, TGA-DTA, XRD, XPS, TEM, TPD of pre-adsorbed pyridine (acidity) and CO_2 (basicity), DRIFT of adsorbed pyridine, and methylbutynol (MBOH) test reaction. ZrO_x showed the highest conversion and selectivity values, which was attributed to the existence of acid–base pair sites (as evidenced by the MBOH test reaction), whereas the introduction of titanium resulted in the drop of both conversion and selectivity probably due to the increase in Brönsted-type acidity. As for MgO_x, it had a predominantly basic character that led to the production of the condensation product of one molecule of furfural and one molecule of acetone, and thus resulted in a lower selectivity to furfuryl alcohol. The TiO_x solid was found to be mainly acidic and exhibited both Lewis and Brönsted acid sites. The presence of the latter could account for the lower selectivity to furfuryl alcohol. All in all, these results seemed to suggest that the MPV reaction is favored on Lewis acid sites and especially on acid–base pair sites. The process was accelerated under microwave irradiation.

Keywords: furfural; MPV reaction; acid–base characterization; methylbutynol test reaction

1. Introduction

The transformation of natural residues from agriculture into platform molecules is one of the promising research lines in obtaining high added value chemical products [1,2]. One of those platform molecules is furfural [3], which can be obtained from lignocellulose [4]. It contains an aromatic ring and an aldehyde group which makes it a versatile molecule to obtain a wide range of chemical compounds [5], and one of the most important ones is furfuryl alcohol. This alcohol is widely used in the production of thermostatic resins, rubbers, fibers, adhesives, and some fine chemicals [5–7]. Furfuryl alcohol is mainly produced by furfural hydrogenation. Approximately 60% of the furfural produced is used to synthesize furfuryl alcohol. The catalytic liquid-phase hydrogenation of furfural to produce furfuryl alcohol has been extensively investigated in the presence of catalysts based on Ni, Co, Cu, Pt, and Pd [8–12]. Cu–Cr-based catalysts are commonly used in the industry, but environmentally friendlier catalysts are required. The transformation of furfural to furfuryl alcohol can also be carried out through the hydrogen transfer from a donor, which are typically secondary alcohols such as propan-2-ol, using the so-called Meerwein–Ponndorf–Verley (MPV) process [13]. This reaction involves the formation of a six-membered ring transition state in which both the reducing alcohol and the carbonyl compound are coordinated to the metal center (Lewis site) [14]. The assistance of the basic sites has also been proposed for the formation of the six-membered ring [15].

A wide range of heterogeneous catalysts has been described for the MPV process such as zirconia [16,17], mesoporous silica [18,19], zeolites [20], and alumina [21,22].

In previous papers, our research group described that zirconium gels calcined at low temperatures (ca. 200 °C) were quite selective to the corresponding unsaturated alcohol [16,17] in the MPV process. In the present work, different gels consisting of pure ZrO_x, TiO_x, and MgO_x, or mixed Mg–Ti and Zr–Ti solids and calcined at 200 °C were synthesized and tested in the MPV reduction of furfural to furfuryl alcohol to try and cast further light on the nature of the active sites responsible for the desired catalytic activity. The possibility of carrying out the reaction with microwave-assisted heating was also evaluated.

2. Results and Discussion

2.1. Textural, Structural, and Acid–Base Characterization of the Solids

The Brunnauer–Emmett–Teller (BET) surface areas as well as Mg/Ti and Zr/Ti atomic ratios (both nominal and experimental) of the synthesized solids are depicted in Table 1. The highest BET areas (in the 219–263 m^2/g range) corresponded to solids consisting of Zr and/or Ti, whereas lower values were found for the systems containing magnesium (42–81 m^2/g). In regard to the chemical composition, experimental results were in general quite similar to the nominal values and was thus evidence of a good precipitation of the metals during the synthesis.

Table 1. Brunnauer–Emmett–Teller (BET) surface area and atomic composition (nominal and experimental) of the different solids synthesized in the present study.

Catalyst	BET Surface Area (m^2/g)	M/Ti Ratio (M = Zr or Mg)		
		Nominal	ICP-MS	XPS
ZrO_x	221	-	-	-
Zr3Ti1	251	3.0	2.43	2.29
ZrTi	263	1.0	0.78	1.03
Zr1Ti3	219	0.33	0.34	0.45
TiO_x	232	-	-	-
Mg1Ti3	42	0.33	0.45	0.39
MgTi	81	1.0	1.06	0.62
Mg3Ti1	68	3.0	3.02	3.5
MgO_x	66	-	-	-

The TGA-DTA profiles of the different solids are shown in Figures 1 and 2 (Zr–Ti and Mg–Ti solids, respectively). The ZrO_x heat flow profile (Figure 1B) exhibited two main peaks centered at ca. 104 °C and 436 °C, respectively. The first endothermal peak corresponded to the loss of water whereas the second peak was the so-called glow exotherm attributed to the crystallization of zirconia [17,23]. For Zr–Ti mixed systems, the glow-exotherm was shifted to higher temperatures (450–466 °C) which suggests that titanium retards zirconium crystallization. In the case of MgO_x solids (Figure 2B), the heat flow profile exhibited two main endothermal peaks centered at ca. 117 °C and 385 °C. The latter peak was assigned to the transformation of $Mg(OH)_2$ into MgO [24]. For Mg–Ti solids, the presence of titanium seemed to favor such a transformation as evidenced by the shift of the peak to lower temperatures (in the 318–357 °C range).

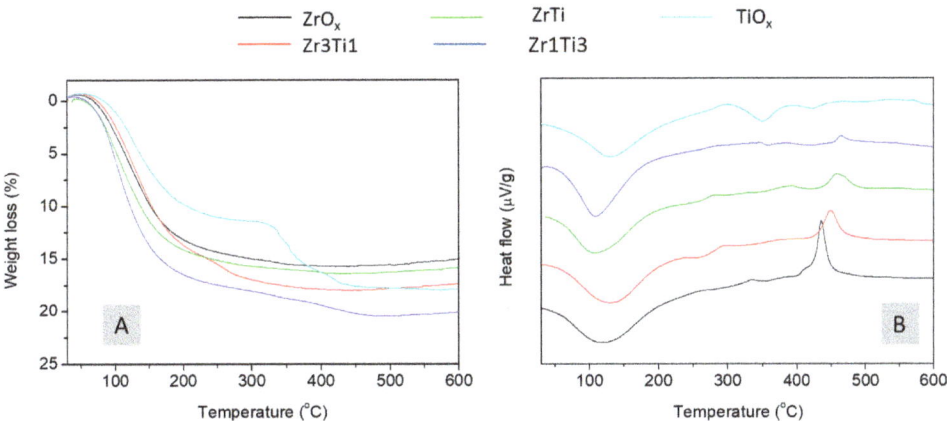

Figure 1. TG-DTA profiles of the precursor gels of the catalysts based on ZrO_x and TiO_x. Weight loss (**A**) and heat flow (**B**) profiles.

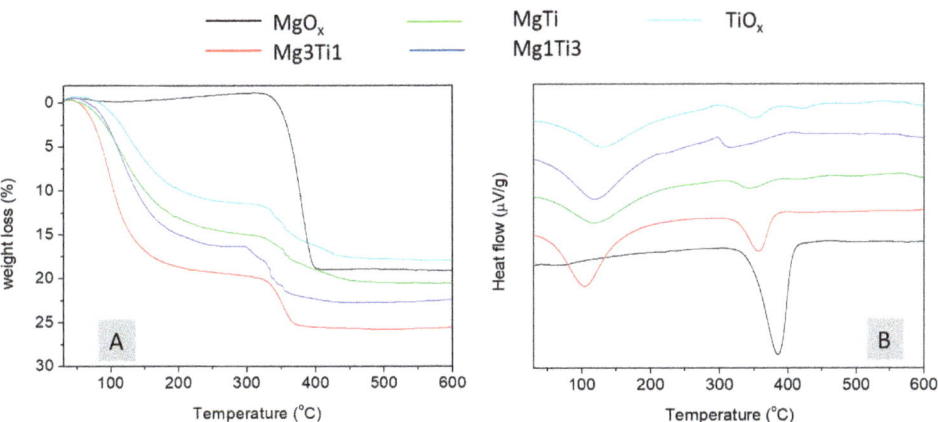

Figure 2. TG-DTA profiles of the precursor gels of catalysts based on MgO_x and TiO_x. Weight loss (**A**) and heat flow (**B**) profiles.

X-ray diffractograms of Zr-containing solids (Figure 3) showed evidence of their amorphous character, which was consistent with the TGA-DTA profiles; crystallization occurred at temperatures above 300 °C. In the case of MgO_x (Figure 3B), there were some peaks present due to the $Mg(OH)_2$ brookite structure. Those peaks were also evident in Mg3Ti1 solid whereas higher titanium contents resulted in the disappearance of brookite signals and the appearance of some new signals which could be assigned to MgO5Ti2 pseudobrookite.

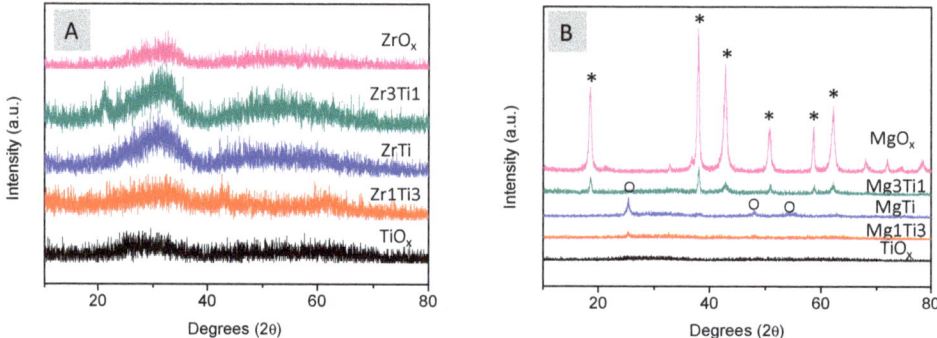

Figure 3. X-ray diffractograms of the different solids synthesized in the present work. Zr-Ti solids (**A**) and Mg-Ti systems (**B**). The corresponding pure compounds have also been included for the sake of comparison. * and ∘ denote brookite and pseudobrookite phases, respectively.

Transmission electron microscopy (TEM) images of all the samples are represented in Figure 4. As can be seen in the central part of Figure 4, MgO_x, ZrO_x, and TiO_x exhibited quite different textures which allowed us to distinguish them in mixed solids.

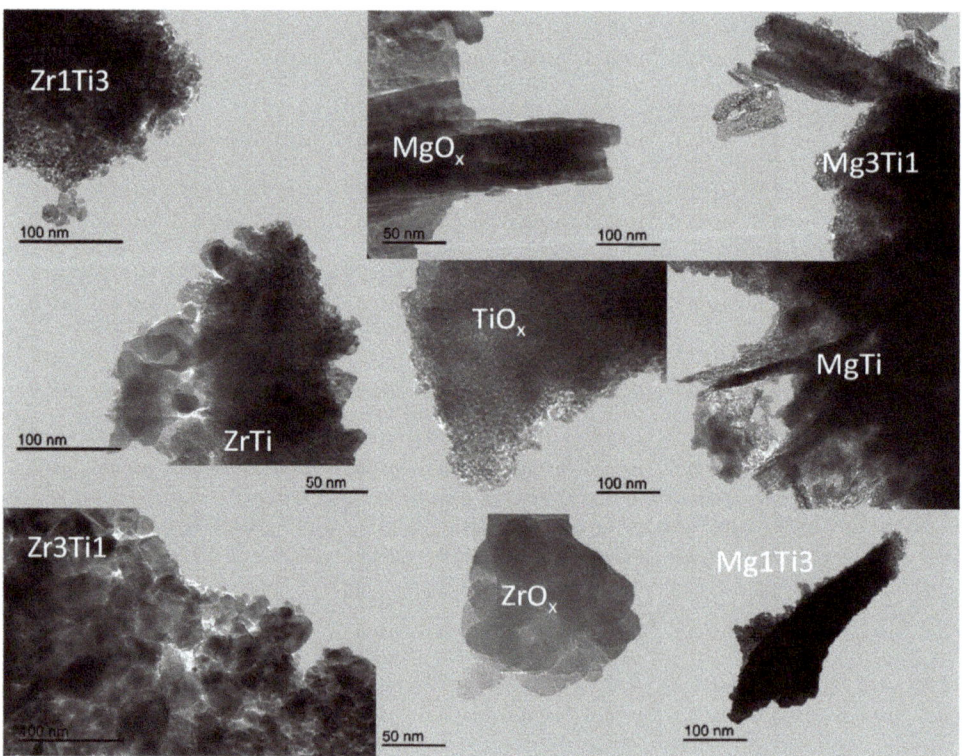

Figure 4. Transmission electron microscopy (TEM) images of the different solids.

X-ray photoelectron spectroscopy (XPS) profiles of Mg–Ti solids are represented in Figure 5. The signal for Mg1 in MgO_x presented two types of magnesium atoms. Moreover, as titanium content increased, there was a shift of signals to higher binding energies (from 1303.0 to 1303.8 eV for MgO_x

and Mg1Ti3, respectively). A similar trend was observed for the Ti(2p) signal, and the Ti2p3/2 signal shifted, in this case, to lower binding energies in the presence of magnesium (Ti2p3/2 signal at 458.4 and 458.1 eV for TiO_x and Ti3Mg1, respectively). These results suggest the existence of some Mg–Ti interaction in Mg–Ti solids.

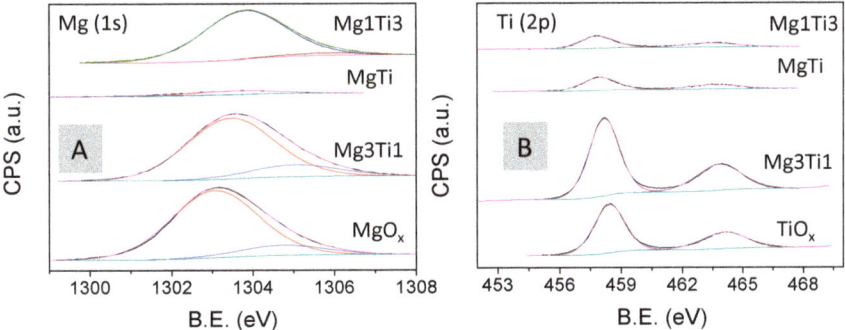

Figure 5. X-ray photoelectron spectroscopy (XPS) profiles of Mg(1s) (**A**) and Ti (2p) (**B**) in Mg–Ti solids.

As far as the Zr–Ti XPS profiles were concerned (Figure 6), there was suggestion of some Zr–Ti interaction as evidenced by the Zr3d and Ti2p signals shifting to higher and lower binding energy values, respectively, as the Ti content increased.

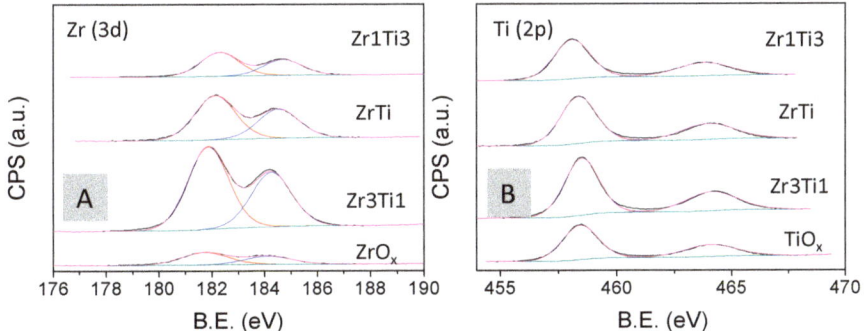

Figure 6. XPS profiles of Zr(3d) (**A**) and Ti (2p) (**B**) in Zr–Ti solids.

Surface acid–base characterization of the solids was carried out by TPD of pre-adsorbed CO_2 (basicity) and pyridine (acidity), and the main results are summarized in Table 2. As can be seen, ZrO_x exhibited a good balance between acid and basic sites (CO_2/py = 1.09), TiO_x was mainly acidic (CO_2/Py = 0.57) and MgO_x was a predominantly basic solid (CO_2/py = 3.39). As for the corresponding mixed solids, they all had an acid–base characteristic between the corresponding pure solids.

Table 2. Acid–base characteristics of the solids as determined by CO_2-TPD and Py-TPD, respectively.

Catalyst	µmol CO_2/g	µmol Py/g	CO_2/Py
ZrO_x	774	707	1.09
Zr3Ti1	728	753	0.97
ZrTi	658	921	0.71
Zr1Ti3	460	635	0.72
TiO_x	371	650	0.57
Mg1Ti3	508	622	0.82
MgTi	1126	616	1.83
Mg3Ti1	1142	354	3.22
MgO_x	1096	323	3.39

Complementary acid–base results could be obtained using the methylbutynol test reaction (Figure 7) which allow us to distinguish between acid, base, and acid–base pair sites.

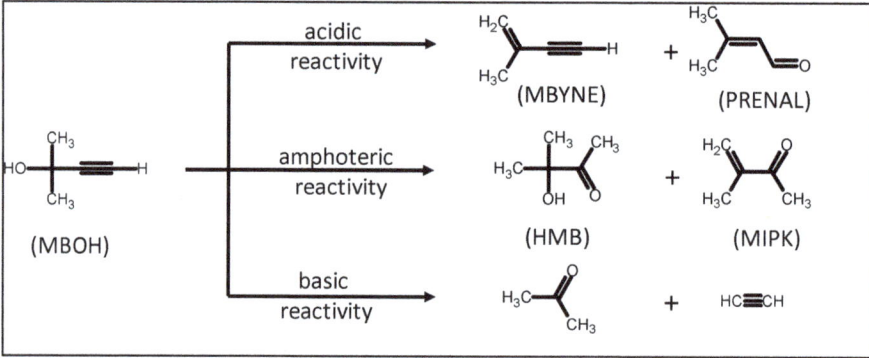

Figure 7. Overall reaction scheme as proposed by Lauron-Pernot et al. [25]. MBOH, 2-methyl-3-butyn-2-ol; MBYNE, 3-methyl-3-buten-1-yne; PRENAL, 3-methyl-2-buten-1-al; HMB, 3-hydroxy-3-methyl-2-butatone; MIPK, 3-methyl-3-buten-2-one.

As can be seen in Table 3, the methylbutynol (MBOH) test reaction confirmed the results found in the TPD studies of pre-adsorbed CO_2 and pyridine. Therefore, MgO_x mainly yielded acetone and acetylene (basic reactivity, 96.2% selectivity), TiO_x was mainly acidic (73.1% selectivity), and ZrO_x was predominantly amphoteric (53.4%) and mainly yielded 3-methyl-3-buten-2-one (MIPK). This was evidence for the presence of acid–base pair sites in ZrO_x.

Table 3. MBOH reaction. Comparison between selectivities of the pure oxides. The reaction conditions were as follows: microcatalytic pulse reactor; 20 mg catalyst, 200 °C, methylbutynol (MBOH) pulses of 0.5 µL (see experimental section).

Catalyst	Conversion (%)	S_{basic} (%)	S_{acid} (%)	$S_{amphoteric}$ (%)
MgO_x	5.6	96.2	3.8	0
TiO_x	1.0	12.3	73.1	14.6
ZrO_x	1.0	26.3	20.3	53.4

Further diffuse reflectance infrared Fourier transform (DRIFT) pyridine studies were performed on Zr–Ti solids to distinguish between Lewis and Brönsted acid sites (Figure 8). Peaks observed at 1443 and 1603 cm^{-1} were attributed to the presence of pyridine adsorbed on Lewis acid sites, whereas the band at ca. 1486 cm^{-1} could be due to adsorbed pyridine on both Lewis and Brönsted sites [26]. The signal at ca. 1534 cm^{-1} corresponded to pyridine on Brönsted acid sites [27]. An estimation of the Lewis/Brönsted acid site ratio can be made by integrating signals using the molar extinction coefficients [28]. Therefore, Lewis/Brönsted values of 15.4 and 4.9 could be obtained for ZrO_x and TiO_x, respectively.

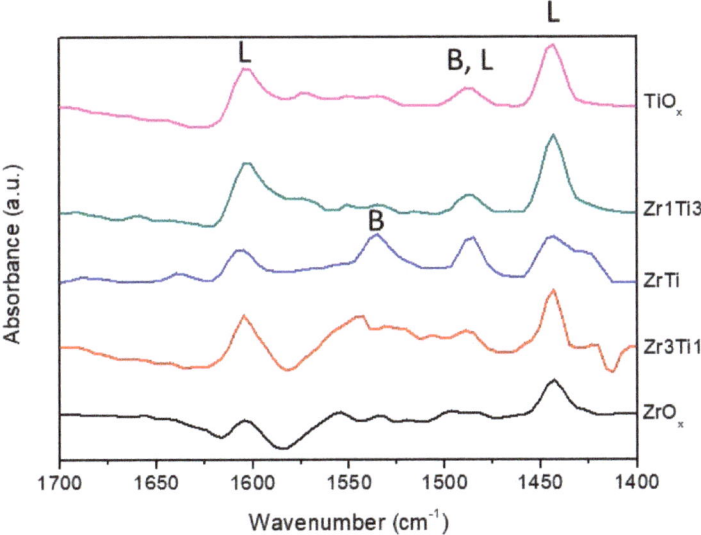

Figure 8. Diffuse reflectance infrared Fourier transform (DRIFT) studies of pyridine chemisorbed on the Zr–Ti solids. B and L stand for Brönsted and Lewis acid sites, respectively.

All in all, the acid–base studies indicated that MgO_x was mainly basic, ZrO_x was amphoteric, and TiO_x was acidic. Moreover, the highest Lewis/Brönsted site ratio corresponded to ZrO_x solids.

2.2. Catalytic Activity in Furfural Hydrogenation into Furfuryl Alcohol

The solids were then tested for liquid-phase MPV reduction of furfural to furfuryl alcohol (FUOL) using propan-2-ol as the hydrogen donor. The main results are summarized in Table 4.

Table 4. Results obtained for experiments under conventional (t = 20 h) and microwave (t = 2 h) heating on the different solids expressed in terms of conversion, selectivity to furfuryl alcohol (FUOL), and FUOL yield. The reaction conditions were as follows: 100 °C, molar propan-2-ol/furfural ratio of 10.8, and furfural/catalyst weight ratio of 5.8. Maximum microwave power was set at 300 W.

Catalyst	Conventional Heating			Microwave Heating		
	Conversion (%)	Selectivity FUOL (%)	Yield FUOL (%)	Conversion (%)	Selectivity FUOL (%)	Yield FUOL (%)
ZrO_x	50.1	90.4	45.3	27.6	96.8	26.7
Zr3Ti1	42.9	88.2	37.8	19.8	97.9	19.4
ZrTi	30.2	79.0	23.9	20.9	75.2	15.7
Zr1Ti3	22.3	80.5	18.0	17.1	79.5	13.6
TiO_x	16.2	68.7	11.1	7.4	53.5	4.0
Mg1Ti3	11.6	35.4	4.1	7.2	31.5	2.3
MgTi	13.4	46.6	6.3	8.6	47.8	3.1
Mg3Ti1	15.8	48.0	7.6	7.4	46.0	3.4
MgO_x	15.2	56.0	8.5	7.9	57.8	4.6

Firstly, the solids were tested under conventional heating. As can be seen, ZrO_x was the most active solid, followed by TiO_x and MgO_x. Both conversion and selectivity to furfuryl alcohol dropped upon the introduction of titanium in Zr–Ti and Mg–Ti solids. This seems to indicate that the interaction evidenced by XPS (and also suggested by TGA-DTA profiles) is detrimental to activity. In the case of MgO$_x$, the main by-product was the condensation product between one molecule of furfural and one molecule of acetone. Microwave heating was also tested, and results for t = 2 h are given in Table 4. The reactions were indeed accelerated under microwave irradiation. For instance, for t = 2 h, conversions of 6.5% (not shown) and 27.6% as well as selectivities of 97.0 and 96.8% to furfuryl alcohol were achieved on ZrO$_x$ under conventional and microwave heating, respectively. The results under microwave irradiation confirmed the observed activity trend in experiments under conventional heating.

The higher selectivity values (over 90%) found for ZrO$_x$ could be ascribed to the existence of acid–base pair sites. As suggested by Komanoya et al. [29], there would be a synergistic effect of acid–base pair sites: base sites could activate methylene groups in propan-2-ol bonded to Lewis sites. A tentative reaction mechanism on those acid–base pair sites is presented in Figure 9. The better catalytic performance of ZrO$_x$ as compared to TiO$_x$ could also be explained as the result of Lewis sites being more active than Brönsted sites in the MPV reaction [30,31].

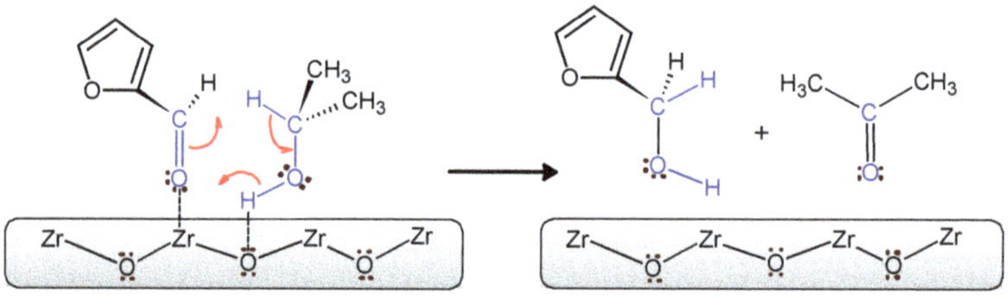

Figure 9. Suggested mechanism for furfural hydrogenation into furfuryl alcohol on acid–base pair sites in ZrO$_x$ through transfer hydrogenation from propan-2-ol.

3. Materials and Methods

For the synthesis of the catalysts, the following compounds were used: aqueous solutions of ammonium hydroxide (5 N) (Fluka 318620-2L, Honeywell, Bucharest, Romania) and hydrochloric acid (1 M) (Fluka 318949-2L); propan-2-ol (Sigma-Aldrich 190764-2.5L, Merck KGaA, Darmstadt, Germany); hydrated zirconium(IV) oxynitrate (Sigma-Aldrich 346462, Merck KGaA, Darmstadt, Germany); magnesium nitrate hexahydrate (Sigma-Aldrich 237175-1KG, Merck KGaA, Darmstadt, Germany); and titanium isopropoxide (Sigma-Aldrich 20527-3, Merck KGaA, Darmstadt, Germany).

The synthesis of the catalysts was carried out by the sol–gel method [32], following previous studies in our research group [17,33–35].

Two types of mixed systems were synthesized: titanium gels with magnesium and titanium gels with zirconium, with different (Mg or Zr)/Ti molar ratios (0%, 25%, 50%, 75%, and 100%). The syntheses were carried out at a constant pH of 10 with magnetic stirring at 700 rpm. The pH was kept constant using a pump (Atlas syringe pump, Syrris, (Hertfordshire, UK), which added 5 N ammonium hydroxide or 1 M hydrochloric acid throughout the synthesis process. The precipitate was filtered, washed with water, dried at 120 °C overnight, and calcined at 200 °C for 8 h. After calcination, the catalysts were sieved (0.149 μm). The nomenclature of the solids was as follows: TiO$_x$, ZrO$_x$, and MgO$_x$ for solids based on pure titanium, zirconium, and magnesium gels, respectively. For mixtures of gels, the nomenclature included the symbol of the metals followed by a number referring to their atomic ratio in the mixture. For instance, Mg3Ti1 indicates a magnesium–titanium system containing 75% Mg and 25% Ti (i.e., Mg/Ti atomic ratio of 3).

Thermogravimetric analyses (TGA) were performed on a Setaram SetSys 12 instrument (Caluire, France). A 20 mg amount of sample (precursor gels of the catalysts) was placed in an alumina crucible and heated at temperatures ranging from 30 to 600 °C (heating rate of 10 °C/min) under a synthetic air stream (50 mL/min) in order to measure weight loss, heat flow, and derivative weight loss.

X-ray photoelectron spectroscopy (XPS) data were recorded at the Central Service for Research Support (SCAI) of the University of Córdoba on 4 mm × 4 mm pellets of 0.5 mm thickness that were obtained by gently pressing the powered materials. The samples were outgassed to a pressure below about 2×10^{-8} Torr at 150 °C in the instrument pre-chamber to remove chemisorbed volatile species. The main chamber of the Leibold-Heraeus LHS10 spectrometer used, which is capable of operating down to less than 2×10^{-9} Torr, was equipped with a EA-200MCD hemispherical electron analyzer with a dual X-ray source using Al Kα (hυ = 1486.6 eV) at 120 W and 30 mA. C (1 s) was used as the energy reference (284.6 eV).

Transmission electron microscopy (TEM) images were obtained at the Central Service for Research Support (SCAI) of the University of Córdoba using a JEOL JEM 1400 microscope available at SCAI. The samples were mounted on 3 mm holey carbon copper grids.

X-ray patterns of the samples in the 10–80° (2θ) range was registered in a D8 Advanced Diffractometer (Bruker AXS) equipped with a Lynxeye detector.

Surface areas of solids were obtained from nitrogen adsorption-desorption isotherms obtained at liquid nitrogen temperature on a Micromeritics ASAP-2010 instrument following the Brunnauer–Emmett–Teller (BET) method. All samples were degassed to 0.1 Pa at 120 °C before measurement.

Surface basicity of the catalysts was determined on a Micromeritics Autochem II instrument by thermal programmed desorption of pre-absorbed CO_2 (TPD-CO_2) with TCD detection. An amount of 100 mg of each catalyst was loaded into a reactor of 10 mm ID and placed in a furnace. Solids were cleaned with an Ar stream (20 mL/min) by heating to 200 °C at a rate of 10 °C/min for 60 min and then cooled down to 40 °C. At that temperature, the catalysts were saturated with the probe molecule using 5% CO_2/Ar flow at 20 mL/min for 60 min. After saturation, physisorbed CO_2 was removed by a flowing Ar stream for 30 min (20 mL/min). Then, the temperature-programmed desorption of chemisorbed CO_2 was carried out by ramping the temperature from 40 to 200 °C (heating rate 5 °C/min) and holding the final temperature for 60 min. The amount of CO_2 adsorbed was determined from a calibration graph constructed from the injection of variable volumes of 5% CO_2/Ar.

The surface acidity of the catalysts was determined by thermal programmed desorption of pre-adsorbed pyridine (TPD-PY) with TCD detection. A 30 mg amount of sample was introduced in a 10 mm ID reactor that was placed inside an oven. The solids were cleaned under a He flow (75 mL/min) by heating to 200 °C at a rate of 10 °C/min and then cooled down to 50 °C. At that temperature, the solids were exposed for 30 min to a pyridine-saturated He flow. After saturation, physisorbed pyridine was removed by flowing a pure He stream for 60 min (75 mL/min). Then, the temperature-programmed desorption of chemisorbed pyridine was carried out by ramping the temperature from 50 to 200 °C (heating rate 10 °C/min) and holding the final temperature for 30 min. Desorbed pyridine was quantified against a calibration graph constructed from variable volumes of pyridine injected.

Complementary studies using diffuse reflectance infrared Fourier transform (DRIFT) spectra of adsorbed pyridine were carried out on a FTIR instrument (Bomem MB-3000, ABB Corporate, Zurich, Switzerland) equipped with an "environmental chamber" (Spectra Tech, Jefferson Court, Oak Ridge, TN, USA) placed in a diffuse reflectance attachment (Spectra Tech, Collector). A resolution of 8 cm^{-1} was used with 256 scans averaged to obtain a spectrum from 4000 to 400 cm^{-1}. In each measurement, the reference was the same sample after heating at 150 °C. Pyridine adsorption was carried out at 150 °C for 45 min to allow the saturation of the catalyst surface. The physisorbed pyridine was then cleaned with a N_2 flow (50 mL/min) and its spectrum was registered.

The MBOH test reaction was carried out as described elsewhere [36]. A microcatalytic pulse reactor (1/8 in i.d. quartz tubular reactor) was placed in the injection port of a gas chromatograph (GC System 7890A, Agilent Technologies, Santa Clara, CA, USA). The reactor was packed with alternating layers of quartz wool with the catalyst (20 mg) placed between them. Prior to each run, the catalyst was pre-treated in the reactor at 200 °C for 2 h under nitrogen (75 mL/min). MBOH pulses of 0.5 µL were then carried out.

The MPV reaction of furfural was carried out under conventional heating in a Carousel 12 Plus™ Reaction Station, Discovery Technologies, and reactions under microwaves were carried out in a CEM-DISCOVER apparatus with PC control. In both cases, temperature was 100 °C. Maximum microwave power was set at 300 W. The reactions by conventional heating were carried out with 100 mg of catalyst, 5 mL of propan-2-ol, and 0.5 mL of furfural over 20 h; the reactions in the microwave oven kept the same catalyst, propan-2-ol, and furfural ratios at a volume of reaction of 2 mL with a reaction time of 2 h. Analysis of reaction products was carried out by gas chromatography (GC-FID System 7890A, Agilent Technologies, equipped with a HP-5 chromatographic column) using the corresponding calibration graphs.

4. Conclusions

Several solids consisting of pure magnesium, zirconium, titanium, and mixed magnesium–titanium as well as zirconium–titanium gels were obtained through the sol–gel process and calcined at 200 °C. The presence of titanium retarded the crystallization of zirconium oxide whereas transformation of $Mg(OH)_2$ into MgO was favored in the presence of titanium. XPS results also suggested the existence of some Mg–Ti and Zr–Ti interaction in mixed gels. In regard to the acid–base properties as determined from the TPD of pre-adsorbed pyridine and CO_2, the ZrO_x system exhibited a good balance between acid and base sites, whereas TiO_x and MgO_x were predominantly acidic and basic, respectively. The MBOH test reaction evidenced the presence of acid–base pair sites in ZrO_x, and pyridine DRIFT studies showed that acid sites in ZrO_x were mainly of the Lewis type whereas both Brönsted and Lewis sites were present in TiO_x and Zr–Ti mixed solids. The most active and selective catalyst in the MPV reduction of furfural to furfuryl alcohol was ZrO_x whereas both parameters decreased in Zr–Ti solids as the titanium content increased. These results suggest that acid–base pair sites are particularly active in MPV reduction and that Lewis acid sites are more active than Brönsted acid ones. The same reactivity order was found for the reactions under microwave irradiation which led to an acceleration of the process as compared to conventional heating.

Author Contributions: A.M., J.H.-C. and F.J.U. conceived and designed the experiments; J.H.-C., A.P. and M.J.C.-R. performed the experiments; J.H.-C., A.P. and M.J.C.-R. analyzed the data; J.H.-C., A.M. and F.J.U. wrote the paper.

Funding: This research was funded by the Ramón Areces Foundation.

Acknowledgments: The scientific support from the Central Service for Research Support (SCAI) at the University of Cordoba is acknowledged.

Conflicts of Interest: The authors declare no conflict of interest.

References

1. Chheda, J.N.; Huber, G.W.; Dumesic, J.A. Liquid-phase catalytic processing of biomass-derived oxygenated hydrocarbons to fuels and chemicals. *Angew. Chem. Int. Ed.* **2007**, *46*, 7164–7183. [CrossRef] [PubMed]
2. Turhollow, A.; Perlack, R.; Eaton, L.; Langholtz, M.; Brandt, C.; Downing, M.; Wright, L.; Skog, K.; Hellwinckel, C.; Stokes, B.; et al. The updated billion-ton resource assessment. *Biomass Bioenergy* **2014**, *70*, 149–164. [CrossRef]
3. Bozell, J.J.; Petersen, G.R. Technology development for the production of biobased products from biorefinery carbohydrates-the US Department of Energy's "Top 10" revisited. *Green Chem.* **2010**, *12*, 539–554. [CrossRef]
4. Mamman, A.S.; Lee, J.-M.; Kim, Y.-C.; Hwang, I.T.; Park, N.-J.; Hwang, Y.K.; Chang, J.-S.; Hwang, J.-S. Furfural: Hemicellulose/xylosederived biochemical. *Biofuels Bioprod. Biorefin.* **2008**, *2*, 438–454. [CrossRef]
5. Chatterjee, C.; Pong, F.; Sen, A. Chemical conversion pathways for carbohydrates. *Green Chem.* **2015**, *17*, 40–71. [CrossRef]
6. Corma, A.; Iborra, S.; Velty, A. Chemical Routes for the Transformation of Biomass into Chemicals. *Chem. Rev.* **2007**, *107*, 2411–2502. [CrossRef] [PubMed]

7. Rothe, M.; Bauer, K.; Garbe, D. *Common Fragrance and Flavour Materials. Preparation, Properties and Uses*; VCH Verlagsgesellschaft mbH: Weinheim, Germany, 1985.
8. Liu, H.; Huang, Z.; Zhao, F.; Cui, F.; Li, X.; Xia, C.; Chen, J. Efficient hydrogenolysis of biomass-derived furfuryl alcohol to 1,2-and 1,5-pentanediols over a non-precious Cu-Mg$_3$AlO$_{4.5}$ bifunctional catalyst. *Catal. Sci. Technol.* **2016**, *6*, 668–671. [CrossRef]
9. Rao, R.; Dandekar, A.; Baker, R.T.K.; Vannice, M.A. Properties of Copper Chromite Catalysts in Hydrogenation Reactions. *J. Catal.* **1997**, *171*, 406–419. [CrossRef]
10. Sang, S.; Wang, Y.; Zhu, W.; Xiao, G. Selective hydrogenation of furfuryl alcohol to tetrahydrofurfuryl alcohol over Ni/γ-Al$_2$O$_3$ catalysts. *Res. Chem. Intermed.* **2017**, *43*, 1179–1195. [CrossRef]
11. Taylor, M.J.; Durndell, L.J.; Isaacs, M.A.; Parlett, C.M.A.; Wilson, K.; Lee, A.F.; Kyriakou, G. Highly selective hydrogenation of furfural over supported Pt nanoparticles under mild conditions. *Appl. Catal. B Environ.* **2016**, *180*, 580–585. [CrossRef]
12. Thompson, S.T.; Lamb, H.H. Palladium–Rhenium Catalysts for Selective Hydrogenation of Furfural: Evidence for an Optimum Surface Composition. *ACS Catal.* **2016**, *6*, 7438–7447. [CrossRef]
13. Meerwein, H.; Schmidt, R. Ein neues Verfahren zur Reduktion von Aldehyden und Ketonen. *Justus Liebigs Annalen der Chemie* **1925**, *444*, 221–238. [CrossRef]
14. Ooi, T.; Ichikawa, H.; Maruoka, K. Practical Approach to the Meerwein–Ponndorf–Verley Reduction of Carbonyl Substrates with New Aluminum Catalysts. *Angew. Chem.* **2001**, *113*, 3722–3724. [CrossRef]
15. Ivanov, V.; Bachelier, J.; Audry, F.; Lavalley, J.C. Study of the Meerwein-Pondorff-Verley Reaction Between Ethanol and Acetone on Various Metal-Oxides. *J. Mol. Catal.* **1994**, *91*, 45–59. [CrossRef]
16. Montes, V.; Miñambres, J.F.; Khalilov, A.N.; Boutonnet, M.; Marinas, J.M.; Urbano, F.J.; Maharramov, A.M.; Marinas, A. Chemoselective hydrogenation of furfural to furfuryl alcohol on ZrO$_2$ systems synthesized through the microemulsion method. *Catal. Today* **2018**, *306*, 89–95. [CrossRef]
17. Axpuac, S.; Aramendía, M.A.; Hidalgo-Carrillo, J.; Marinas, A.; Marinas, J.M.; Montes-Jiménez, V.; Urbano, F.J.; Borau, V. Study of structure-performance relationships in Meerwein-Ponndorf-Verley reduction of crotonaldehyde on several magnesium and zirconium-based systems. *Catal. Today* **2012**, *187*, 183–190. [CrossRef]
18. Antunes, M.M.; Lima, S.; Neves, P.; Magalhães, A.L.; Fazio, E.; Neri, F.; Pereira, M.T.; Silva, A.F.; Silva, C.M.; Rocha, S.M.; et al. Integrated reduction and acid-catalysed conversion of furfural in alcohol medium using Zr,Al-containing ordered micro/mesoporous silicates. *Appl. Catal. B Environ* **2016**, *182*, 485–503. [CrossRef]
19. Iglesias, J.; Melero, J.A.; Morales, G.; Moreno, J.; Segura, Y.; Paniagua, M.; Cambra, A.; Hernandez, B. Zr-SBA-15 Lewis Acid Catalyst: Activity in Meerwein Ponndorf Verley Reduction. *Catalysts* **2015**, *5*, 1911–1927. [CrossRef]
20. Iglesias, J.; Melero, J.A.; Morales, G.; Paniagua, M.; Hernández, B. Dehydration of Xylose to Furfural in Alcohol Media in the Presence of Solid Acid Catalysts. *ChemCatChem* **2016**, *8*, 2089–2099. [CrossRef]
21. Kim, M.S.; Simanjuntak, F.S.H.; Lim, S.; Jae, J.; Ha, J.-M.; Lee, H. Synthesis of alumina–carbon composite material for the catalytic conversion of furfural to furfuryl alcohol. *J. Ind. Eng. Chem.* **2017**, *52*, 59–65. [CrossRef]
22. Lopez-Asensio, R.; Cecilia, J.A.; Jimenez-Gomez, C.P.; Garcia-Sancho, C.; Moreno-Tost, R.; Maireles-Torres, P. Selective production of furfuryl alcohol from furfural by catalytic transfer hydrogenation over commercial aluminas. *Appl. Catal. A Gen.* **2018**, *556*, 1–9. [CrossRef]
23. Stefanic, G.; Music, S.; Popovic, S.; Sekulic, A. FT-IR and laser Raman spectroscopic investigation of the formation and stability of low temperature t-ZrO$_2$. *J. Mol. Struct.* **1997**, *408–409*, 391–394. [CrossRef]
24. Formosa, J.; Chimenos, J.M.; Lacasta, A.M.; Haurie, L. Thermal study of low-grade magnesium hydroxide used as fire retardant and in passive fire protection. *Thermochim. Acta* **2011**, *515*, 43–50. [CrossRef]
25. Lauron-Pernot, H.; Luck, F.; Popa, J.M. Methylbutynol: A new and simple diagnostic tool for acidic and basic sites of solids. *Appl. Catal.* **1991**, *78*, 213. [CrossRef]
26. Osman, A.I.; Abu-Dahrieh, J.K.; Rooney, D.W.; Halawy, S.A.; Mohamed, M.A.; Abdelkader, A. Effect of precursor on the performance of alumina for the dehydration of methanol to dimethyl ether. *Appl. Catal. B Environ.* **2012**, *127*, 307–315. [CrossRef]
27. Lu, J.; Kosuda, K.M.; Van Duyne, R.P.; Stair, P.C. Surface Acidity and Properties of TiO$_2$/SiO$_2$ Catalysts Prepared by Atomic Layer Deposition: UV-visible Diffuse Reflectance, DRIFTS, and Visible Raman Spectroscopy Studies. *J. Phys. Chem. C* **2009**, *113*, 12412–12418. [CrossRef]

28. Emeis, C.A. Determination of Integrated Molar Extinction Coefficients for Infrared Absorption Bands of Pyridine Adsorbed on solid Acid Catalysts. *J. Catal.* **1993**, *141*, 347–354. [CrossRef]
29. Komanoya, T.; Nakajima, K.; Kitano, M.; Hara, M. Synergistic Catalysis by Lewis Acid and Base Sites on ZrO$_2$ for Meerwein-Ponndorf-Verley Reduction. *J. Phys. Chem. C* **2015**, *119*, 26540–26546. [CrossRef]
30. Guo, Z.-K.; Hong, Y.-C.; Xu, B.-Q. Transfer hydrogenation of cinnamaldehyde with 2-propanol on Al$_2$O$_3$ and SiO$_2$-Al$_2$O$_3$ catalysts: Role of Lewis and Brönsted acidic sites. *Catal. Sci. Technol.* **2017**, *7*, 4511–4519. [CrossRef]
31. Miñambres, J.F.; Aramendía, M.A.; Marinas, A.; Marinas, J.M.; Urbano, F.J. Liquid and gas-phase Meerwein–Ponndorf–Verley reduction of crotonaldehyde on ZrO$_2$ catalysts modified with Al$_2$O$_3$, Ga$_2$O$_3$ and In$_2$O$_3$. *J. Mol. Catal. A Chem.* **2011**, *338*, 121–129. [CrossRef]
32. Debecker, D.P.; Mutin, P.H. Non-hydrolytic sol-gel routes to heterogeneous catalysts. *Chem. Soc. Rev.* **2012**, *41*, 3624–3650. [CrossRef] [PubMed]
33. Minambres, J.F.; Marinas, A.; Marinas, J.M.; Urbano, F.J. Activity and deactivation of catalysts based on zirconium oxide modified with metal chlorides in the MPV reduction of crotonaldehyde. *Appl. Catal. B Environ.* **2013**, *140*, 386–395. [CrossRef]
34. Minambres, J.F.; Marinas, A.; Marinas, J.M.; Urbano, F.J. Chemoselective crotonaldehyde hydrogen transfer reduction over pure and supported metal nitrates. *J. Catal.* **2012**, *295*, 242–253. [CrossRef]
35. Aramendia, M.A.; Borau, V.; Jimenez, C.; Marinas, J.M.; Ruiz, J.R.; Urbano, F.J. Influence of the preparation method on the structural and surface properties of various magnesium oxides and their catalytic activity in the Meerwein-Ponndorf-Verley reaction. *Appl. Catal. A Gen.* **2003**, *244*, 207–215. [CrossRef]
36. Aramendia, M.A.; Borau, V.; Garcia, I.M.; Jimenez, C.; Marinas, A.; Marinas, J.M.; Porras, A.; Urbano, F.J. Comparison of Different Organic Test Reaction over Acid-Base Catalysts. *Appl. Catal. A Gen.* **1999**, *184*, 115–125. [CrossRef]

© 2018 by the authors. Licensee MDPI, Basel, Switzerland. This article is an open access article distributed under the terms and conditions of the Creative Commons Attribution (CC BY) license (http://creativecommons.org/licenses/by/4.0/).

Review

Zeolites as Acid/Basic Solid Catalysts: Recent Synthetic Developments

Valentina Verdoliva [1], Michele Saviano [2] and Stefania De Luca [1,*]

1. Institute of Biostructures and Bioimaging, National Research Council, 80134 Naples, Italy; valentina.verdoliva@gmail.com
2. Institute of Crystallography, National Research Council, 70126 Bari, Italy; msaviano@unina.it
* Correspondence: stefania.deluca@cnr.it; Tel.: +39-081-253-4514

Received: 13 February 2019; Accepted: 1 March 2019; Published: 8 March 2019

Abstract: The zeolites are porous solid structures characterized by a particular framework of aluminosilicates, in which the incorporation of the Al^{+3} ions generates an excess of negative charge compensated by cations (usually alkali or alkali earth) or protons. In the latter case, they are employed as catalysts for a wide variety of reactions, such as dehydration, skeletal isomerization and cracking, while the catalytic activity of basic zeolites has not found, up to now, any industrial or whatever relevant application in chemical processes. In the present review, we firstly intend to give an overview of the fundamental chemical composition, as well as the structural features of the zeolite framework. The purpose of this paper is to analyze their key properties as acid, both Lewis and Brønsted, and basic solid support. Their application as catalysts is discussed by reviewing the already published works in that field, and a final remark of their still unexplored potential as green, mild, and selective catalyst is also reported.

Keywords: zeolites; catalysis; solid acid; solid base; chemical modification; alkylation; glycosidation

---

1. Introduction

Zeolites, both naturally occurring or synthetic, are solids of aluminosilicates with crystalline structure and pore size in the microporous range (<10 Å). They can be described as a framework of SiO_4 and AlO_4 tetrahedrons cross-linked at their corners by a common oxygen ion. The isomorphic substitution of Si^{+4} with a trivalent ion Al^{+3} generates a negative charge in the lattice, which needs to be compensated. Inclusion in the crystal structure of cations, such as alkali or alkali earth metals, compensates the electrovalence of the tetrahedrons containing aluminum. The spaces between the tetrahedrons, which constitute their channels, are naturally occupied by water molecules, so that hydrated zeolites can be generally represented by the formula [1–3]

$$M^{+2}{}_{x/n}[(AlO_2)_x(SiO_2)_y] \cdot wH_2O$$

Dehydrated zeolites are able to act as molecular sieves, since the dimension of their pores is such as to accept for absorption only molecules of a certain size, shape, and polarity. The available void volume of the zeolites is directly connected to their sieving properties. There are a variety of ways in which they are employed in order to take advantage of these properties. Indeed, alkali cationic zeolites are largely employed as adsorbents for industrial gas purification, such as CO_2 capture [4]. They have also a high tendency to capture water and other polar molecules, such as NH_3 and H_2S [5].

The charge-balancing cations can be partially or totally exchanged by other cations, since they are bonded to the lattice through Coulombic interactions, or either by the ammonium ion, which can be later decomposed, or by heat treatment in gaseous ammonia and a proton, thus generating a Brønsted acid form of the zeolite [6]. Recently, protonated zeolites have also been synthesized by employing templating agents, in fact their decomposition leads to having a residual proton, which neutralizes the net negative charge on the framework aluminum atoms (Figure 1). Brønsted acid sites in zeolites are the bridging hydroxyl groups between Al and Si tetrahedral, while the second source of acidity, Lewis acidity, is ascribed to both framework and extra-framework aluminum species (EFAL) [3].

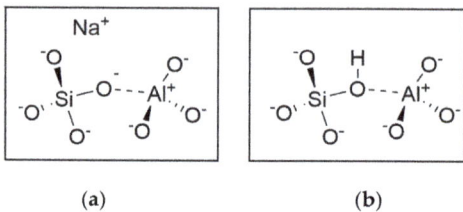

(a) (b)

Figure 1. Representation of the basic (**a**) and acid (**b**) sites in Na- zeolites.

In the last thirty years, zeolites, as solid acids, have become extremely successful as catalysts and have been employed for a wide variety of reactions, such as the cracking of carbon–carbon bonds, skeletal isomerizations, polymerization, aromatic alkylation with alkenes or alcohols, and other acid catalyzed reactions [7]. The reason for their success in catalysis is related to several specific features. First of all, zeolites, because of their cage-like porous structure, have a very high surface area and adsorption capacity. In addition, the sizes of their channels and cavities are in the range typical for many molecules of interest (5–12 Å) and offer at the same time, shape selectivity, which can be used to direct a given catalytic reaction toward the desired product, avoiding undesired side reactions. It is also worth mentioning that for the hydrogenation/dehydrogenation, the activity of the acid zeolites is implemented by metals like Pt, Pd, Ni and, in this case, the solid catalyst acts as a support providing its large surface as a deposit of the metals listed above [4].

The Brønsted acidity of zeolites, when the negative charge present on the Al-substituted framework is compensated by alkaline cations, can be generated by exchanging the synthesis cations with protons. In other words, the number of the Brønsted acid sites of a zeolite depends on the framework Si/Al ratio. On the other hand, it has been experimentally demonstrated that, while the density of these acid sites increases with the Al^{+3} content, a low strength of them is observed, which means the Brønsted acidity of zeolites is also influenced by the presence of the Lewis acidity. More in general, several factors are involved with the catalytic performance of the zeolites, the ability of the substrate to sieve through their pores, the topology of the crystal framework and extra-framework and, also, the electrostatic field generated by the framework Si/Al ratio [8,9].

Among the solid catalysts employed for the Friedel–Craft aromatic acylation, the zeolites have successfully replaced the Lewis acids $AlCl_3$, $FeCl_3$, $SnCl_4$, and $TiCl_4$ for academic and industrial applications, due to the shape selectivity offered by this material. In some cases several metals, in particular, metal triflates such as $La(OTf)_3$, $Ce(OTf)_4$, $Y(OTf)_3$, and $Zn(OTf)_2$, were incorporated in their framework and more stable, efficient and recyclable catalysts were obtained. The results in this field allowed advances toward the employment of more eco-compatible methodologies and are greatly reviewed by Sartori and Maggi [10]. The same authors also provided an exhaustive overview about the heterogeneous catalysis applied in the field of the protection and deprotection reactions of functional groups during the organic synthesis. Zeolites are also extensively employed in this field, and have the advantage of providing an easier reaction handling and final purification step, together with an improved selectivity obtained for the protective chemistry [11].

After all, the application of zeolites as acid catalysts has been basically linked with the development of refinery and petrochemistry. In oil refinery, they play a key role, since they result highly active, selective, and durable catalysts, and, as a matter of fact, have replaced mineral acids as the new environmentally friendly solid catalyst. Moreover, it has been proved that these materials would have a key role also for the biomass conversion [12,13]. In fact, zeolites, as solid catalysts, offer the advantage of being tunable and recyclable, and are easily separated from reaction mixtures. For all the listed reasons, they are particularly useful for clean and economical production processes of fine chemicals, which is becoming an area of growing interest [14].

In all the cited applications, the zeolites are used for their acid–base properties; however, the chemical discussion for defying this key role is still an active debate [4]. Both acidic and basic zeolites are available, but acidic zeolites have found broad application in important industrial processes as well as in organic chemistry reactions. In comparison, much less attention has been paid, in chemical technology, to the application of the zeolites as "basic catalysts" for academic reactions, while it results totally absent for industrial processes. Indeed, the structure–activity relationship for the microporous basic properties of zeolites is still an open discussion. What is clear is that the zeolite acid–base pair is different from that of the homogeneous acid–base compounds, since their strength is rather influenced by the type of exchangeable cations, aluminum content, structural dimension, and porosity [4,15,16].

Herein, we do not intend to give a comprehensive review, but instead an overview of the most recent organic reactions that have employed zeolites as mild acid or basic solid catalysts for efficient and selective chemical transformation of quite sensitive substrates. In particular, the present paper focuses on the recent results dealing with zeolites as basic/acid solids for organic transformations of peptide and carbohydrate molecules, since wide applications were found on these substrates. Moreover, in recent studies of the modulation activity of zeolites, for both purposes, acid and basic catalysts are reported, in order to prove its valuable application in some of the very useful organic reactions. For all the reported organic process, the employment of zeolites as solid catalysts could avoid the formation of by-products and waste, more usually associated with traditional homogeneous catalysts, besides being easily separated from the final products and also regenerated, if it was necessary, in order to be reused more times in the same process. The final aim is to exploit the potential and the additional advantages of replacing the currently used homogenous catalyst for standard organic reactions with heterogeneous catalysts such as the zeolites.

The structural and composition description of the several types of zeolites, natural or synthetic, is not reported in this review, it has been extensively reviewed in a recently published article [17].

We deal in particular with the Zeolites A (Si/Al ratio usually 1) structurally denoted with the code LTA (Linde Type A), in particular with Na-LTA, which contains sodium as cation (Figure 2), it counterbalances the negative charge of the oxygen atoms and has a pore diameter of 4 Å: $Na_{12}[(AlO_2)\cdot(SiO_2)]_{12}\cdot 27H_2O$.

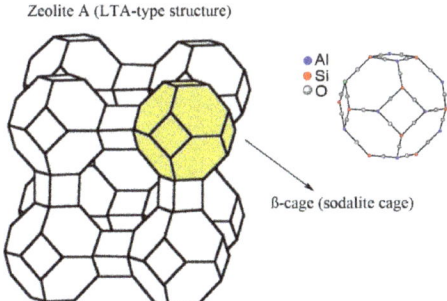

Figure 2. The structure of zeolite Na-LTA.

When 75% of its sodium is replaced by potassium, it is referred to as a zeolite of 3 Å pore size, alternatively, the replacement of sodium by calcium gives rise to the type of 5 Å, with an increased diameter.

Other zeolites taken under consideration in this review (for them catalytic properties) are Zeolite X (FAU-type structure, Si/Al ratio usually 1.2), Zeolite Y (FAU-type structure, Si/Al ratio usually 2.4),

Zeolite ZSM-5 (MFI-type structure, Si/Al ratio usually 15), and Natrolite (Na$_2$[Al$_2$Si$_3$O$_{10}$]·2H$_2$O) [9,18] (Figure 3).

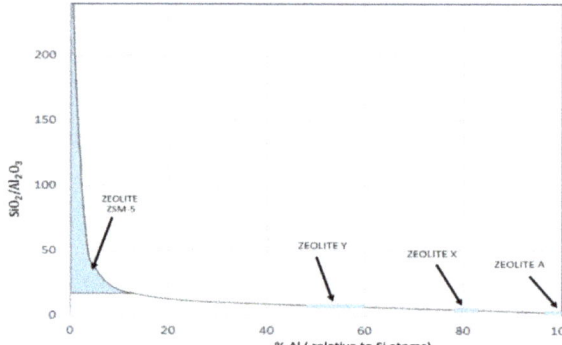

Figure 3. Al atom concentration and SiO$_2$/Al$_2$O$_3$ ratio for various zeolites (adaptation from [9], Nature 1984).

Moreover we also report on several examples of recently published works performed by tuning the acid–base properties of the zeolite catalysts. They depend on the aluminum content in the zeolite framework, as well as on the nonframework charge balancing constituents that can be adequately replaced. Some other references are only cited [19–23].

2. Glycosidation Reactions Promoted by Molecular Sieves

2.1. Glycosidation Reaction via Activation of Glycosyl Trichloro- and N-Phenyltrifluoroacetimidates

Iadonisi and coauthors performed the glycosidation of primary and secondary saccharides by using acid-washed 4 Å molecular sieves as catalyst that successfully activated glycosyl imidates, acting at the same time as promoters and as drying agents [24]. In major detail, a donor such as trichloro- or trifluoroacetimidate, and an acceptor, employed in a slight stoichiometric excess, were dissolved under an Ar atmosphere in toluene and in presence of activated (T = 200 °C overnight under vacuum) 4 Å aw (acid washed) MS (Scheme 1). The reaction was heated to a temperature of 70 °C and then stirred for two hours. Although common acidic activators like lanthanide triflates avoid the high temperature, the obtained yields with molecular sieves as catalyst resulted quite competitive. In order to exclude that residual acids of the acid washed zeolite could have catalyzed the glycosidation reaction, the 4 Å aw MS were washed several times with distillated water before their use and the efficiency of the reaction was not affected.

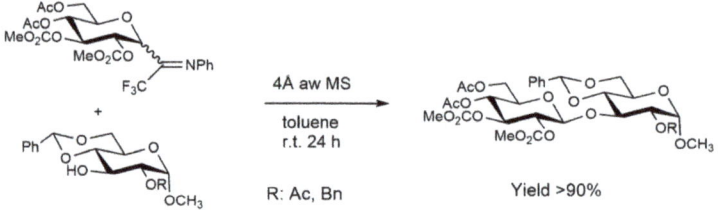

Scheme 1. Activation of glycosyl imidates with activated 4 Å aw MS in a glycosidation reaction.

2.2. Carbinol Glycosidation Reaction of the 17-β-Estradiol

The same authors, by only using the same commercially available 4 Å aw MS, performed a regioselective carbinol glycosidation of 17β-Estradiol [25]. The reaction proceeded overnight via a mild activation of a glycosyl trichloroacetimidate, in 1,2-dichloroethane as solvent; the final glycosidated estradiol was obtained in a satisfactory yield (Scheme 2). The illustrated procedure conferred water solubility to a steroid derivative that has a wide interest for its pharmacokinetic properties.

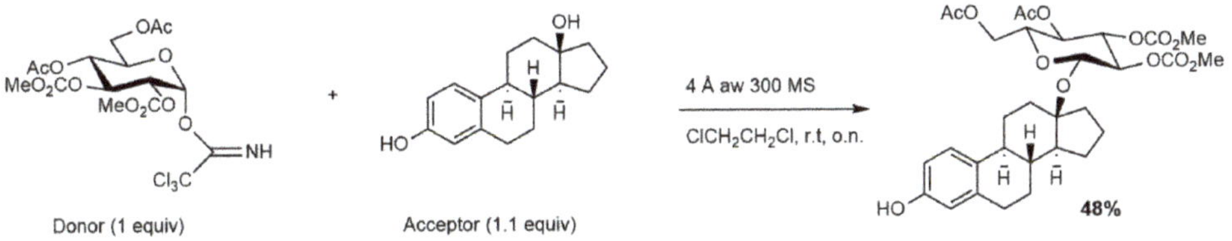

Scheme 2. Regioselective carbinol glycosidation of Estradiol in presence of activated 4 Å aw MS.

2.3. Selective α-Fucosylation for Preparing the Antigenic Trisaccharide Sequence of Lewis X

Since the α-anomer of the L-fucose is widely present in active oligosaccharide sequences, Iadonisi and coauthors have devoted big attention to the development of synthetic strategies to introduce fucose, with its peculiar α-linkage, into oligosaccharide domains [26] (Scheme 3).

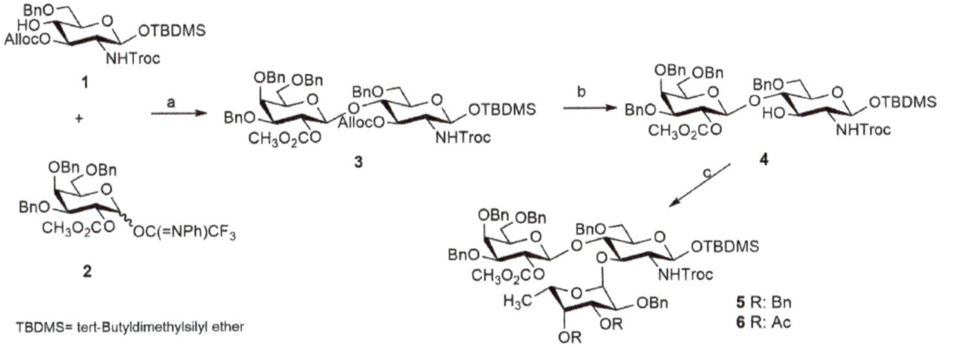

Reaction conditions: a) 4 Å aw MS, dichloroethane, 24 h at 5 °C, 24 h at r.t., 76%; b) Pd(PPh$_3$)$_4$, dimedone, THF, 1.5 h at r.t., 70%; c) for **5**: donor (3 equiv), procedure A: Yb(OTf)$_3$ (0.1 equiv), 4 Å aw MS, −30 °C, 1-3 h, 81%; solvent: 4:1 CH$_2$Cl$_2$/ Et$_2$O/ dioxane; for **6**: donor (2.5 equiv), procedure B: 4 Å aw MS, from 0 °C to r.t., 24-36 h, 42%; solvent: toluene or ClCH$_2$CH$_2$Cl

Scheme 3. α-fucosylation reaction promoted by activated 4 Å aw MS to synthesize a trisaccharide antigenic fragment.

In this regard, the investigation must take into account two important aspects: fucosyl donors are rather reactive and so prone to decomposition and that the α-fucosylation reaction is efficiently assisted by the presence of acetyl groups on the donor. A mild synthetic protocol for preparing the antigenic trisaccharide sequence of Lewis X has proven that 4 Å aw MS are able to promote a highly stereocontrolled α-fucosylation, having the advantage of long-range participation of the fucose acetyl protective groups. In major detail, following a procedure previously illustrated, glucosamine **1** was reacted with a galactosyl imidate **2** providing that disaccharide **3** was quickly deprotected alcohol **4** for the subsequent α-fucosylation reaction. In case of a perbenzylated donor such as the fucosyl *N*-(phenyl)trifluoroacetoimidate, the reaction took place in a short time (1–3 h) in the presence of a catalytic amount of Yb(OTf)$_3$. The final yield and the stereoselectivity were totally satisfying for the trisaccharide **5**. For a partially acetylated fucose derivative, only activated 4 Å aw MS (T = 200 °C overnight under vacuum) were used as catalyst and an excess of two equivalent of the donor (**6**). The reaction time was prolonged to 24–36 h, the yield was quite satisfying, and complete stereocontrol was achieved. The developed strategy disclosed a remarkable mildness that can characterize the synthetic procedure for preparing important biological saccharidic targets.

2.4. Synthesis of a Tetrasaccharide Sequence of Globo H

Globo H (Figure 4) is an antigen oligosaccharide widely employed for the development of vaccines against some cancers, the terminal portion of it, a tetrasaccharide that was proven to be endowed with biological activity, was nimbly synthesized by Iadonisi and coauthors combining mild reaction strategies [27].

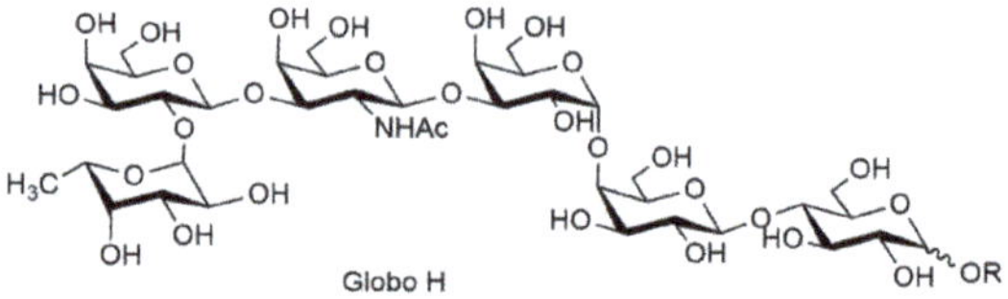

Figure 4. Chemical structure of Globo H.

As shown in Scheme 4, by coupling the azido-functionalized donor **1** with the acceptor **2** in acetonitrile as solvent and Yb(OTf)$_3$ as catalyst, the β-linked disaccharide **3** was prepared thanks to the β-directing solvent effect. After deacetylation and successive benzylidene formation, acceptor **4** was coupled with galactose donor **5**. For this reaction only, activated 5 Å aw MS were used as catalyst, they gave a higher yield of the final trisaccharide product in a shorter reaction time if compared with the results obtained by using the smaller pore size 4 Å aw MS. Basic deprotection performed with a solution of potassium bicarbonate in methanol allowed the preparation of the acceptor **7** that was subsequently reacted, at low temperature, with fucosyl donor **8**. The reaction is performed in an α-directing system solvent (DCM/diethyl ether/dioxane 4:1:1) and with the efficient activation of Yb(OTf)$_3$. Then, by a reduction procedure, protecting groups such as benzyl and benzylidene were removed, and the azido-group was converted in an amino group. The obtained final product (**10**) was considered a suitable precursor for developing novel analogs of the original biological target.

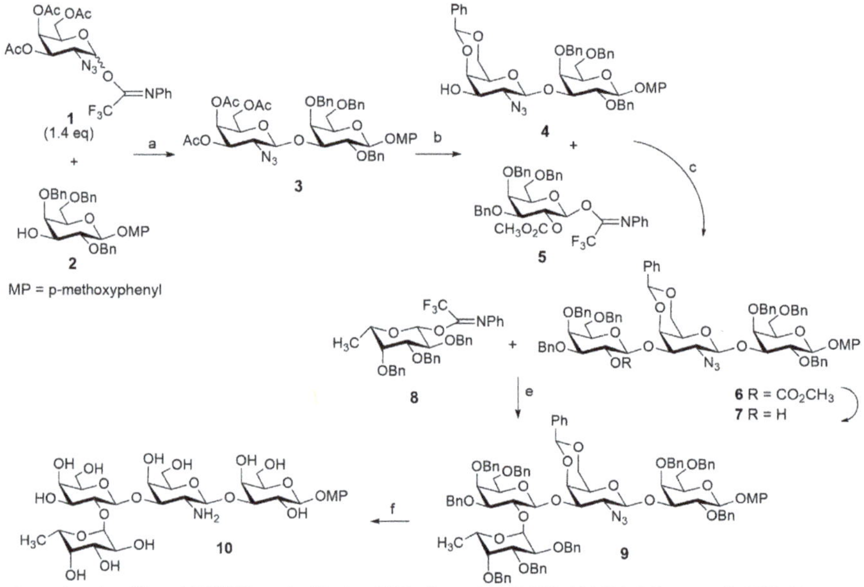

Reagents and conditions: (a) Yb(OTf)$_3$, acetonitrile, from 0 °C to rt, overnight, 70-77%; (b) 9:1 MeOH/aq ammonia (32%), 3 h, then acetonitrile, benzaldehyde dimethyl acetal, camphorsulfonic acid (cat.) 70 °C, 3 h, 80% overall yield; (c) 5Å aw MS, dichloroethane/cyclohexane 5:1, rt, overnight, 75%; (d) K$_2$CO$_3$, methanol, 40 °C, 8 h, 89%; (e) Yb(OTf)$_3$, DCM/diethyl ether/dioxane 4:1:1, from -30 °C to rt, 66%; (f) Pd(OH)$_2$, H$_2$, 3:3:1 DCM/MeOH/H$_2$O, rt, 90%.

Scheme 4. Synthesis of the tetrasaccharide.

3. O-Protection of Sugars Promoted by Molecular Sieves

3.1. Protection Procedure of Sugar Alcoholic Functions

Iadonisi and coauthors achieved interesting results by employing the acid-washed 4 Å molecular sieves (300 aw MS) to install protecting groups on primary and secondary alcoholic functions of saccharides [28]. In particular, benzhydrylation and tritylation were performed under the following conditions. To a mixture of a model saccharide dissolved in anhydrous toluene and in presence of activated 300 aw MS (T = 200 °C overnight under vacuum), several portions of diphenylmethanol were added during a time-lapse of several hours at room temperature. Concerning the tritylation, it was performed by a single addition of a slight stoichiometric excess of triphenyl methanol (Scheme 5). The procedure followed a dehydration mechanism avoiding the employment of strong protic or Lewis acid, it also allowed the protection of the less reactive secondary alcoholic functionalities under mild conditions that result compatible with both, acid and base labile protecting groups. Molecular sieves were thoroughly washed and then reused for another reaction. The protection procedure resulted totally ineffective when performed in presence of standard, not acid washed, 4 Å MS.

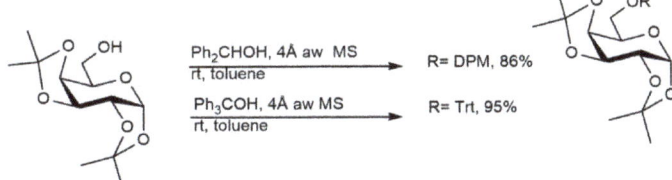

Scheme 5. Benzhydrylation and tritylation of a saccharide in presence of activated 4 A aw MS.

3.2. O-Acetylation and De-O-Acetylation of Carbohydrates Activated by Molecular Sieves

Iadonisi and coauthors have also employed 4 Å MS, without any previous activation, for promoting the selective acetylation of primary and secondary carbohydrate alcoholic groups [29]. The reaction was performed in neat acetic anhydride at 0 °C and provided the final monoacetylated primary product when continued at room temperature, while the heating of the reaction mixture promoted a di-acetylated carbohydrate synthesis (Scheme 6). In both cases, the final products were obtained in high yields after few hours.

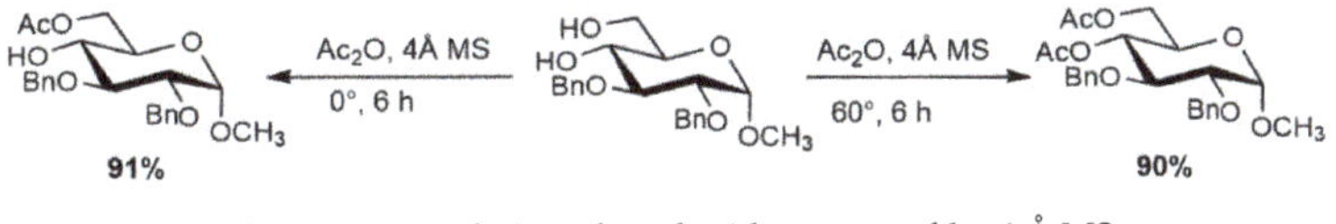

Scheme 6. Acetylation of saccharide promoted by 4 Å MS.

Ravindranathan Kartha and coauthors proved that activated 4 Å molecular sieves are able to promote the selective deacetylation of several O-acetylated and O-benzylated carbohydrate derivatives [30]. The procedure consist of using methanol as solvent over activated 4 Å molecular sieves, thus the methanolic sodium methoxide was hypothesized to be generated. The reaction allowed the deprotection almost quantitatively after a variable time, from minutes to several hours (Scheme 7).

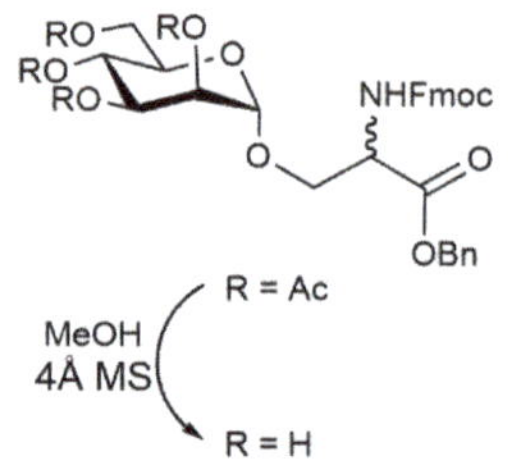

Scheme 7. De-O-acetylation of saccharide promoted by activated 4 Å MS.

4. Alkylation Reactions Promoted by Molecular Sieves

4.1. Side Chain Mono-N-Alkylation of Fmoc-Amino Acids and Peptide Sequences

Our group reported on an *N*-alkylation as the first reaction promoted by activated 4 Å molecular sieves (280 °C, T = 4 h under vacuum) [31]. The reaction was performed on *ortho*-Ns-protected basic Fmoc-amino acids (Fmoc-Lys(Ns)-OH; Fmoc-Orn(Ns)-OH; Fmoc-Dab(Ns)-OH; Fmoc-Dap(Ns)-OH) by using different alkyl halides (Scheme 8). One of the obtained mono-*N*-alkylated amino acids was introduced as a building block into a peptide sequence.

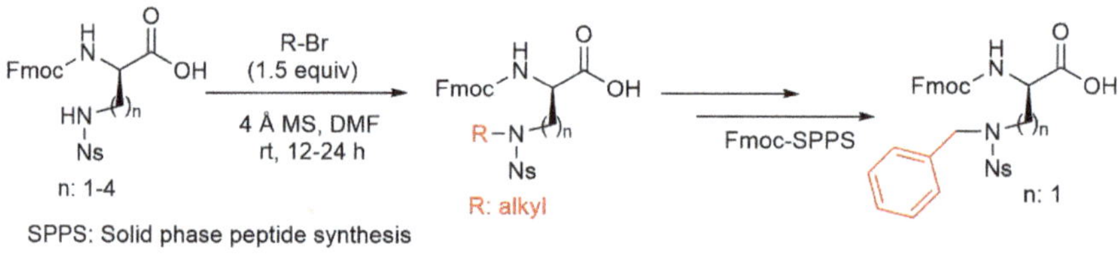

Scheme 8. N-alkylation of nosyl-protected Fmoc-amino acid promoted by activated 4 Å MS.

The success of the obtained results prompted us to explore the new *N*-alkylation strategy on fully deprotected peptide sequences as postsynthetic modification strategy [32]. The reaction was performed in DMF under an Ar atmosphere and employed under very mild conditions, only activated 4 Å MS as basic catalyst and the appropriate alkyl halide. The nosyl-protected amino group of the amino acid residue performed the nucleophilic substitution, so that a mono-*N*-alkylation was performed on the peptide side chain (Scheme 9). The strategy allowed the introduction of different substituents on model peptides and provided high chemoselectivity and excellent conversion yields.

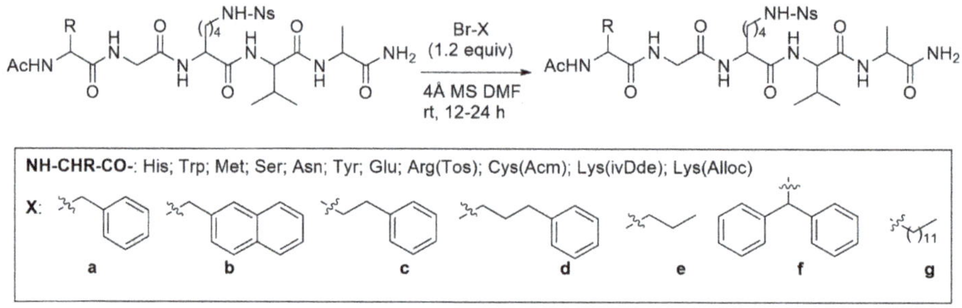

Scheme 9. Chemoselective *N*-alkylation of a peptide sequence promoted by activated 4 Å MS.

4.2. S-Alkylation Reaction for Introducing Peptide Modification

Considering the strong nucleophilicity exhibited by the cysteine sulfhydryl group [33], our group explored and tuned the alkylation reaction conditions on peptide sequences containing a cysteine residue [34]. In analogy with the *N*-alkylation protocol, postsynthetic *S*-alkylation was promoted by activated 4 Å MS that acts as a base, the mild conditions employed could avoid the racemization of the cysteine residue. The reactivity of the thiol group was explored by performing the reaction on the same model peptide with different alkyl bromides, while the chemoselectivity was assessed by adding at the *N*-extremity of a peptide model the amino acid residues representative of the range of functional groups usually present on the side chains of natural peptides (Scheme 10). The reaction time was increased from 1 to 3 h, when less reactive alkyl bromides were employed; however, in all cases a high yield of the final product was obtained. The reaction time was drastically reduced when the protocol was further implemented by a microwave irradiation (40 °C for 5 min) [35]. In addition, the methodology allowed the discrimination of the cysteine thiol group reactivity upon that of other sensitive nucleophilic peptide functionalities. Even in the presence of a lysine residue a polyalkylated product was obtained in a low yield, which was farther reduced when MW-activation was applied.

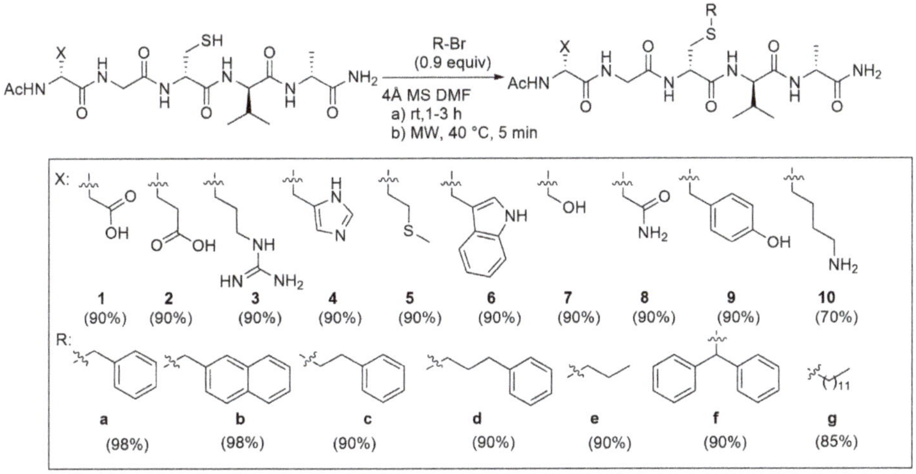

Scheme 10. Chemoselective *S*-alkylation of a peptide sequence promoted by activated 4 Å MS.

We also succeeded in functionalizing a peptide with polyethylene glycol, a widespread substituent present on bioactive peptides in order to increase their in vivo stability and solubility, as well as with a fluorophore such as 5-(bromo-methyl)fluorescein (Figure 5).

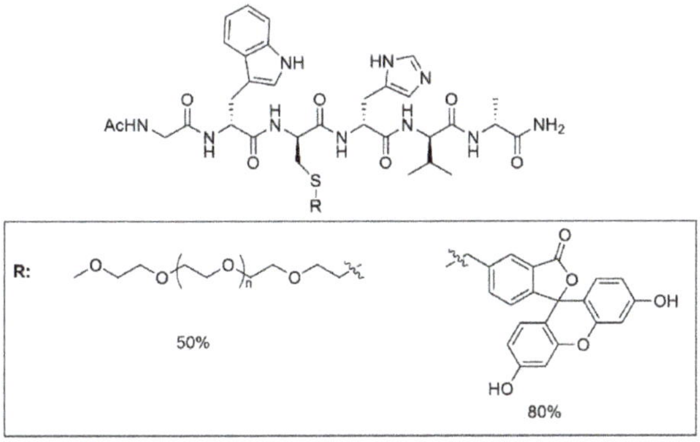

Figure 5. *S*-alkylation promoted by activated 4 Å MS of peptide with useful substituents.

The most interesting results were obtained by introducing lipid functionalities in several tailor-made peptides, such as farnesyl and hexadecyl groups (Figure 6), which represent the bioactive portions of their Ras parent proteins [36].

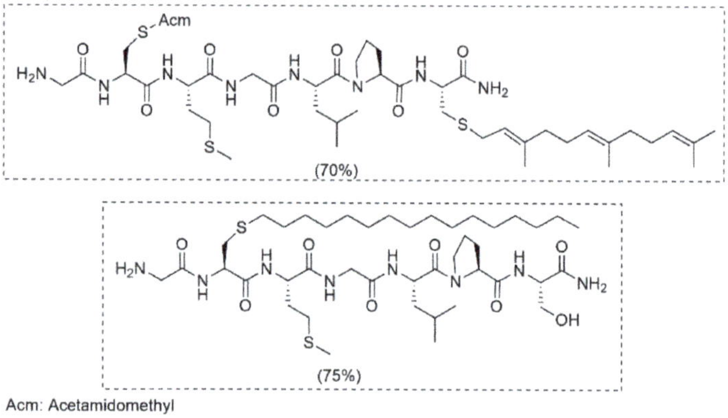

Figure 6. Chemical structure of S-lipidated peptides.

4.3. Solid-Phase S-Alkylation Reaction for Introducing Peptide Modification

Our group has implemented the developed a postsynthetic S-alkylation procedure promoted by 4 Å MS in solid phase, with the aim of considering it a routinely solid-phase peptide synthetic step [37]. The reaction is performed on a peptidyl resin and the chemical process can be considered a solid-state procedure, since both the catalyst (molecular sieves) and the substrate (the peptide anchored on the resin) are provided in solid-state. After a completion of the peptide sequence, the N-terminus was acetylated and then the cysteine thiol group was freed from the highly acid labile protecting group (Mmt = 4-methoxytrityl) for the subsequent S-alkylation reaction that was performed in DMF under an Ar atmosphere, in presence of activated molecular sieves. The alkyl bromide was employed in a stoichiometric excess and the reaction mixture was kept under stirring for 15 h at room temperature. The efficiency of the solid phase S-alkylation was investigated by reacting a peptide model with different alkyl bromides (Scheme 11).

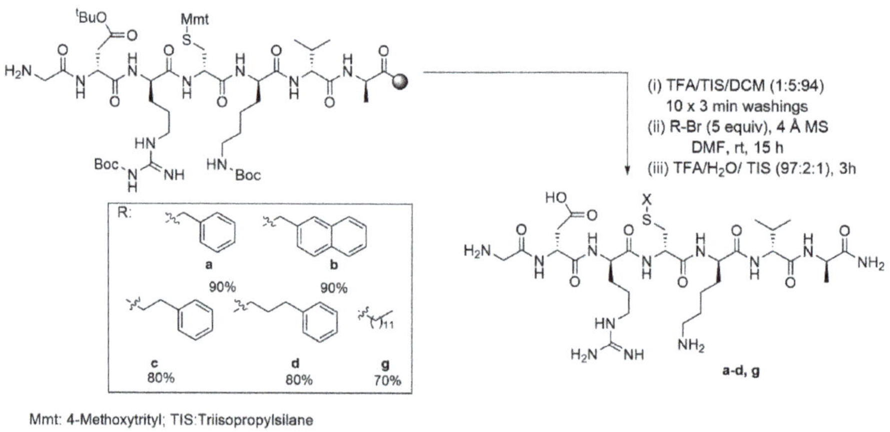

Scheme 11. Solid-phase S-alkylation of peptide sequences anchored on resin promoted by activated 4 Å MS.

By combining both protocols—the 4 Å MS solution and solid-phase strategy—it was possible to prepare a multialkylated peptide, such as the C-terminal N-RAS decapeptide shown in Scheme 12. It contains the palmityl group introduced in solid phase and the farnesyl that was necessarily introduced on the fully deprotected functionalized peptide **2**. In fact, the farnesyl could avoid the final acid treatment for the deprotection and the cleavage of the peptide, which can affect its stereochemistry.

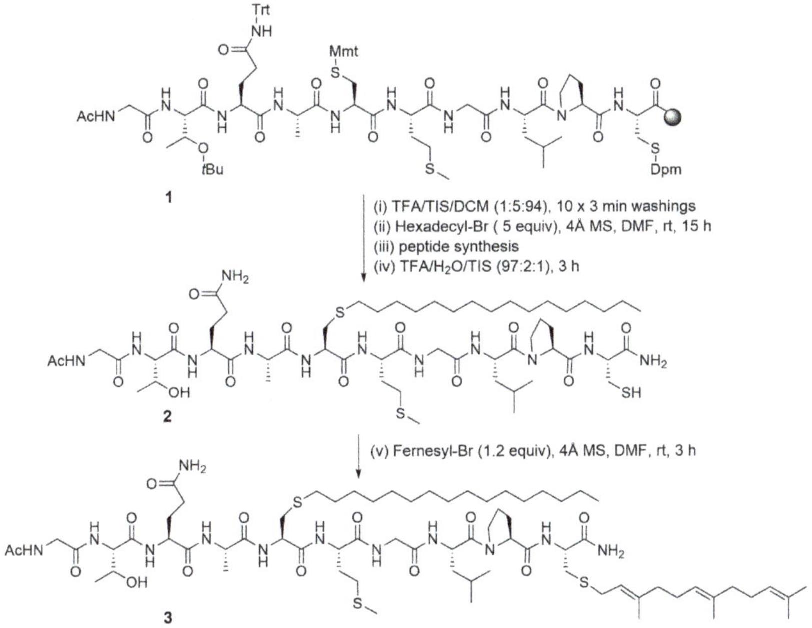

Scheme 12. Solution and solid-phase approach promoted by activated 4 Å MS to perform a multialkylation.

4.4. Chemoselective Glycosylation of Peptides through S-Alkylation Reaction

The S-alkylation procedure promoted by activated 4 Å molecular sieves was also employed by our group to prepare S-linked glycopeptides [38]. The reaction was performed on peptide sequences containing the sensitive cysteine residue that could react with the chosen glycosyl halide, in DMF, under an Ar atmosphere, and in the presence of activated molecular sieves as the basic catalyst. A good chemoselectivity for this site-directed functionalization of the cysteine thiol group was obtained, even in presence of other nucleophiles. In addition, the reactivity of the sulfhydryl was explored toward a certain number of monosaccharides (acetobromo-α-D-glucose; acetobromo-α-D-galactose; 1-chloro-N-acetyl-glucosamine) and also toward a disaccharide such as the acetobromo-maltose. The reaction occurred through a S_N2 mechanism and allowed the formation of S-β-linked glycopeptide (Scheme 13). The final compounds were prepared in moderate yields; however, they were obtained without any manipulation of the sugar derivative that was in all cases commercially available, neither a peculiar activation of the thiol group was necessary in order to achieve chemo and stereoselective cysteine functionalization.

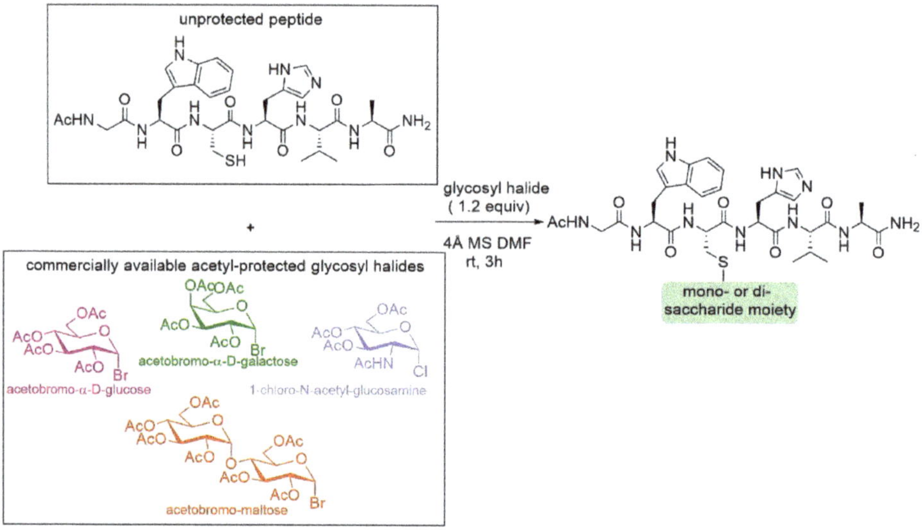

Scheme 13. *S*-glycosylation of peptide sequences promoted by activated 4 Å MS using different carbohydrate derivatives.

4.5. Lanthionine-Containing Peptide Obtained via Cysteine S-Alkylation on Cyclic Sulfamidates

Our group has also developed a 4 Å MS promoted strategy to prepare lantipeptides, peptide molecules that contain the unusual amino acid lanthionine [39]. The reaction consists of the *S*-alkylation of a cysteine, already inserted in a peptide sequence, on a cyclic sulfamidate. The obtained modified peptide contains a stereochemically pure lanthionine residue (Scheme 14). The mild reaction conditions allowed for the employment of orthogonal and peptide-compatible protecting groups on the sulfamidate derivatives, which is a key objective of the proposed procedure, since they can be selectively removed in order to perform the subsequent cyclization and introduce the thioether ring, typical structural motif of the natural lantibiotics. The successful outcome of the new methodology, both in terms of efficiency and chemoselectivity, was proved by the excellent yield values (80–95%) found for the obtained monoalkylated peptides.

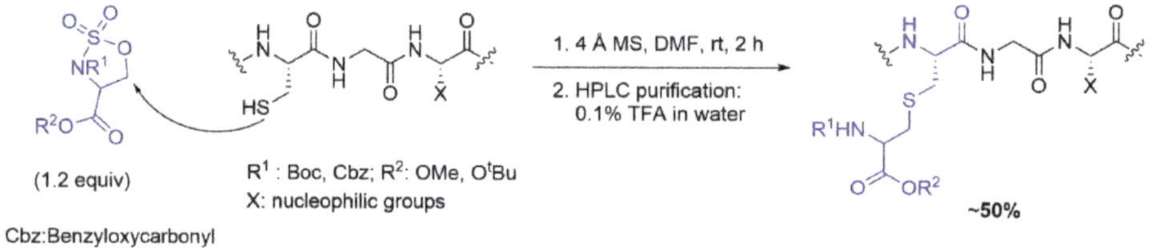

Scheme 14. *S*-alkylation of peptide sequences performed on cyclic sulfamidates and promoted by activated 4 Å MS.

As an application of the new methodology, the synthesis of Halβ ring B was successful carried out, it serves as a template for future preparation of lantibiotic analogs (Scheme 15).

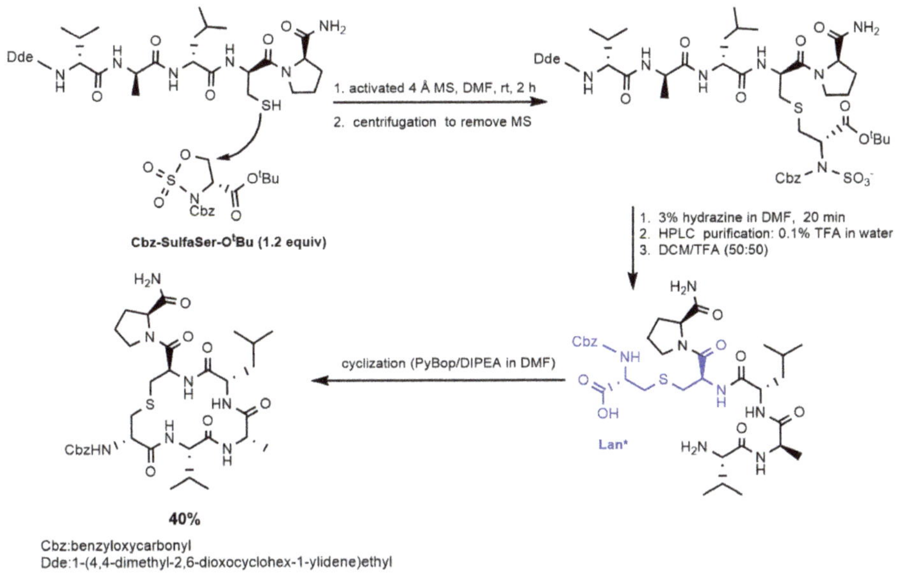

Scheme 15. *S*-alkylation of peptide sequence promoted by activated 4 Å MS to prepare bioactive lantibiotic fragment.

4.6. Benzylation of Arylcyanamides Performed via Acid Catalysis of Zeolites

Azarifar and coauthors performed a substitution reaction of cyanamides by employing benzyl bromide and several arylcyanamides in acetonitrile, at 65 °C and in presence of acid zeolite as catalyst. It is worth remembering that substituted cyanamides are useful and widespread employed building blocks for the preparation of biologically active substrates, such as antiviral and anticancer agents. Several zeolites have been employed, some naturally occurring in the Lewis acid form (Na-Y and Na-ZSM-5) and others obtained by preparing their Brønsted H-form (H-Y and H-ZSM-5). The exchange of Na^+ with the proton was performed by treatment of the zeolite with NH_4NO_3 followed by a drying and calcination process (Scheme 16) [40].

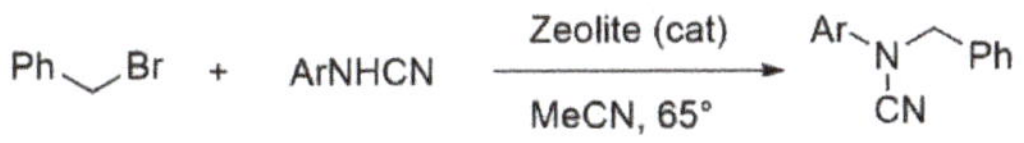

Scheme 16. *N*-benzylation of arylcyanamide promoted by acidic zeolites.

The zeolite H-Y exhibited the best result in terms of yield, probably due to the high number of acidic sites and the dimension of the zeolite pores together with the presence of the super-cages inside the aluminosilicate solid.

The hypothesized reaction path is reported in Scheme 17 and suggests a $S_N 2$ mechanism into the zeolite channels by acidic catalysis assistance.

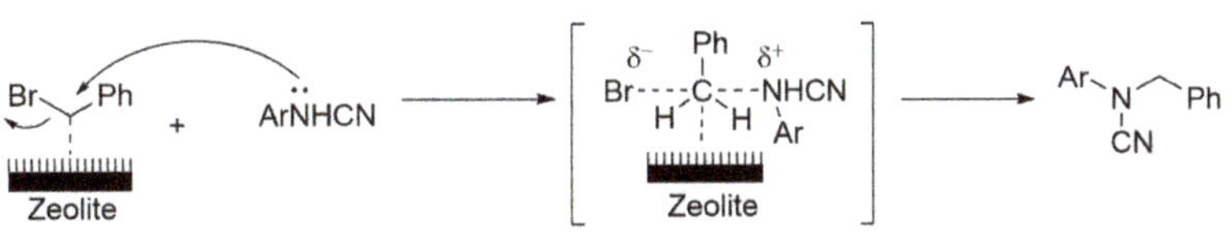

Scheme 17. Hypothesized mechanism for *N*-benzylation of arylcyanamides promoted by acidic zeolite.

It is important to refer that the catalyst was recovered by filtration and reused, keeping the original activity. Moreover, the reaction was totally unsuccessful in absence of the zeolite.

4.7. Microwave Irradiation Promotes the N-Alkylation of Imidazole in Presence of a Basic Zeolite as Catalyst

The alkylation of heterocycles is an important synthetic step for the preparation of biological active compounds such as fungicides and anticonvulsants. Costarrosa and coauthors reported on an alkylation reaction performed on imidazole by using different alkylating agents, alkaline-exchanged X-type zeolite as catalyst, and microwave irradiation to promote the chemical process [41].

The sodium ions of 13X zeolites were exchanged with some other alkaline earth ions by employing 2M solutions of the corresponding salts, $CaCl_2$, $BaCl_2$, and $Sr(NO_3)_2$. The procedure was performed as follows (Scheme 18). The imidazole and the chosen zeolite, previously dried at 353 K, were blended in a Teflon vessel, then an excess of 9 equivalent of alkylating agent (1-Br-butane, 2-Br-butane, and 1-Br-hexane) was added. The reactor vessel was placed in a microwave oven and irradiated at 300 and 700 W during different time-lapses (1, 3, and 5 min). The reaction products were extracted in acetone.

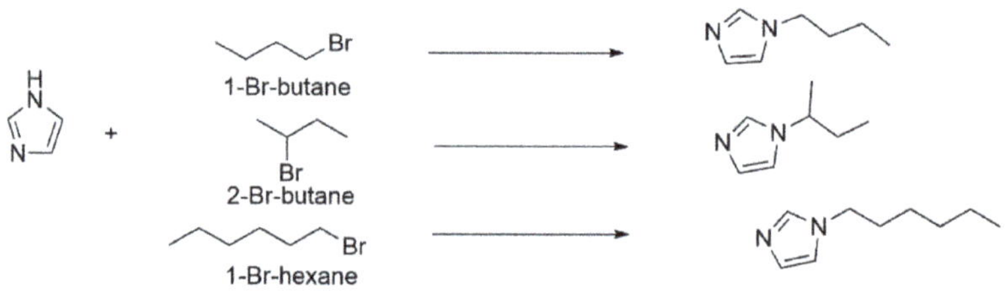

Scheme 18. N-Alkylation of imidazole with different alkylating agents in presence of basic zeolites.

The obtained results indicated the following trend of basicity NaX > CaX > BaX > SrX, which is not related to the basic character of the alkaline earth ions exchanged zeolites (BaX > SrX > CaX). In general, the comprehensive analysis of the performed experiments assessed that the catalytic activity depends on the basicity of the employed zeolite, the polarizing ability of the exchanged cation, and the power of the employed microwave irradiation. For 1-Br-butane and 2-Br-butane, the best catalyst was revealed to be NaX when microwave irradiation was produced at 300 W. An increase in this value would affect the activity only of the less basic catalysts. In the case of 1-Br-hexane, the found trend was different. In fact, at the highest power of 700 W, the best catalyst resulted to be SrX with the shortest time of irradiation, which is in agreement with the strongest basic character of SrX zeolite.

5. Useful Examples of Broad Scope Reactions Promoted by Molecular Sieves

5.1. Synthesis of α-Aminophosphonates by Using Natural Natrolite as Reusable and Efficient Catalyst

Bahari and coauthors reported on a novel high yielding procedure for synthesizing α-aminophosphonates [42]. The one-pot reaction uses aromatic aldehydes or ketones, substituted anilines, and trialkyl phosphites under solvent-free conditions and Natrolite zeolite as catalyst. The optimized reaction conditions were found by employing 4-methylbenzaldehyde (1 mmol), aniline (1 mmol), and triethyl phosphine (1 mmol); concerning the catalyst, the optimum amount of natural Iranian natrolite was found to be 0.05 g (Scheme 19). It is important to underline that the reaction does not occur in absence of the catalyst and that it could be easily recovered by filtration after addition of ethylacetate to the mixture. As final step, the catalyst was washed of water and ethanol and dried in order to be reused without any loss of efficiency.

In conclusion, the developed procedure revealed to be efficient (high yield in a short reaction time), environmentally friendly (little waste was produced), and economically convenient (a natural and reusable zeolite was employed as catalyst).

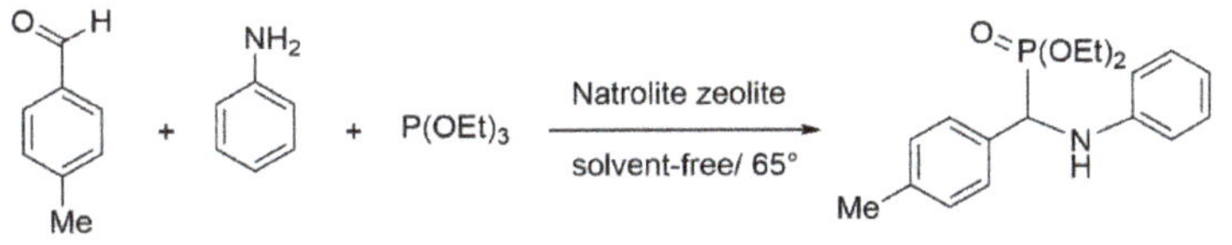

Scheme 19. α-aminophosphonate catalyzed by Natrolite zeolite.

5.2. Metal-Catalyzed Enolization of Nucleophiles Precursors

Hasegawa and coauthors developed an efficient catalytic activation protocol for generating metal enolate, by employing nitromethane as nucleophile and Lewis acid, such as chiral Ni(II) ions, in combination with 4 Å MS [43]. The reaction is performed in alcohol as solvent and provides the active enolate that can be subsequently involved in a Michael addition reaction with α,β-unsaturated carbonyl electrophiles. Following the hypothesized mechanism, the catalytic activity of the molecular sieves, employed without any preactivation, consist of capturing a proton (Scheme 20). In other words, the molecular sieves behaved an effective base, even with a smaller pore size of 3 Å, while the 5 Å MS resulted much less efficient

Scheme 20. Ni(II) nitronate generated by the catalyst activity of basic 4 Å MS that captures the α-proton of a nucleophile precursor.

5.3. Different Base-Catalyzed Reactions Promoted by Methylammonium- Faujasite

It is well-known that the basicity of the zeolites increases with the decrease of the electronegativity of the compensating cations, since it enhances the electron density on the framework oxygens. Therefore, the basic strength of the cationic zeolites follows the order $Li^+ < Na^+ < K^+ < Rb^+ < Cs^+$. With this idea in mind, Martins and coauthors prepared exchanged Cs-FAU and also exchanged methylammonium-FAU, which showed the highest ion exchange degree.

$(CH_3)NH_3^+$-FAU catalytic power for several reactions, such as Knoevenagel and Claisen-Shmidt condensation as well as alcoholysis of propylene oxide (Scheme 21), gave the best results, and is six times more active than Cs-FAU [44].

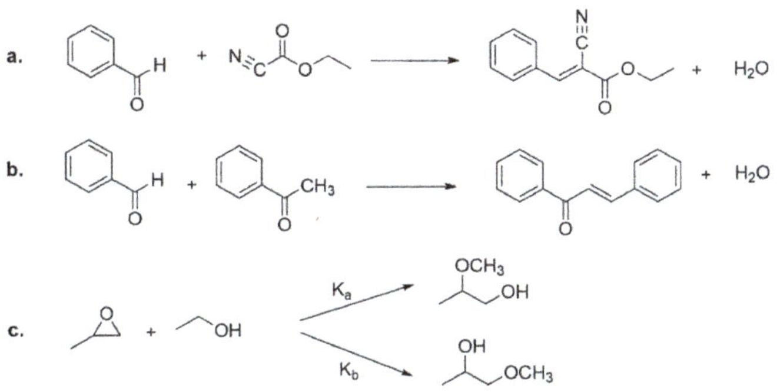

a. Knoevenagel condensation of benzaldehyde with ethyl cyanoacetate; b. Claisen-Schmidt condensation of benzaldehyde with acetophenone; c. Propylene oxide alcoholysis with methanol.

Scheme 21. Three different base-catalyzed reactions promoted by methylammonium FAU zeolites.

The experimental studies were performed on both modified FAU types, X (Si/Al = 1.4) and Y (Si/Al = 2.5). The TPD-CO_2 test (temperature-programmed desorption of CO2) found that the zeolites containing the highest content of aluminum present also have a higher basicity character. On the other hand the O1s-XPS measurements (X-ray photoelectron spectroscopy) proved that the presence of the organic cation methyl-ammonium increases the basic character of the zeolite, since it exhibited the highest ion exchange degree, so that the highest number of catalytic sites. It was proved that the acid–base pair of the alkyl-ammonium-FAU is more efficient in providing basic catalysis for all the performed reactions, which confirm the great potential of the new low-cost developed catalyst.

5.4. Benzene Ethylation Performed with Ethanol in Presence of Modified Acidic Zeolite

Ethylbenzene is one of the most used chemical intermediate for several synthetic processes; in fact, it is the main feedstock for the production of styrene, while it is also utilized for the manufacture of cellulose acetate, acetophenone, diethylbenzene, propylene oxide, and other substances.

Emana and Chand reported on the ethylation of benzene by an electrophilic substitution performed upon catalysis of an acidic zeolite that activates the alkene obtained by dehydration of ethanol [45]. In details, all reagents were placed in a continuous down flow tubular quartz reactor allocated in a controlled furnace. Two different kinds of zeolite, previously activated for 1 h in nitrogen atmosphere, were employed. One of them was the Brønsted acidic H-ZSM-5 zeolite, obtained from the Na-ZSM-5 upon exchange of the cations with protons by using a 1 M solution of NH_4NO_3. Subsequently, the impregnation of H-ZSM-5 with magnesium and boron was performed by using a solution of boric acid and magnesium nitrate as source of the desired elements. The best results, in terms of both selective and high-yield production, of ethylbenzene were obtained by employing Mg (5%)-B (4%)-HZSM-5 zeolite; in fact, other by-products were obtained in a negligible yield (Scheme 22).

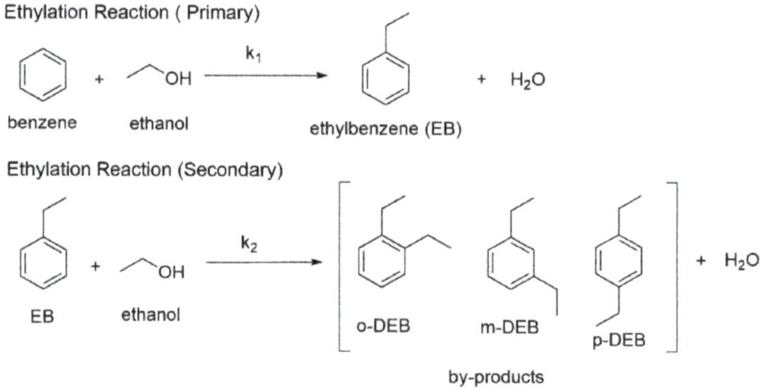

Scheme 22. Reactions occurring during the ethylation of benzene with ethanol in presence of modified acidic zeolite.

5.5. Basic Mesoporous Zeolite Proved to Be an Efficent Catalyst for Several Reactions: Condensation, Hydroxylation, and Cycloaddition Reactions

Mesoporous ZSM-5 zeolite was prepared under basic conditions and its calcined form was treated with NH_4OH. Sarmah and coauthors studied the catalytic activity of the obtained basic zeolite by performing several base catalyzed reactions such as the synthesis of substituted styrenes, carbinolamides, naphthopyrans, and cyclic carbonates via Knovenagal condensation, amide hydroxylation, one-pot multicomponent reaction, and cycloaddition reactions (Scheme 23) [46].

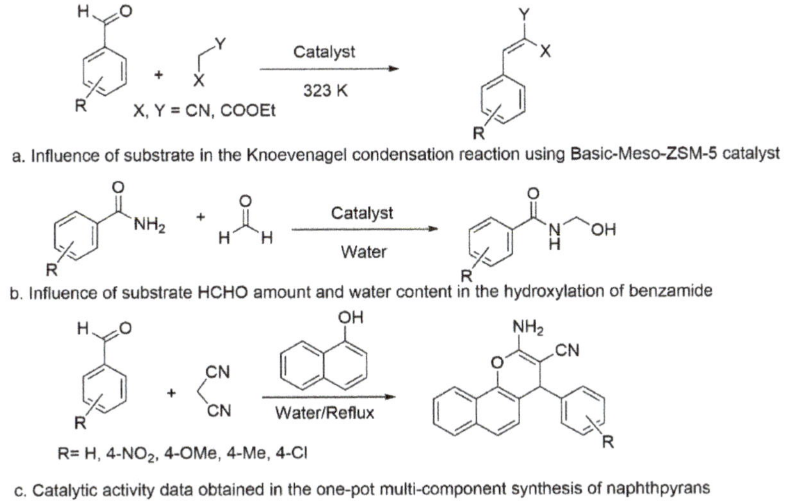
Catalyst
323 K
X, Y = CN, COOEt
a. Influence of substrate in the Knoevenagel condensation reaction using Basic-Meso-ZSM-5 catalyst
Catalyst
Water
b. Influence of substrate HCHO amount and water content in the hydroxylation of benzamide
Water/Reflux
R= H, 4-NO2, 4-OMe, 4-Me, 4-Cl
c. Catalytic activity data obtained in the one-pot multi-component synthesis of naphthpyrans

Scheme 23. Basic mesoporous zeolite catalyzes the synthesis of substituted styrenes, carbinolamides, naphthopyrans, and cyclic carbonates.

5.6. *Synthesis of Polysubstituted Cyclopropanes Catalyzed by Basic Zeolite*

Rama and coauthors reported on a very convenient protocol to promote the one-pot two-step synthesis of substituted cyclopropane. The reaction utilizes substituted benzyl halides, aromatic aldehydes, pyridine, acetonitrile derivatives, and potassium-exchanged Y zeolites as basic catalyst (Scheme 24). The same reaction performed in absence of the zeolite gave no product, since the pyridine basicity was insufficient for catalyzing the final addition that provides the cyclopropane, while it could only promote the Knoevenagel condensation [47].

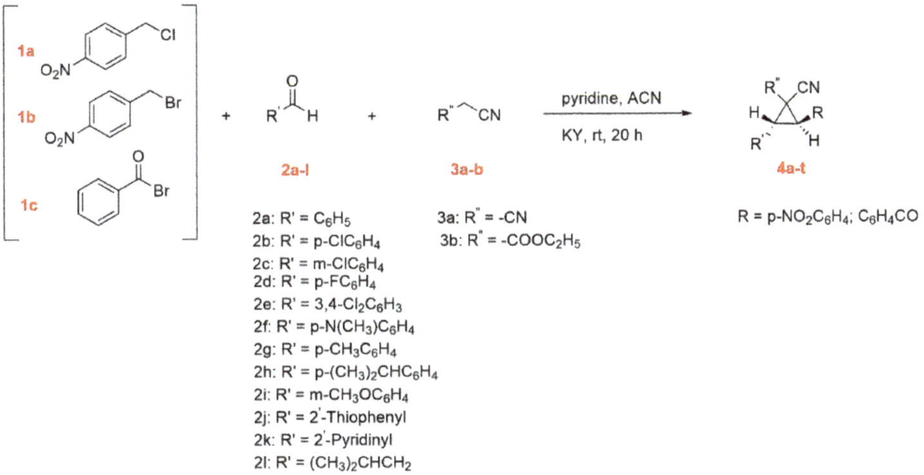
1a
1b
1c
2a-l
3a-b
pyridine, ACN
KY, rt, 20 h
4a-t
2a: R' = C6H5
2b: R' = p-ClC6H4
2c: R' = m-ClC6H4
2d: R' = p-FC6H4
2e: R' = 3,4-Cl2C6H3
2f: R' = p-N(CH3)C6H4
2g: R' = p-CH3C6H4
2h: R' = p-(CH3)2CHC6H4
2i: R' = m-CH3OC6H4
2j: R' = 2'-Thiophenyl
2k: R' = 2'-Pyridinyl
2l: R' = (CH3)2CHCH2
3a: R" = -CN
3b: R" = -COOC2H5
R = p-NO2C6H4; C6H4CO

Scheme 24. Synthesis of polysubstituted cyclopropanes promoted by basic KY zeolites.

The catalyst was prepared by a cation substitution procedure performed on NaY with nitrate solution of potassium, lithium and cesium. The yield of the catalyzed reactions followed the basicity degree of the zeolites (CsY > KY > NaY > LiY). Indeed, KY was preferred to CsY for cost effectiveness and easy handling reasons. The exchanged cation increased the negative charges on the aluminum center, which was transferred to the adjacent oxygen atom. It is believed that the basic cites are localized on the surface-bound metal oxides and the plausible catalytic mechanism is reported in Scheme 25.

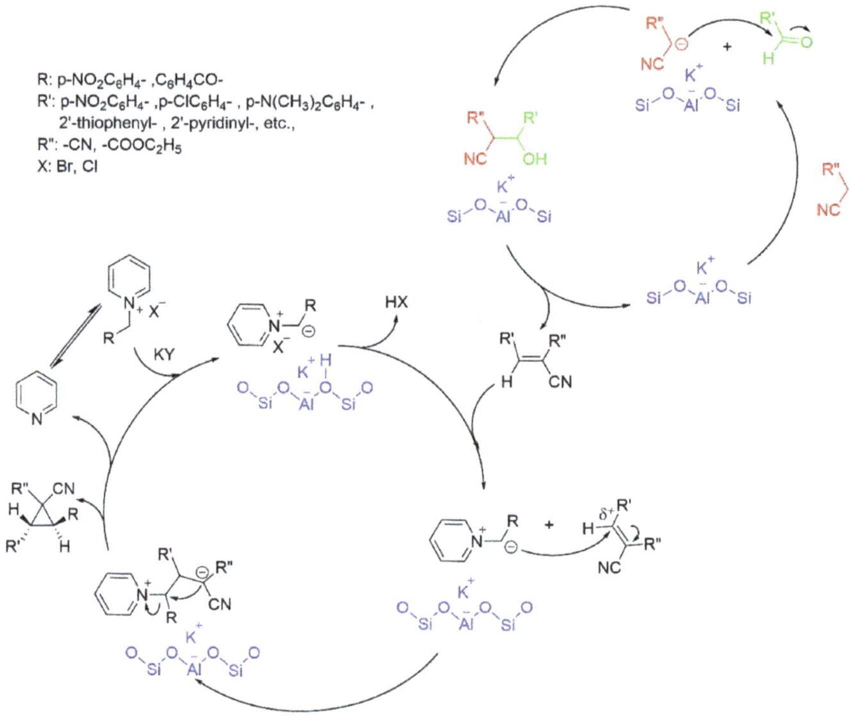

Scheme 25. Plausible mechanism for synthesis of polysubstituted cyclopropanes catalyzed by basic KY zeolites.

In conclusion, the developed protocol requires very mild conditions and is characterized by good yields, a simple work-up, and reusability of the catalyst. It was actually used for three consecutive reactions and retained its full activity.

6. Final Remarks

The acid–base sites of zeolites are related to the different charge of the two ions Al^{+3} and Si^{+4} present in the crystal structure, so (AlO_4) tetrahedrons have a net negative charge that requires balancing. As previously discussed, cations or protons can ensure the framework electroneutrality; in the latter case the zeolite is considered a solid acid that can give heterogeneous catalysis based on Brønsted acidity of bridging Si-(OH)-Al sites [48,49]. On the other hand, in the cationic zeolite the oxygen atoms that possess a negative charge are characterized by a certain basicity/nucleophilicity, also being the conjugate base of the corresponding acid form (Figure 1). After all, the basicity of the zeolites is a documented property, considering its deprotonating catalytic activity reported in several publications, as well as all the spectroscopic studies performed to ensure this behavior [4].

We have already discussed into the introduction the influence of the framework and extra-framework chemical composition on the number and the strength of the acidic sites of protonated zeolites and so of their catalytic activity. Since the reactions occur into the zeolite pores, it must be taken into account also of the zeolite geometry that causes the shape selectivity and the confinement effect of the well-defined crystal pore sizes, where the catalytic active sites are located. This molecular sieving property of the zeolites can select the desired reaction path, which gives rise to the so-called "transition state shape selectivity", that means only certain configurations of the transition state are allowed. In this regard, the modulation of the acid site strength has been considered a big challenge: it seems fairly certain that the electric field present in zeolites is crucial in stabilizing ionic pairs of the transition state [6]. On the other hand, nonuniformity of the zeolites in the proton transfer process can affect the selectivity of the chemical process, thus generating unwanted by-products. However, the attention for tuning the catalytic performance of the protonated acidic zeolites as well as the cationic basic zeolites has been reported for several chemical transformations [50,51], some of them illustrated in the present review.

Concerning the cationic zeolites, such as alkaline-LTA, it has been proven that they are naturally endowed with mild basic properties, which is a great advantage compared to the basic strength of the standardly employed organic bases. The exchange of Na^+ ion with larger alkali metals (Cs^+ for example), which are weaker Lewis acids due to their larger size, increases to some extent the basicity of the zeolite oxides, however the increase of the steric hindrance compensates this effect, since the access to the basic sites or to the overall cavity is reduced. In any case, the basic catalytic activity is regularly modulated and these materials result in not very strong solid bases. It likely represents the main reason why cationic zeolites have been considered of less practical importance for industrial applications. In our opinion, the moderate catalytic activity as basic solid should be considered a strong point for the implementation in organic reactions of these cationic zeolites, since they are able to provide great selectivity, due to the mild condition offered, together with ease handling of the reaction mixture and a simple final work up.

In addition to these features, economic feasibility and reusability make zeolites in general attractive heterogeneous benign catalysts, which can accomplish the need to develop new environmentally friendly chemical processes and products, avoiding the use and generation of hazardous substances.

Funding: This research received no external funding.

Conflicts of Interest: The authors declare no conflict of interest.

References

1. Pines, H. Acid-Catalyzed Reactions. In *The Chemistry of Catalytic Hydrocarbon Conversions*, 1st ed.; Academic Press: Evanston, IL, USA, 1981; pp. 1–304. ISBN 9780323155922.
2. Weitkamp, J. Zeolites and catalysis. *Solid State Ion.* **2000**, *131*, 175–188. [CrossRef]
3. Primo, A.; Garcia, H. Zeolites as catalysts in oil refining. *Chem. Soc. Rev.* **2014**, *43*, 7548–7561. [CrossRef] [PubMed]
4. Busca, G. Acidity and basicity of zeolites: A fundamental approach. *Microporous Mesoporous Mater.* **2017**, *254*, 3–16. [CrossRef]
5. Ugal, J.R.; Hassan, K.H.; Ali, I.H. Preparation of type 4Å zeolite from Iraqi Kaolin: Characterization and properties measurements. *J. Assoc. Arab Univ. Basic Appl. Sci.* **2010**, *9*, 2–5. [CrossRef]
6. Corma, A. Inorganic Solid Acids and Their Use in Acid-Catalyzed Hydrocarbon Reactions. *Chem. Rev.* **1995**, *95*, 559–614. [CrossRef]
7. Corma, A. From Microporous to Mesoporous Molecular Sieves Materials and Their Use in catalysis. *Chem. Rev.* **1997**, *97*, 2373–2420. [CrossRef] [PubMed]
8. Busca, G. Base and Basic Materials in Chemical and Environmental Processes. Liquid Versus Solid Basicity. *Chem. Rev.* **2010**, *110*, 2217–2249. [CrossRef]
9. Haag, W.O.; Lago, R.M.; Weisz, P.B. The active site of acidic aluminosilicate catalysts. *Nature* **1984**, *309*, 589–591. [CrossRef]
10. Sartori, G.; Maggi, R. Update 1 of: Use of Solid Catalysts in Friedel-Crafts Acylation Reactions. *Chem. Rev.* **2011**, *111*, PR181–PR214. [CrossRef]

11. Sartori, G.; Maggi, R. Protection (and Deprotection) of Functional Groups in Organic Synthesis by Heterogeneous Catalysis. *Chem. Rev.* **2010**, *113*, PR1–PR54. [CrossRef]
12. Ennaert, T.; Van Aelst, J.; Dijkmans, J.; De Clercq, R.; Schutyser, W.; Dusselier, M.; Verboekend, D.; Sels, B.F. Potential and challenges of zeolite chemistry in the catalytic conversion of biomass. *Chem. Soc. Rev.* **2016**, *45*, 584–611. [CrossRef] [PubMed]
13. Louis, B.; Gomes, E.S.; Losch, P.; Lutzweiler, G.; Coelho, T.; Faro, A., Jr.; Pinto, J.F.; Cardoso, C.S.; Silva, A.V.; Pereira, M.M. Biomass-assisted Zeolite Syntheses as a Tool for Designing New Acid Catalysts. *ChemCatChem* **2017**, *9*, 2065–2079. [CrossRef]
14. Kloetstra, K.R.; van Bekkum, H. Base and Acid Catalysis by the Alkali-containing MCM-41 Mesoporous Molecular Sieve. *J. Chem. Soc. Chem. Commun.* **1995**, 1005–1006. [CrossRef]
15. Barthomeuf, D. Conjugate Acid-Base Pairs in Zeolites. *J. Phys. Chem.* **1984**, *88*, 42–45. [CrossRef]
16. Romero, M.D.; Rodrìguez, A.; Gòmez, J.M. Basicity in zeolite. In *New Topics in Catalysis Research*; McReynolds, D.K., Ed.; Nova Science Pub Inc.: Madrid, Spain, 2007; pp. 197–220. ISBN 1-60021-286-7.
17. Guo, P.; Yan, N.; Wang, L.; Zou, X. Database Mining of Zeolite Structures. *Cryst. Growth Des.* **2017**, *17*, 6821–6835. [CrossRef]
18. Titiloye, J.O.; Tschaufeser, P.; Parker, S.C. Recent Advances in Computational Studies of Zeolites. In *Spectroscopic and Computational Studies of Supramolecular Systems. Topics in Inclusion Science*; Davies, J.E.D., Ed.; Springer: Dordrecht, The Netherlands, 1992; Volume 4, pp. 137–185. ISBN 978-94-015-7989-6.
19. Lima, C.G.S.; Moreina, N.M.; Paixão, M.W.; Corrêa, A.G. Heterogenous green catalysis: Application of zeolites on multicomponent reactions. *Nanocatalysis* **2019**, *15*, 7–12. [CrossRef]
20. Park, S.; Biligetu, T.; Wang, Y.; Nishitoba, T.; Kondo, J.N.; Yokoi, T. Acidic and catalytic properties of ZSM-5 zeolites with different Al distributions. *Catal. Today* **2018**, *303*, 64–70. [CrossRef]
21. Arya, K.; Rajesh, U.R.; Rawat, D.S. Proline confined FAU zeolite: Heterogeneous hybrid catalyst for the synthesis of spiroheterocycles via a Mannich type reaction. *Green Chem.* **2012**, *14*, 3344–3351. [CrossRef]
22. Narasimharao, K.; Hartmann, M.; Thiel, H.H.; Ernst, S. Novel solid basic catalysts by nitridation of zeolite beta at low temperature. *Microporous Mesoporous Mater.* **2006**, *90*, 377–383. [CrossRef]
23. Saravanamurugan, S.; Riisager, A. Zeolite catalized transformation of carbohydrates to alkyl levulinates. *ChemCatChem* **2013**, *5*, 1754–1757. [CrossRef]
24. Adinolfi, M.; Barone, G.; Iadonisi, A.; Schiattarella, M. Activation of Glycoyl Trihaloacetimidates with Acid-washed Molecular Sieves in the Glycosidation Reaction. *Org. Lett.* **2003**, *5*, 987–989. [CrossRef] [PubMed]
25. Adinolfi, M.; Iadonisi, A.; Pezzella, A.; Ravidà, A. Regioselective Phenol or Carbinol Glycosidation of 17p-Estradiol and Derivatives Thereof. *Synlett* **2005**, *12*, 1848–1852. [CrossRef]
26. Adinolfi, M.; Iadonisi, A.; Ravidà, A.; Schiattarella, M. Moisture Stable Promoters for Selective α-Fucosylation Reactions: Synthesis of Antigen Fragments. *Synlett* **2004**, *2*, 0275–0278. [CrossRef]
27. Adinolfi, M.; Iadonisi, A.; Ravidà, A.; Schiattarella, M. Versatile Use of Ytterbium(III) Triflate and Acid Washed Molecular Sieves in the Activation of Glycosyl Trifluoroacetamide Donors. Assemblage of a Biologically Relevant Tetrasaccharide Sequence of Globo H. *J. Org. Chem.* **2005**, *70*, 5316–5319. [CrossRef] [PubMed]
28. Adinolfi, M.; Barone, G.; Iadonisi, A.; Schiattarella, M. Mild benzhydrylation and tritylation of saccharidic hydroxyls promoted by acid washed molecular sieves. *Tetrahedron Lett.* **2003**, *44*, 3733–3735. [CrossRef]
29. Adinolfi, M.; Barone, G.; Iadonisi, A.; Schiattarella, M. An easy approach for the acetylation of saccharidic alcohols. Applicability for regioselective protections. *Tetrahedron Lett.* **2003**, *44*, 4661–4663. [CrossRef]
30. Kartha, R.K.P.; Mukhopadhyay, B.; Field, R.A. Practical de-O-acylation reactions promoted by molecular sieves. *Carbohydr. Res.* **2004**, *339*, 729–732. [CrossRef] [PubMed]
31. Monfregola, L.; De Luca, S. Synthetic strategy for side chain mono-N-alkylation of Fmoc-amino acids promoted by molecular sieves. *Amino Acids* **2011**, *41*, 981–990. [CrossRef]
32. Monfregola, L.; Leone, M.; Calce, E.; De Luca, S. Postsynthetic Modification of Peptide via Chemoselective N-Alkylation of Their Side Chains. *Org. Lett.* **2012**, *14*, 1664–1667. [CrossRef]
33. Calce, E.; De Luca, S. The Cysteine S-Alkylation Reaction as a Synthetic Method to Covalently Modify Peptide Sequences. *Chem. Eur. J.* **2016**, *22*, 1–11. [CrossRef]
34. Calce, E.; Leone, M.; Monfregola, L.; De Luca, S. Chemical Modifications of Peptide Sequences via S-Alkylation Reaction. *Org. Lett.* **2013**, *15*, 5354–5357. [CrossRef] [PubMed]

35. Calce, E.; De Luca, S. Microwave heating in peptide side chain modification via cysteine alkylation. *Amino Acids* **2016**, *48*, 2267–2271. [CrossRef]
36. Calce, E.; Leone, M.; Monfregola, L.; De Luca, S. Lipidated peptides via post-synthetic thioalkylation promoted by molecular sieves. *Amino Acids* **2014**, *46*, 1899–1905. [CrossRef] [PubMed]
37. Calce, E.; Leone, M.; Mercurio, F.A.; Monfregola, L.; De Luca, S. Solid-Phase S-Alkylaion Promoted by Molecular Sieves. *Org. Lett.* **2015**, *17*, 5646–5649. [CrossRef] [PubMed]
38. Calce, E.; Digilio, G.; Menchise, V.; Saviano, M.; De Luca, S. Chemoselective Glycosylation of Peptides through S-Alkylation Reaction. *Chem. Eur. J.* **2018**, *24*, 1–9. [CrossRef] [PubMed]
39. De Luca, S.; Digilio, G.; Verdoliva, V.; Saviano, M.; Menchise, V.; Tovillas, P.; Jiménez-Osés, G.; Peregrina, J.M. A Late-Stage synthetic approach to lanthionine-containing peptides via S-alkylation on cyclic sulfamidates promoted by molecular sieves. *Org. Lett.* **2018**, *20*, 7478–7482. [CrossRef] [PubMed]
40. Azarifar, D.; Soleimanei, F.; Aliani, F. Benzylation of arylcyanamides catalyzed by acidic zeolites. *J. Mol. Catal. A Chem.* **2013**, *377*, 7–15. [CrossRef]
41. Costarrosa, L.; Ruiz-Martínez, J.; Rios, R.V.R.A.; Silvestre-Albero, J.; Rojas-Cervantes, M.L.; Sepúlveda-Escribano, A. Basic zeolites as catalysts in the N-alkylation of imidazole: Activation by microwave irradiation. *Microporous Mesoporous Mater.* **2009**, *120*, 115–121. [CrossRef]
42. Bahari, S.; Sajadi, S.M. Natrolite zeolite: A natural and reusable catalyst for one-pot synthesis of α-aminophosphonates under solvent-free conditions. *Arabian J. Chem.* **2017**, *10*, S700–S704. [CrossRef]
43. Hasegawa, M.; Ono, F.; Kanemasa, S. Molecular sieves 4Å work to mediate the catalytic metal enolization of nucleophile precursors: Application to catalyzed enantioselective Michael addition reactions. *Tetrahedron Lett.* **2008**, *49*, 5220–5223. [CrossRef]
44. Martins, L.; Hölderich, W.; Cardoso, D. Methylammonium-FAU zeolite: Investigation of the basic sites in base catalyzed reactions and its performance. *J. Catal.* **2008**, *258*, 14–24. [CrossRef]
45. Emana, A.N.; Chand, S. Alkylation of benzene with ethanol over modified HZSM-5 zeolite catalysts. *Appl. Petrochem. Res.* **2015**, *5*, 121–134. [CrossRef]
46. Sarmah, B.; Satpati, B.; Srivastava, R. Highly efficient and recyclable basic mesoporous zeolite catalyzed condensation, hydroxylation, and cycloaddition reactions. *J. Colloid Interface Sci.* **2017**, *493*, 307–316. [CrossRef] [PubMed]
47. Rama, V.; Kanagaraj, K.; Subramanian, T.; Suresh, P.; Pitchumani, K. Pyridinium ylide-assisted KY zeolite catalyzed tandem synthesis of polysubstituted cyclopropanes. *Catal. Commun.* **2012**, *26*, 39–43. [CrossRef]
48. Jiao, J.; Kanellopoulos, J.; Wang, W.; Ray, S.S.; Foerster, H.; Freude, D.; Hunger, M. Characterization of framework and extra-framework aluminum species in non-hydrated zeolites Y by ^{27}Al spin-echo, high-speed MAS, and MQMAS NMR spectroscopy at B_0 = 9.4 to 17.6 T. *Phys. Chem. Chem. Phys.* **2005**, *7*, 3221–3226. [CrossRef] [PubMed]
49. Abraham, A.; Lee, S.; Shin, A.; Hong, S.B.; Prins, R.; van Bokhoven, J.A. Influence of framework silicon to aluminium ratio on aluminium coordination and distribution in zeolite Beta investigated by ^{27}Al MAS and ^{27}Al MQ MAS NMR. *Phys. Chem. Chem. Phys.* **2004**, *6*, 3031–3036. [CrossRef]
50. Kim, H.; Cho, H.S.; Kim, C.; Choi, M. Gradual Disordering of LTA Zeolite for Continuous Tuning of the Molecular Sieving Effect. *J. Phys. Chem. C* **2017**, *121*, 6807–6812. [CrossRef]
51. Rimer, J.D. Rational design of zeolite catalysts. *Nat. Catal.* **2018**, *1*, 488–489. [CrossRef]

© 2019 by the authors. Licensee MDPI, Basel, Switzerland. This article is an open access article distributed under the terms and conditions of the Creative Commons Attribution (CC BY) license (http://creativecommons.org/licenses/by/4.0/).

MDPI
St. Alban-Anlage 66
4052 Basel
Switzerland
Tel. +41 61 683 77 34
Fax +41 61 302 89 18
www.mdpi.com

Catalysts Editorial Office
E-mail: catalysts@mdpi.com
www.mdpi.com/journal/catalysts

www.ingramcontent.com/pod-product-compliance
Lightning Source LLC
LaVergne TN
LVHW070437100526
838202LV00014B/1616